Nonlinear Ultrasonic Guided Waves

Online at: https://doi.org/10.1088/978-0-7503-4911-6

Nonlinear Ultrasonic Guided Waves

Cliff J Lissenden

Department of Engineering Science and Mechanics, The Pennsylvania State University, University Park, PA, USA

IOP Publishing, Bristol, UK

ISBN   978-0-7503-4911-6 (ebook)
ISBN   978-0-7503-4909-3 (print)
ISBN   978-0-7503-4912-3 (myPrint)
ISBN   978-0-7503-4910-9 (mobi)

DOI   10.1088/978-0-7503-4911-6

Multimedia content is available for this book from https://doi.org/10.1088/978-0-7503-4911-6.

Version: 20240601

IOP ebooks

British Library Cataloguing-in-Publication Data: A catalogue record for this book is available from the British Library.

Published by IOP Publishing, wholly owned by The Institute of Physics, London

IOP Publishing, No.2 The Distillery, Glassfields, Avon Street, Bristol, BS2 0GR, UK

US Office: IOP Publishing, Inc., 190 North Independence Mall West, Suite 601, Philadelphia, PA 19106, USA

Cover image: Jessie Lissenden.

Dedicated to Debbie, the most amazing woman in my world.
Thanks for putting up with me.

Contents

9 Finite amplitude pulse loading

Part III Applications

10 Numerical simulations

11 Making measurements

Preface

The last few years have brought a proliferation of journal articles that advance the state of knowledge regarding nonlinear ultrasonic guided waves for nondestructive evaluation of structural components. However, the foundational knowledge underlying this progress has yet to be consolidated into a single source that makes the knowledge accessible to newcomers to the field. Having one consistent source will make it easier for graduate students and others entering into research and development in this field to understand the fundamental concepts and know how to apply them. The book enables readers to design and test their own measurement system numerically and experimentally. The general aims of the book are to:

- Provide an entry point for those interested in the research and development of nondestructive evaluation using nonlinear ultrasonic guided waves.
- Be self-contained in terms of the background necessary for effective analysis of nonlinear ultrasonic guided waves and emphasize the differences between nonlinear guided waves and nonlinear bulk waves.
- Empower readers to become doers by providing sufficient details and examples.
- Summarize experimental results in the literature for a broad range of wave types, frequencies, and types of interaction.

Acknowledgments

This book testifies to the curiosity, proficiency, creativity, and persistence of the graduate students that I have been fortunate enough to work alongside. I would like to specifically mention the contributions to this subject by Yang Liu, Vamshi Chillara, Baiyang Ren, Gloria Choi, Hwanjeong Cho, Mostafa Hasanian, Christopher Hakoda, Anurup Guha, Chaitanya Bakre, Lalith Pillarisetti, and Hamidreza Afzalimir (who did all the simulations in chapter 10 and more). Thanks to Chung Seok Kim and Sungho Choi for introducing me to nonlinear ultrasonics and laser ultrasonics, respectively. Thanks to Francesco Costanzo for shedding light on the continuum mechanics content. Thanks to Joe Rose for teaching me about guided waves and life. Last but not least, thanks to my colleague Akhlesh Lakhtakia for inspiring me to write.

Author biography

Cliff J Lissenden

Cliff Lissenden is a professor of engineering science and mechanics at Penn State. His academic training is BSCE from Virginia Tech, MSCE from the University of Virginia, and PhD in Civil Engineering/Applied Mechanics from the University of Virginia. His expertise is in ultrasonic guided waves for nondestructive evaluation and the mechanical behavior of materials.

Part I

Analysis techniques

IOP Publishing

Nonlinear Ultrasonic Guided Waves

Cliff J Lissenden

Chapter 1

Introduction

'Nothing ventured, nothing gained' is an old proverb that encourages us to expand our reach. This chapter motivates us to venture into applying nonlinear ultrasonic guided waves to nondestructively evaluate material degradation to gain awareness of the material state, and thereby provide sharp tools for managing the safe operation and maintenance of assets used for energy production, ground and aerospace vehicles, civil structures, manufacturing equipment and elsewhere. We use a brief foray into fracture mechanics to provide the requisite motivation for utilizing nonlinear guided waves, although the elegance of nonlinear wave mechanics is itself compelling.

Image credit: Jessie Lissenden

1.1 Motivation

This book is motivated by the need to monitor material degradation at very early stages. Nonlinear ultrasonic guided waves have both strong potential and numerous challenges for this task. As nonlinear ultrasonic guided waves are not presently used commercially to any significant extent, we need to examine the knowledge gaps that must be filled for widespread application to become reality. Moreover, there are fascinating physics behind the propagation of nonlinear ultrasonic guided waves. Consider that acoustic wave amplitudes as small as tens of nanometers from a locomotive can propagate kilometers in the rail. Consider that secondary wave amplitudes can increase linearly when internal resonance occurs. Consider that a pulsed laser can actuate Rayleigh waves where shock fronts develop and may even initiate cracking. The primary objective of the book is to articulate the background that should be mastered to effectively investigate and apply nonlinear ultrasonic guided waves for nondestructive evaluation applications.

Imagine a future where electrical energy is generated in nuclear power plants that have no planned outages in contrast to today when planned outages make it convenient and relatively inexpensive to conduct nondestructive inspections. How can safety be ensured within reasonable economic cost? Knowing that outages for the sole purpose of inspection would incur unreasonable economic costs and huge lost-opportunity costs, we must turn to online structural health, or condition, monitoring. What information should the condition monitoring provide? Is it sufficient to detect the presence of flaws that influence strength and fracture properties? This line of questions leads us to ponder how flaws progress to the point of fracture, which is the purview of damage mechanics and fracture mechanics. Online condition monitoring is most effective for damage processes that occur on a relatively long timescale because there is ample time to track the progression of damage. Our premise is that the sooner the damage mechanisms can be detected, the more effectively the structural health can be managed. This chapter expounds upon our premise. As a simple example we present the oft-used concept of a normalized damage index (DI). When the material is in its pristine state DI = 0, and failure occurs when DI = 1. The health index (HI) is just the opposite. In the pristine state HI = 1 and at failure HI = 0. Figure 1.1 depicts two contrived cases of damage detection in a reactor pressure vessel. In this imagined future world, a digital twin is used to manage the reactor pressure vessel operations. At the worst hot spot, damage is detected at HI = 0.8 for case 1 and at HI = 0.9 for case 2. However, the damage is detected at life fractions of 0.95 and 0.5 for cases 1 and 2, respectively. Case 2 is clearly preferred based on the early warning it provides, which can be leveraged to perform logistics of maintenance for safe and economical operations. The nonlinear damage evolution, which is common, exacerbates the problem and demonstrates the importance of early damage detection.

The physics of nonlinear wave propagation are fascinating. Nonlinear acoustic waves in fluids exhibit increasing amplitudes with propagation distance and shock formation, for example from supersonic aircraft. Solitons having a balance between nonlinearity and attenuation occur naturally as tidal bores and tsunamis. Likewise, nonlinear electromagnetic waves play a variety of roles in optical fibers. The nonlinear ultrasonic

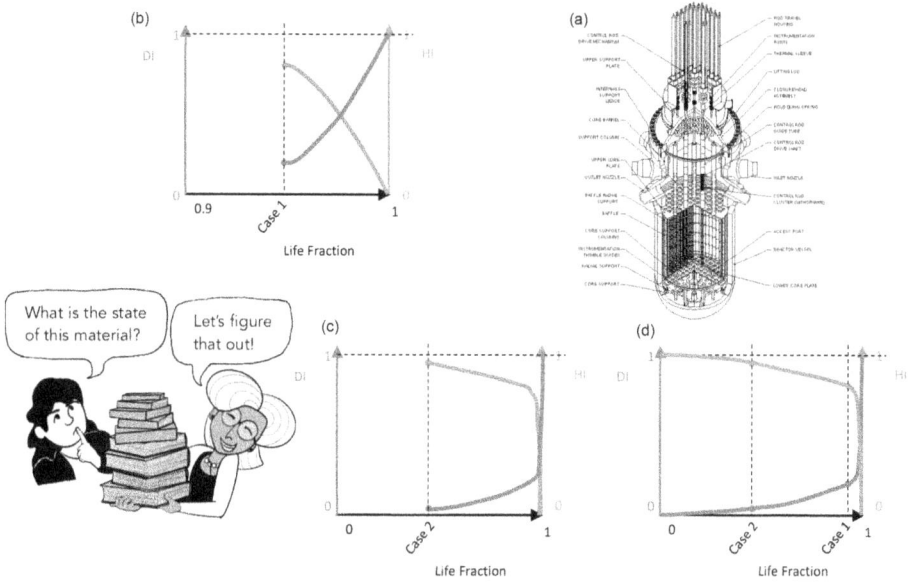

Figure 1.1. (a) Illustration of RPV digital twin from which we plot HI and DI versus life fraction for (b) case 1, (c) case 2, and (d) the entire life curve. Bottom left image credit: Jessie Lissenden.

guided wave physics demands much of mathematics to provide understanding. We impose boundary conditions on the wave equation to determine the dispersion curves and wavestructures for the freely propagating waves in a waveguide, which is solved as an eigenproblem. The nonlinear wave equation requires special attention to the frame of reference when defining the tensor variables. Continuum mechanics provides the necessary framework to define strain and stress tensors precisely. For weak nonlinearity a perturbation methodology can be implemented, whose solution relies upon a complex reciprocity theorem to determine the orthogonality relation necessary to implement a normal mode expansion. More sophisticated solution methods, such as the method of multiple scales, are required to deal with strong nonlinearity that leads to steep discontinuities or even shock formation. All of these solutions require idealizations that need to be checked by computational analyses, typically using finite element methods. But in the end, it is only real measurements that determine the utility of the nonlinear wave mechanics for acquiring useful information about the material state.

In linear ultrasonics all wave features are measured at the excitation frequency, while in nonlinear ultrasonics the features are measured at frequencies different from the excitation frequency or frequencies. This has been demonstrated to provide improved sensitivity and early detection of material degradation—for which there is a tremendous interest for operating and maintaining fleets of aircraft, vehicles, ships, and power plants as well as manufacturing and industrial equipment and structures. The interest in guided waves is due primarily to their penetration power and utility for otherwise inaccessible structures, and because so many components of structures and mechanical systems are natural waveguides. The advantages of penetration power and access to inaccessible structures add to the advantage of using a significantly higher frequency relative to vibration monitoring, and the improved sensitivity that comes with it.

Nonlinear ultrasonic guided waves have strong potential for nondestructive evaluation of structures, in particular for detecting material degradation at an early stage. Once detected, action can then be taken to repair or replace the component based on its condition. This enables a paradigm shift in maintenance philosophy from schedule-based or upon failure, to condition-based with sufficient warning time and specificity to allow for logistics. The schematic in figure 1.2 illustrates one benefit of early detection of material degradation due to fatigue. Linear ultrasonics is quite good at detecting the presence of open cracks based on the discontinuity associated with the traction-free surfaces created by the crack. Unfortunately, it is common for only 5%–10% of the fatigue life to remain once the crack is large enough to open—leaving little time for maintenance. Nonlinear ultrasonics has the potential to shift the detection threshold to a much earlier point in the service life. Nonlinear ultrasonics shifting the detection threshold is not unique to fatigue, but applies to creep and thermal aging processes that also occur over long timescales. Figure 1.2 is a schematic, we need to develop the ability for such plots to be data-based. This is a critical knowledge gap.

Figure 1.2 illustrates that fatigue and fracture provide powerful motivation to develop nonlinear ultrasonic guided waves for early detection of material degradation. Without going deep into fracture mechanics (there are many good sources available, e.g. [1]), let's highlight some key concepts that provide a driving force for nonlinear ultrasonic nondestructive evaluation. Fracture mechanics enables analysts to predict remaining fatigue life based on crack growth, which is the root cause of fatigue failures, by presuming the existence of a crack or crack-like flaw. Essential terminology is summarized below. To provide some clarity, this section on fracture mechanics is intended to motivate the subject of nonlinear ultrasonic guided waves, but it is very different to the remainder of the book.

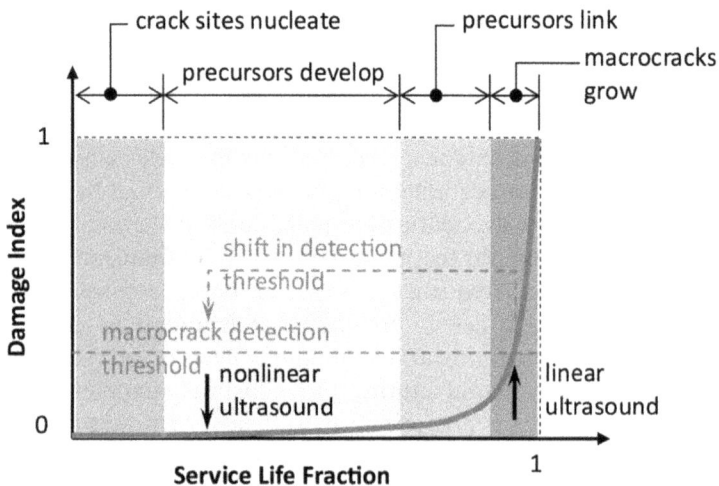

Figure 1.2. Shifting detection to earlier in the service life using nonlinear ultrasound.

Defects and flaws. Although it is atypical to distinguish between defects and flaws in the fracture mechanics literature, it is useful here to concisely delineate the appropriate terminology. Defects are present in all materials, however, their presence does not necessarily prevent the material from performing at its optimal level. The presence of dislocations (line defects in the crystal lattice) is a perfect example of defects; they exist in newly processed crystalline materials and the evolution of their number and distribution could affect the material behavior. On the other hand, flaws definitely affect material behavior. Examples of flaws are fatigue cracks, creep-induced voids, corrosion pits, and delaminations. Defects may grow to become flaws or not. Additional discussion with slightly different terminology is provided by Farrar and Worden [2] and Worden *et al* [3] with respect to structural health monitoring. We repeat for emphasis that fracture mechanics posits the existence of a crack-like flaw in the material. There can be no fracture mechanics analysis without a flaw present.

Stress intensity factor. This term is used to describe the stress state in the vicinity of a crack tip to avoid the dilemma associated with the stress field singularity at the tip of a sharp crack. For crack growth due to normal separation of the crack faces, the stress intensity factor can be written as

$$K_I = FS\sqrt{\pi a}, \tag{1.1}$$

where F is the geometry factor that depends on crack size, loading, and structure geometry, S is the nominal stress in the far-field, and a is a measure of the crack size. The subscript I delineates normal separation from shear modes of crack growth denoted by II and III.

Fracture toughness. Sudden unstable crack growth occurs when the stress intensity factor attains a critical value, K_{Ic}, which may also be thought of as the crack size becoming critical $a \rightarrow a_c$. K_{Ic} is a material parameter determined by standardized mechanical testing techniques and is known as the fracture toughness. As fracture toughness depends on restraint at the crack tip, K_{Ic} is defined to be the minimum value, which occurs for the plane strain condition. Thus, K_{Ic} represents the plane strain fracture toughness.

Crack growth. As just described, crack growth can be unstable, which is when catastrophic fracture occurs. Stable, or subcritical, crack growth can occur for cracks less than the critical size due to repeated loads (i.e. fatigue) or if there is environmental assistance. Figure 1.3 is a scanning electron microscope image of a fracture surface in an aluminum alloy showing the striations associated with crack front advancement in each loading cycle. It is a shame that this kind of image is only available after fracture. Example 1.1 provides fatigue crack growth data for a stainless steel alloy.

Paris power law. This applies to steady state crack growth, where Paul Paris found that there exists a power law relationship between crack growth rate *da/dN* and the range of the stress intensity factor ΔK_I,

$$\frac{da}{dN} = C(\Delta K_I)^m, \tag{1.2}$$

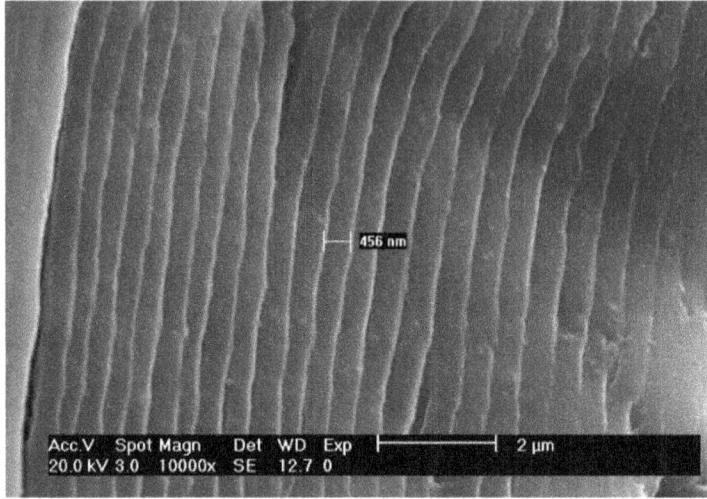

Figure 1.3. Fatigue striations in a 7075 aluminum single-edge notched sample.

where C and m are material parameters fit to experimental data, as shown in example 1.2. Variants to this basic empirical relationship account for the effect of mean stress and other complications. Remaining fatigue life can be estimated by integrating the power law relationship,

$$\int_{N_i}^{N_f} dN = \int_{a_i}^{a_f} \frac{da}{C(\Delta K_I)^m}, \tag{1.3}$$

$$\Delta N = N_f - N_i = \int_{a_i}^{a_f} \frac{da}{C(F\Delta S\sqrt{\pi a})^m}. \tag{1.4}$$

The bounds on the integral are the initial and final crack sizes. The final crack size may come from the fracture condition given by equation (1.1) based on critical values, or some other service-based requirement. The initial crack size typically comes from a nondestructive measurement technique, or that technique's resolution if no crack is actually detected. The geometry factor F is often a complicated relation dependent upon the crack size, which typically requires equation (1.4) to be solved by numerical integration. Example 1.2 is a counter example where the geometry factor is taken to be constant, making it straight-forward to illustrate the role of nondestructive evaluation, and by extension, nonlinear ultrasonic guided waves.

Example 1.1. Fatigue crack growth experiments were conducted on type 347 stainless steel and a calibrated crack opening displacement gage was used to determine the crack length as a function of the cycle number as shown in figure 1.4(a). These data can be manipulated to determine the crack growth rate, da/dN as a function of the number of cycles N and the number of cycles to grow the crack a unit amount dN/da as a function of crack length a. The $dN/da - a$ plot is shown in figure 1.4(b). The dN/da plot is particularly informative for nondestructive evaluation. Since the area under the curve, $\int_{a_i}^{a_f} \frac{dN}{da} da = \Delta N$, is the number of cycles to grow the crack from a_i to a_f, we see that the accuracy to which the initial crack size, a_i, is known has a tremendous impact on the remaining life estimate; most of the area under the curve is skewed towards a_i. It is also important to point out that this approach only predicts the remaining life associated with crack growth from a_i to a_f and has nothing to do with the initiation of the crack and its growth up to a_i.

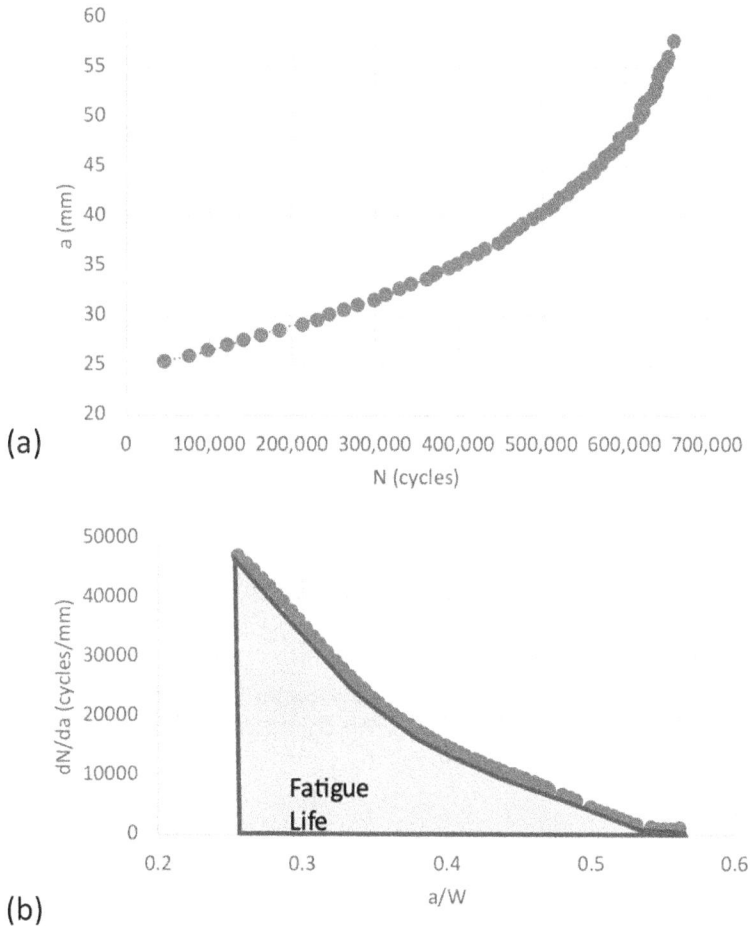

(a)

(b)

Figure 1.4. Fatigue crack growth in a 347 stainless steel compact tension sample: (a) crack size as a function of the number of cycles and (b) the number of cycles to grow the crack a unit amount as a function of nondimensional crack size (W = sample width).

Example 1.2. Predict the number of cycles before crack growth becomes unstable and fracture occurs for a stainless steel vessel containing a fluctuating internal pressure. Post-fabrication quality assurance (QA) testing of the vessel did not detect any cracks. The ultrasonic test method used is known to reliably detect surface cracks greater than 2 mm in length. We will apply the integrated form of the Paris power law (equation (1.4)) to predict the fatigue life.

Analysis inputs are:
1) Inner radius $r = 1$ m
2) Wall thickness $t = 25$ mm
3) Maximum internal pressure $p_{max} = 8$ MPa
4) Minimum internal pressure $p_{min} = 0$ MPa

The stresses in a thin-walled cylindrical pressure are:
1) Hoop stress $\sigma_h = pr/t$
2) Axial stress $\sigma_a = pr/2t$

Although the pressure vessel passed the QA tests, to apply fracture mechanics we must posit the existence of a crack or crack-like flaw. Therefore, we assume that a 2 mm surface breaking crack exists in the pressure vessel. The worst-case scenario is that the crack faces are normal to the largest tensile stress, in this case the hoop stress. We further assume that the crack is semi-circular and located within a transverse cross-section of the cylinder (i.e. normal to the hoop stress). The crack size, $a = 1$ mm, for a surface crack length of 2 mm and crack depth of 1 mm, and the geometry factor $F = 0.728$ (with 10% accuracy provided $a/t < 0.4$ according to figure 8.17(b) in Dowling [4]).

Fatigue crack growth parameters, C and m, can be determined from the test results shown in figure 1.4 if the far-field stresses and stress intensity factor in the compact tension sample are known (they are!). The crack growth rate da/dN is plotted as a function of the stress intensity factor range ΔK_I on log axes in figure 1.5, from which C and m are regressed:
1) $C = 1 \times 10^{-11}$ in units of m/cycle and MPa$\sqrt{m}$
2) $m = 2.8$.

The fracture toughness is also needed to determine the critical crack size; $K_{Ic} = 38$ MPa$\sqrt{m}$ is a reasonable value. We compute the critical crack size from the maximum stress using equation (1.1):

$$K_{Ic} = FS_{max}\sqrt{\pi a_c}$$

$$38 = 0.728\left(\frac{8 \cdot 1000}{25}\right)\sqrt{\pi a_c}$$

$$a_c = 0.0085 \text{ m},$$

and we check that $a_c/t < 0.4$ to ensure that the geometry factor is valid [4]. Analyse equation (1.4) as follows:

$$\Delta N = N_f - N_i = \int_{a_i}^{a_f} \frac{da}{C(F \Delta S \sqrt{\pi a})^m}$$

$$N_f - N_i = \frac{1}{C(F \Delta S \sqrt{\pi})^m} \int_{a_i}^{a_f} \frac{da}{a^{m/2}}$$

$$\Delta S = \frac{\Delta p \cdot r}{t} = 320 \text{ MPa}$$

$$N_f - N_i = \frac{1}{C(F \Delta S \sqrt{\pi})^m} \frac{a_f^\xi - a_i^\xi}{\xi}$$

$$\xi = 1 - \frac{m}{2} = -0.4$$

$$N_f - N_i = \frac{1}{10^{-11}(0.728 \cdot 320 \sqrt{\pi})^{2.8}} \frac{(0.0085)^{-0.4} - (0.001)^{-0.4}}{-0.4}$$

$$N_f - N_i = 107\,976 \text{ cycles.}$$

Therefore, the crack is expected to reach critical size in less than 108 000 cycles. But remember that no crack was detected by the QA tests and that we had to assume it to be the largest size crack that could go undetected. Also remember that the number of cycles to grow a crack a unit amount is skewed toward small initial cracks. The point here is that more accurate life prediction would result from earlier crack detection.

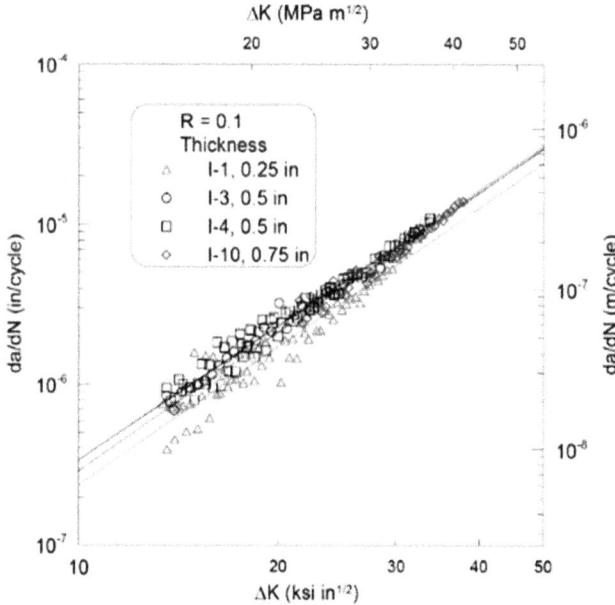

Figure 1.5. Fatigue crack growth in a 347 stainless steel compact tension sample.

It is also important to realize that the Paris power law only applies for the steady state crack growth portion of the crack growth spectrum, it does not address initiation. Unfortunately for accurate predictions, as the fatigue life increases as the stress amplitude decreases, the life fraction spent in crack initiation increases. Thorough treatment of fatigue degradation in metals is provided by Suresh [5] and McDowell [6] and many others. Kulkarni and Achenbach [7] proposed that fatigue life estimation could be a two-stage process; first use damage mechanics to represent the initiation of a macroscale crack followed by fracture mechanics (as in example 1.2) to predict the fracture event. They proposed nonlinear ultrasonics as a means to track material state evolution up to the point of macroscale crack initiation. The nonlinear ultrasonic guided wave techniques described in this book fit nicely into the damage mechanics framework.

There are several challenges associated with deploying nonlinear ultrasonic guided waves for nondestructive evaluation of materials and structures. The intent of this book is to carefully describe those challenges that have already been overcome in the first two decades of study such that future efforts are focused on connecting the measurements of weak nonlinearity to material degradation through wave-material interaction. For emphasis, I want to reiterate that the primary objective of the book is to articulate the background that should be mastered to effectively investigate and apply nonlinear ultrasonic guided waves for nondestructive evaluation applications. It provides a comprehensive foundation for continued research and development of nonlinear ultrasonic guided waves for nondestructive evaluation.

1.2 Brief perspective on nonlinear ultrasonic guided waves

For the purposes of this book, acoustic waves can propagate in both fluid and solid media, ultrasonic waves propagate at frequencies above what a typical human can hear (~20 kHz), and elastic waves propagate in solid media. While acoustic waves in fluids can be guided, here we use the term ultrasonic guided waves to mean waves at frequencies above about 20 kHz that propagate as elastic waves in a solid waveguide. The boundaries and interfaces in the waveguide channel or guide the transport of elastic energy in specific directions, which typically enable it to travel longer distances as well as through otherwise inaccessible material.

The principle of linear superposition is one of the pillars of linear ultrasonics, as such it is used to mathematically describe the many interference phenomena that occur for bulk and guided waves (e.g. phased arrays). Interference phenomena are characterized by wave fields that do not interact. However, as the wave amplitude increases, and becomes known as a finite amplitude wave, its dependence on particle velocity starts to become appreciable and weak interactions occur. Weak interactions resulting in progressive evolution of the waveform can be adequately represented by perturbation solutions to the governing differential equations. In exceptional situations, the wave amplitude can become sufficiently large as to form shocks. The presence of waveform discontinuities invalidates the regular perturbation solutions, and it becomes necessary to use more advanced techniques to solve the boundary value problem. Chapter 9 discusses the exceptional case of finite amplitude surface waves generated by short laser pulses. We adopt the premise that there are problems for which linear ultrasound measurements are insufficient. It is in these cases that nonlinear ultrasound measurements are most helpful. Aside from the previous section, where we allude to early material degradation as a clear area of interest, selecting the applications where nonlinear ultrasonic guided waves should be used is left to the reader. Comparing the capabilities of nonlinear methods with linear methods is a good way to demonstrate the need to apply the nonlinear method, which is typically more demanding. These comparisons are not made in this book simply because the focus is on the nonlinear methods themselves.

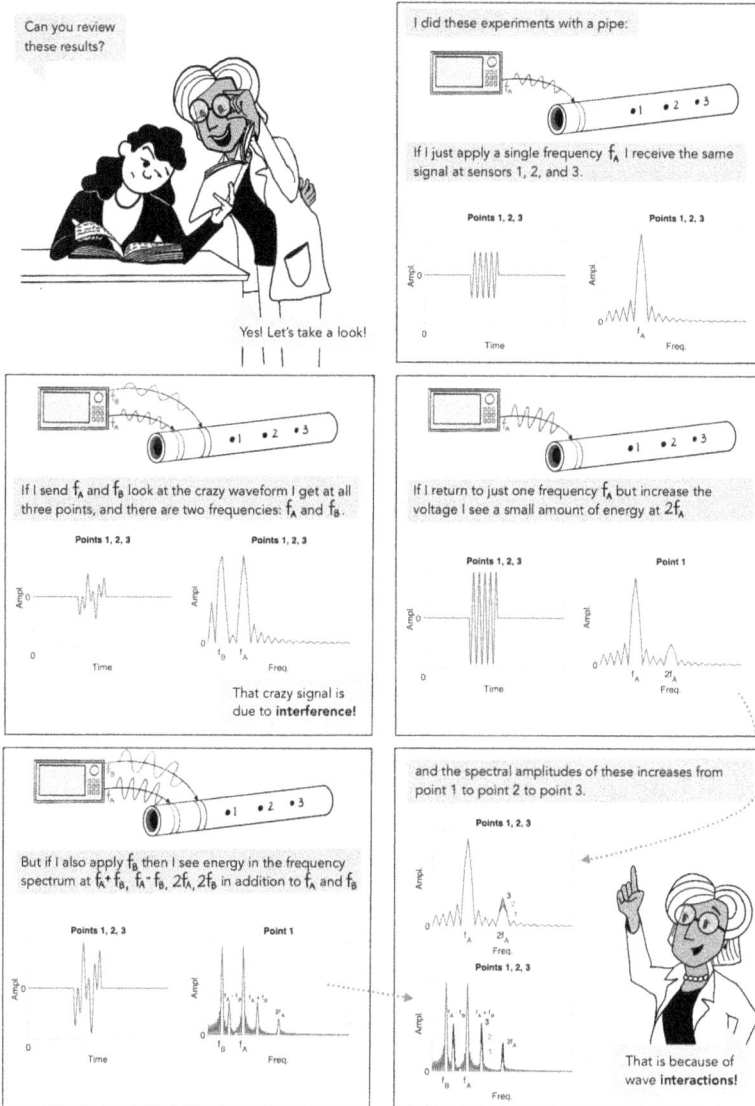

Image credit: Jessie Lissenden

Nonlinear acoustics in fluid media is a well-developed field of study [8–11]. Hamilton and Blackstock [11] are careful to define nonlinear parameters and nonlinear coefficients. However, in the nonlinear acoustics literature on solids their terminology is not strictly followed. I try to clarify the terminology discontinuity in chapter 3. Study of nonlinear ultrasound in solids began in the early 1960s with respect to higher harmonic generation and wave mixing. Applications of acoustoelasticity (i.e. wave speed dependency on stress state) date back to the 1950s (e.g. Hughes and Kelly [12]) for measuring residual stress and determining third order elastic constants. The study of nonlinear ultrasonic guided waves began in the late 1990s with Deng [13], followed by de Lima and Hamilton [14] and Bermes *et al* [15]. There have been a number of review articles [16–19] and books on nonlinear ultrasonics [20, 21], and a tutorial on nonlinear ultrasonic guided waves for nondestructive evaluation [22], but the elevated interest in the subject makes a book dedicated to nonlinear ultrasonic guided waves valuable.

1.3 Approach

This book analyses nonlinear elastic waves, but they are analysed with respect to applications for ultrasonics, therefore both terms are used with the term *elastic waves* emphasizing how they are analysed and the term *ultrasonic waves* emphasizing how they are applied. As is almost always the case with nonlinear ultrasonic guided wave applications, a combination of modeling, simulation, and physical measurement is provided. Even if the objective is to use nonlinear methods to make measurements connected to the microstructure of the material, a continuum mechanics approach is taken for the modeling. Simulations typically use commercially available finite element analysis software. Physical measurements on the other hand, employ a broad spectrum of instruments, transducers, and methods based on the needs of the various applications. Each of these subjects is given appropriate coverage in the book.

In the analysis of linear ultrasonic guided waves, we realize that interference phenomena dominate and are described by the principle of linear superposition, as already noted. When the analysis is extended to nonlinear ultrasonic guided waves, the wave equation becomes nonlinear and linear superposition no longer applies. Therefore, wave interactions occur that distort the waves. We choose to approach the nonlinear problems as:

  i. self-interaction, when a wave interacts with itself, which generates higher harmonics at integer multiples of the primary frequency; and
  ii. mutual interaction, when waves at different frequencies interact, which generates combinational harmonics at frequencies $\pm m\omega_a \pm n\omega_b$ if the primary frequencies are ω_a and ω_b, where m and n are integers.

In fact, we will see that self-interaction is a special case of the mutual interaction problem. This approach enables us to analyse higher harmonic generation problems and wave mixing problems with the same framework.

1.4 Content

Part I of this book is on *analysis techniques* and sets the stage and prepares the toolkit for analysing nonlinear ultrasonic guided waves by introducing the continuum mechanics concepts necessary for nonlinear elastic waves and the mathematical formulation for elastodynamic problems. It presents the 1D nonlinear wave propagation problem to introduce the analysis methods and physical features in a more simplified fashion. The boundary value problems for various guided wave problems are set up for linear problems and extrapolated to nonlinear problems. The first part of the book ends by solving the linear guided wave boundary value problems for free surfaces, plates, pipes, and other waveguides, with a focus on the dispersion relations. Frequency–wavenumber, phase velocity, and group velocity dispersion curves are shown. Focus is on isotropic materials, but transversely isotropic materials are included for generality.

Part II of the book, on *modeling nonlinearity*, uses the toolkit to obtain the solution for nonlinear ultrasonic guided wave problems. The internal resonance conditions are determined by using a reciprocity relation to prove that wave modes in a plate are orthonormal and a normal mode expansion is feasible. The importance of selecting primary waves that generate internally resonant secondary modes is emphasized. Chapter 8 is more comprehensive than what is published in journal articles to date. Nonlinear features of dispersive guided waves are shown for self-interaction and mutual interactions. Part II ends with finite amplitude pulse loading, which has been shown to provide very interesting results for surface waves, including shock formation and crack initiation.

Part III, the final part of the book, highlights *applications* for nonlinear ultrasonic guided waves and hopefully leads to future expansion of the toolkit for nonlinear ultrasonic guided waves. Numerical simulations are an important tool because they allow the analyst to dictate the nonlinearity in the problem. Specifically, the physical measurement system nonlinearity can be eliminated. The order of the strain energy function can be adjusted, and geometric nonlinearity may be excluded. Sensing systems are described, and some sample experimental results are discussed. The book concludes with the author's perspective on current capabilities and future possibilities.

1.5 Closure

I believe that nonlinear ultrasonic guided waves have strong potential to detect material degradation at an early stage, thus enabling maintenance and safe operations of assets to be efficiently managed. A fracture mechanics-based motivation was introduced in this chapter. Readers are encouraged to preview the compelling experimental results highlighted in chapter 12 before proceeding through the analysis techniques described in the remainder of part I.

References

[1] Hertzberg R W, Vinci R P and Hertzberg J L 2020 *Deformation and Fracture Mechanics of Engineering Materials* 6th edn (Hoboken, NJ: Wiley)
[2] Farrar C R and Worden K 2007 An introduction to structural health monitoring *Philos. Trans. R. Soc. Math. Phys. Eng. Sci.* **365** 303–15

[3] Worden K, Farrar C R, Manson G and Park G 2007 The fundamental axioms of structural health monitoring *Proc. R. Soc. Math. Phys. Eng. Sci.* **463** 1639–64

[4] Dowling N E 2013 *Mechanical Behaivior of Materials* 4th edn (London: Pearson)

[5] Suresh S 1991 *Fatigue of Materials* (Cambridge: Cambridge University Press)

[6] McDowell D L 1996 Basic issues in the mechanics of high cycle metal fatigue *Int. J. Fract.* **80** 103–45

[7] Kulkarni S S and Achenbach J D 2008 Structural health monitoring and damage prognosis in fatigue *Struct. Health Monit.* **7** 37–49

[8] Zarembo L K and Krasil'nikov V A 1966 *Introduction to Nonlinear Acoustics* (Moscow: Nauka) (in Russian)

[9] Rudenko O V and Soluyan S I 1977 Theoretical foundations of nonlinear acoustics *Studies in Soviet Science* (New York: Consultants Bureau)

[10] Beyer R T 1984 *Nonlinear Acoutics in Fluids* (New York: Van Nostrand Reinhold)

[11] Hamilton M F and Blackstock D T 1997 *Nonlinear Acoustics* 1st edn (San Diego, CA: Academic)

[12] Hughes D S and Kelly J L 1953 Second-order elastic deformation of solids *Phys. Rev.* **92** 1145–9

[13] Deng M 1998 Cumulative second-harmonic generation accompanying nonlinear shear horizontal mode propagation in a solid plate *J. Appl. Phys.* **84** 3500–5

[14] de Lima W J N and Hamilton M F 2003 Finite-amplitude waves in isotropic elastic plates *J. Sound Vib.* **265** 819–39

[15] Bermes C, Kim J-Y, Qu J and Jacobs L J 2007 Experimental characterization of material nonlinearity using Lamb waves *Appl. Phys. Lett.* **90** 021901

[16] Jhang K-Y 2009 Nonlinear ultrasonic techniques for nondestructive assessment of micro damage in material: a review *Int. J. Precis. Eng. Manuf.* **10** 123–35

[17] Zheng Y, Maev R G and Solodov I Y 1999 Nonlinear acoustic applications for material characterization: a review *Can. J. Phys.* **77** 927–67

[18] Matlack K H, Kim J-Y, Jacobs L J and Qu J 2015 Review of second harmonic generation measurement techniques for material state determination in metals *J. Nondestruct. Eval.* **34** 273

[19] Chillara V K and Lissenden C J 2016 Review of nonlinear ultrasonic guided wave nondestructive evaluation: theory, numerics, and experiments *Opt. Eng.* **55** 011002

[20] 2019 *Nonlinear Ultrasonic and Vibro-Acoustical Techniques for Nondestructive Evaluation* ed T Kundu (Cham: Springer International)

[21] 2020 *Measurement of Nonlinear Ultrasonic Characteristics* (Springer Series in Measurement Science and Technology) ed K-Y Jhang, C Lissenden, I Solodov, Y Ohara and V Gusev (Singapore: Springer)

[22] Lissenden C J 2021 Nonlinear ultrasonic guided waves—principles for nondestructive evaluation *J. Appl. Phys.* **129** 021101

IOP Publishing

Nonlinear Ultrasonic Guided Waves

Cliff J Lissenden

Chapter 2

Preliminaries

Readers interested in ultrasonics might wonder why they should spend their valuable time on concepts from continuum mechanics knowing that ultrasonic wave displacements are typically only tens of nanometers. The answer is in the nonlinearity! The ultrasonic nonlinearity from material nonlinearity needs precisely defined stress and strain quantities to describe the nonlinearities with consistency and accuracy. Continuum mechanics provides a consistent analysis framework suitable for precise mathematical modeling of finite amplitude waves. Fundamental concepts in elastodynamics are introduced.

Image credit: Jessie Lissenden

2.1 Notation

We make combined use of direct tensor (symbolic) notation and index notation in this book, depending upon which is more convenient or revealing for that part of the analysis, although flipping back and forth between notations is minimized to avoid creating headaches. Bold face variables represent vectors or higher rank tensors, and italic variables represent scalars (no indices) or components of vectors and higher order tensors (having one or more free indices). When using Einstein's index notation, summation over 1–3 is implied for repeated indices unless otherwise noted, i.e. $\varepsilon_{kk} = \sum_{k=1}^{3} \varepsilon_{kk}$. Physical quantities given by Cartesian tensors of rank n are represented by 3^n real (or complex) numbers, where rank zero is a scalar, rank one is a vector, and so on. A tensor of rank two can be represented by a 3×3 matrix whose entries are the components of the tensor relative to an underlying (tensor) basis. Modeling nonlinear ultrasonic guided waves requires us to satisfy balance laws by careful connection to a prescribed coordinate system. For the case of cylindrical and hollow cylindrical waveguides the analysis is most readily performed in a curvilinear coordinate system. However, because the calculus in curvilinear coordinates is more complicated and the notation more cumbersome, the analysis in this book is limited to Cartesian coordinates for simplicity. Interested readers are referred to more advanced sections of continuum mechanics texts (e.g. Malvern [1] appendices I and II, Fung and Tong [2] chapters 2 and 16) for details on the necessary covariant, contravariant, and metric tensors. Liu et al's work on nonlinear guided waves in pipes provides a direct application [3–5].

Superscripts T and -1 and are used to denote 'transpose' and 'inverse' (when it exists). Spatial derivatives in the reference Cartesian coordinate system are often written in index notation with a comma, e.g. $u_{i,j} = \frac{\partial u_i}{\partial X_j}$. Given rank-2 tensors $\mathbf{A}$ and $\mathbf{B}$, the inner and outer products can be written directly as $\mathbf{A} \cdot \mathbf{B}$ and $\mathbf{A} \otimes \mathbf{B}$, respectively, or in component form using indices as $A_{ik}B_{kj}$ and $A_{ij}B_{kl}$, respectively. All component expressions in index notation are understood to stem from a global Cartesian coordinate system covering all possible configurations.

Because we use multiple stress and strain definitions in different parts of the book it is important to be clear which definition is being used. The stress tensors used are:

Cauchy, $\mathbf{T}_c = \boldsymbol{\sigma}$
first Piola–Kirchhoff, $\mathbf{T}_{PK1} = \mathbf{S}$
second Piola–Kirchhoff, $\mathbf{T}_{PK2}$

and the strain tensors are:

Lagrangian $\mathbf{E}$ and
infinitesimal $\boldsymbol{\varepsilon}$.

In linear problems we use the Cauchy stress and the infinitesimal strain, but in nonlinear problems we use the Cauchy stress, both Piola–Kirchhoff stresses, and the Lagrangian strain.

2.2 Continuum mechanics

Although our interest is in understanding the propagation characteristics of elastic waves, which transfer energy through discrete packets (phonons), a continuum mechanics approach that ignores the underlying structure of the elastic material is extremely useful because the wavelength is substantially larger than the underlying structure (i.e. grains and lattice defects). Later we will see a paradox, in that nonlinear ultrasonic (guided) waves can provide valuable information about the (ignored) underlying material structure that is very difficult to obtain through any other nondestructive means. Assuming the domain to be continuous enables the rigorous application of calculus to problems where the elastic material is either homogeneous or heterogeneous as well as either isotropic or anisotropic. Our particular interest is in nonlinear elastic materials, for which hyperelastic constitutive equations will be applied. Even though particle displacements associated with elastic wave motion are typically quite small, the problem formulation must be consistent, which means that a finite strain measure will be adopted. Moreover, elastic waves are rapid disturbances whose displacement gradients are large relative to the small displacements, as shown in example 2.1. Therefore, this section provides a concise summary of the key concepts and definitions from continuum mechanics that will be used in the remainder of the book. Readers are referred to several excellent continuum mechanics textbooks for more depth [1, 2, 6, 7].

Example 2.1. Consider an elastic longitudinal wave having a wave speed of 6300 m s^{-1} driven at a frequency f of 1 MHz. For a planar bulk wave, the displacement can be written as

$$u = A \sin(kx_1 - 2\pi ft),$$

where k is the wavenumber. The displacement gradient is

$$\frac{du}{dx_1} = Ak \cos(kx_1 - 2\pi ft).$$

Therefore, the peak displacement gradient value is Ak. Let's take the amplitude A to be 10 nm. Using the fundamental wave mechanics relation that $k = 2\pi f/c$ we obtain

$$Ak = (10^{-8})\left(\frac{2\pi}{6300} \frac{10^6}{}\right) \approx 10^{-5}.$$

In this case the strain is approximately equal to the displacement gradient. Thus, a displacement amplitude of 10 nm results a peak strain of $\sim 10^{-5}$, or 10 $\mu\varepsilon$.

2.2.1 Kinematics

Under the action of applied forces a body can translate, rotate, and deform. We will write the field equations describing nonlinear ultrasonic wave propagation using a Lagrangian formulation, i.e. a referential description. The coordinates used to identify points in the reference and current configurations will be denoted as (X_1, X_2, X_3) and (x_1, x_2, x_3), respectively, and will be referred to as material and spatial coordinates, respectively. Note that depending on the circumstances, (X_1, X_2, X_3) and (x_1, x_2, x_3) might be expressed using a single coordinate system as shown in figure 2.1. The displacement vector **u** is defined in figure 2.1 to be

$$\mathbf{u} = \mathbf{x} - \mathbf{X}, \tag{2.1}$$

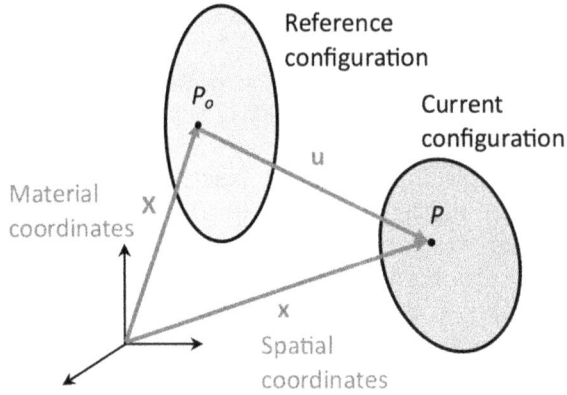

Figure 2.1. Reference and current configurations.

while the deformation gradient tensor $\mathbf{F}$ is defined by $d\mathbf{x} = \mathbf{F} \cdot d\mathbf{X}$, and its components are computed as $F_{ij} = \frac{\partial x_i}{\partial X_j}$. Taking the derivatives with respect to the material coordinates in equation (2.1) gives the displacement gradient $\mathbf{H}$, whose components are:

$$H_{ij} = \frac{\partial u_i}{\partial X_j} = \frac{\partial x_i}{\partial X_j} - \frac{\partial X_i}{\partial X_j} = F_{ij} - \delta_{ij}, \tag{2.2}$$

where δ_{ij} is the Kronecker delta. We should recognize that the relationship between the technically incompatible quantities $\mathbf{H}$ and $\mathbf{F}$ is more delicate than alluded to here. It is, however, nominally valid in an Euclidean context.

Define the right Cauchy–Green strain tensor as

$$\mathbf{C} = \mathbf{F}^T \cdot \mathbf{F} \tag{2.3}$$

and the left Cauchy–Green strain tensor as

$$\mathbf{B} = \mathbf{F} \cdot \mathbf{F}^T. \tag{2.4}$$

Image credit: Jessie Lissenden

Then twice the Lagrangian strain tensor is defined with respect to the right Cauchy–Green strain tensor,

$$2\mathbf{E} = \mathbf{C} - \mathbf{I} \tag{2.5}$$

$$2\mathbf{E} = \mathbf{F}^T \cdot \mathbf{F} - \mathbf{I}.$$

Finally, the component form of the strain–displacement relation is:

$$E_{ij} = \frac{1}{2}\left[\left(\frac{\partial x_k}{\partial X_i}\right)\left(\frac{\partial x_k}{\partial X_j}\right) - \delta_{ij}\right] \tag{2.6}$$

$$E_{ij} = \frac{1}{2}\left[\left(\frac{\partial(u_k + X_k)}{\partial X_i}\right)\left(\frac{\partial(u_k + X_k)}{\partial X_j}\right) - \delta_{ij}\right]$$

$$E_{ij} = \frac{1}{2}\left[(u_{k,i} + \delta_{ki})(u_{k,j} + \delta_{kj}) - \delta_{ij}\right]$$

$$E_{ij} = \frac{1}{2}\left[(u_{k,i}u_{k,j} + u_{i,j} + u_{j,i} + \delta_{ij}) - \delta_{ij}\right]$$

$$E_{ij} = \frac{1}{2}\left[u_{i,j} + u_{j,i} + u_{k,i}u_{k,j}\right]$$

or

$$\mathbf{E} = \frac{1}{2}[\mathbf{H} + \mathbf{H}^T + \mathbf{H}^T\mathbf{H}], \tag{2.7}$$

where $\mathbf{H}$ is the displacement gradient. Therefore, the Lagrangian finite strain tensor is quadratic in the displacement gradient. The infinitesimal strain tensor ε is simply a linearization of $\mathbf{E}$ with respect to $\mathbf{u}$ when $\mathbf{E}$ is viewed as a function of $\mathbf{u}$ and $\mathbf{H}$ is taken as the smallness parameter:

$$\varepsilon = \frac{1}{2}[\mathbf{H} + \mathbf{H}^T], \quad \varepsilon_{ij} = \frac{1}{2}\left[u_{i,j} + u_{j,i}\right]. \tag{2.8}$$

Examples 2.2 and 2.3 demonstrate the kinematic relations for the important cases of uniaxial normal stress and antiplane shear.

Example 2.2. Normal strains.

Consider for example the deformation given by

$$x_1 = (1 + k)X_1$$
$$x_2 = (1 - \nu k)X_2$$
$$x_3 = (1 - \nu k)X_3,$$

where k is a constant and ν is Poisson's ratio.

Compute the deformation gradient $\mathbf{F}$ ($F_{ij} = \frac{\partial x_i}{\partial X_j}$), displacement gradient $\mathbf{H}$ ($H_{ij} = \frac{\partial u_i}{\partial X_j}$), and displacement $\mathbf{u}$ ($u_i = x_i - X_i$):

$$\mathbf{F} = \begin{bmatrix} 1+k & 0 & 0 \\ 0 & 1-\nu k & 0 \\ 0 & 0 & 1-\nu k \end{bmatrix} \quad \mathbf{H} = \begin{bmatrix} k & 0 & 0 \\ 0 & -\nu k & 0 \\ 0 & 0 & -\nu k \end{bmatrix} \quad \mathbf{u} = \begin{Bmatrix} kX_1 \\ -\nu k X_2 \\ -\nu k X_3 \end{Bmatrix}$$

Compute the strain tensors $\mathbf{C}$, $\mathbf{E}$, and ε:

$$\mathbf{C} = \begin{bmatrix} (1+k)^2 & 0 & 0 \\ 0 & (1-\nu k)^2 & 0 \\ 0 & 0 & (1-\nu k)^2 \end{bmatrix}$$

$$\mathbf{E} = \frac{1}{2}\begin{bmatrix} (1+k)^2 - 1 & 0 & 0 \\ 0 & (1-\nu k)^2 - 1 & 0 \\ 0 & 0 & (1-\nu k)^2 - 1 \end{bmatrix}$$

$$\mathbf{E} = \begin{bmatrix} k + \frac{1}{2}k^2 & 0 & 0 \\ 0 & -\nu k + \frac{1}{2}(\nu k)^2 & 0 \\ 0 & 0 & -\nu k + \frac{1}{2}(\nu k)^2 \end{bmatrix}$$

$$\varepsilon = \begin{bmatrix} k & 0 & 0 \\ 0 & -\nu k & 0 \\ 0 & 0 & -\nu k \end{bmatrix}.$$

This kinematic state is associated with a uniaxial normal stress state.

Example 2.3. Shear strain.

Consider for example shear deformation given by

$$x_1 = X_1 + kX_2$$
$$x_2 = X_2$$
$$x_3 = X_3,$$

where k is a constant.

Compute the deformation gradient $\mathbf{F}$ ($F_{ij} = \frac{\partial x_i}{\partial X_j}$), displacement gradient $\mathbf{H}$ ($H_{ij} = \frac{\partial u_i}{\partial X_j}$), and displacement $\mathbf{u}$ ($u_i = x_i - X_i$):

$$\mathbf{F} = \begin{bmatrix} 1 & k & 0 \\ 0 & 1 & 0 \\ 0 & 0 & 1 \end{bmatrix} \quad \mathbf{H} = \begin{bmatrix} 0 & k & 0 \\ 0 & 0 & 0 \\ 0 & 0 & 0 \end{bmatrix} \quad \mathbf{u} = \begin{Bmatrix} kX_2 \\ 0 \\ 0 \end{Bmatrix}$$

Compute the strain tensors $\mathbf{C}$, $\mathbf{E}$, and ε:

$$\mathbf{C} = \begin{bmatrix} 1 & k & 0 \\ k & k^2 + 1 & 0 \\ 0 & 0 & 1 \end{bmatrix}$$

$$\mathbf{E} = \frac{1}{2} \begin{bmatrix} 1-1 & k & 0 \\ k & k^2+1-1 & 0 \\ 0 & 0 & 1-1 \end{bmatrix}$$

$$\mathbf{E} = \frac{1}{2} \begin{bmatrix} 0 & k & 0 \\ k & k^2 & 0 \\ 0 & 0 & 0 \end{bmatrix}$$

$$\varepsilon = \frac{1}{2} \begin{bmatrix} 0 & k & 0 \\ k & 0 & 0 \\ 0 & 0 & 0 \end{bmatrix}.$$

This kinematic state is associated with an antiplane shear stress state. The Lagrangian strain tensor has a normal strain $E_{22} = \frac{1}{2}k^2$, but the infinitesimal strain tensor does not.

2.2.2 Balance laws

It is necessary and sufficient for the satisfaction of the balance of momentum laws that there exists a physical quantity, known as the Cauchy stress σ, such that

$$\nabla \cdot \sigma + \rho \mathbf{b} = \rho \dot{\mathbf{v}}, \quad \sigma_{ij,j} + \rho b_i = \rho \dot{v}_i \tag{2.9}$$

in the interior of the body, where ρ is the mass density, $\mathbf{b}$ is the volumetric body force, and $\mathbf{v}$ is the velocity,

$$\sigma = \sigma^T, \quad \sigma_{ij} = \sigma_{ji} \tag{2.10}$$

and

$$\mathbf{t}^{(n)} = \sigma^T \cdot \mathbf{n}, \quad t_i^{(n)} = \sigma_{ji} n_j \tag{2.11}$$

on the boundary of the body.

2.2.3 Stress

Applied forces, whether surface tractions or body forces, cause internal forces, which can be quantified as force intensities in different ways. Most common is the Cauchy stress tensor, whose components can be interpreted as tractions acting on a cube, as shown in figure 2.2, which is defined in the current (spatial) configuration. Due to its symmetry, the Cauchy stress tensor σ_{ij} is describable by six scalar components, three normal stresses ($i = j$) and three shear stresses ($i \neq j$).

However, when using the Lagrangian formulation and Lagrangian strain, the balance laws are simpler to formulate when stress is defined relative to the reference

$$\boldsymbol{\sigma} = \begin{bmatrix} \sigma_{11} & \sigma_{12} & \sigma_{13} \\ \sigma_{21} & \sigma_{22} & \sigma_{23} \\ \sigma_{31} & \sigma_{32} & \sigma_{33} \end{bmatrix}$$

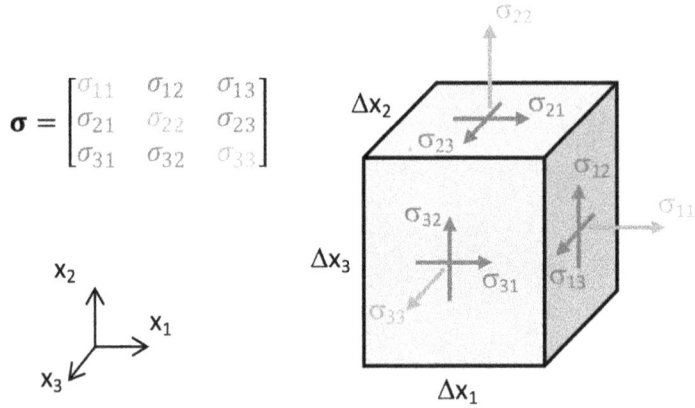

Figure 2.2. Tractions representing Cauchy stress components (hidden faces not shown).

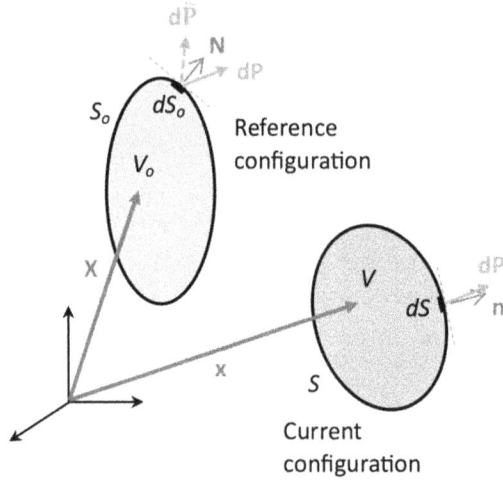

Figure 2.3. Force vectors.

configuration. Therefore, the first Piola–Kirchhoff stress is used. Suppose that the force $d\mathbf{P}$ acts on the surface dS, which was dS_o in the reference configuration as shown in figure 2.3. The force can be written in terms of the Cauchy and first Piola–Kirchhoff stresses as

$$d\mathbf{P} = (\mathbf{n} \cdot \boldsymbol{\sigma})dS \tag{2.12}$$

$$d\mathbf{P} = (\mathbf{N} \cdot \mathbf{T}_{PK1})dS_o, \tag{2.13}$$

where $\mathbf{n}$ and $\mathbf{N}$ are the unit normals to the surface in the current and reference configurations, respectively, at $\mathbf{X}$. $\mathbf{T}_{PK1}$ is a two-point tensor associating the force components in the current configuration with an oriented area in the reference configuration. The second Piola–Kirchhoff stress is different in that it gives a force $d\tilde{\mathbf{P}}$ related to the force $d\mathbf{P}$ in the same way that a material vector $d\mathbf{X}$ at $\mathbf{X}$ is related by the deformation to the corresponding spatial vector $d\mathbf{x}$ at $\mathbf{x}$. In other words, by direct analogy with $d\mathbf{X} = \mathbf{F}^{-1} \cdot d\mathbf{x}$ we have $d\widetilde{\mathbf{P}} = \mathbf{F}^{-1} \cdot d\mathbf{P}$. Therefore,

$$d\widetilde{\mathbf{P}} = (\mathbf{N} \cdot \mathbf{T}_{PK2})dS_o \tag{2.14}$$

$$d\widetilde{\mathbf{P}} = \mathbf{F}^{-1} \cdot (\mathbf{N} \cdot \mathbf{T}_{PK1})dS_o \tag{2.15}$$

$$d\widetilde{\mathbf{P}} = \mathbf{F}^{-1} \cdot (\mathbf{n} \cdot \boldsymbol{\sigma})dS. \tag{2.16}$$

Equations (2.14) and (2.15) premultiplied by $\mathbf{F}$ provide the relationship between $\mathbf{T}_{PK1}$ and $\mathbf{T}_{PK2}$

$$\mathbf{F} \cdot \mathbf{F}^{-1} \cdot (\mathbf{N} \cdot \mathbf{T}_{PK1})dS_o = \mathbf{F} \cdot (\mathbf{N} \cdot \mathbf{T}_{PK2})dS_o$$

$$\mathbf{T}_{PK1} = \mathbf{T}_{PK2} \cdot \mathbf{F}^T. \tag{2.17}$$

To relate the Cauchy stress to the first Piola–Kirchhoff stress we use the kinematic relation for area change between the reference and current configurations

$$\mathbf{n}dS = \det(\mathbf{F})\mathbf{N} \cdot \mathbf{F}^{-1}dS_o. \tag{2.18}$$

Equating the right-hand sides of equations (2.12) and (2.13) and using equation (2.18)

$$\det(\mathbf{F})\mathbf{N} \cdot \mathbf{F}^{-1} \cdot \boldsymbol{\sigma}dS_o = \mathbf{N} \cdot \mathbf{T}_{PK1}dS_o \tag{2.19}$$

$$\mathbf{N} \cdot [\det(\mathbf{F})\mathbf{F}^{-1} \cdot \boldsymbol{\sigma} - \mathbf{T}_{PK1}] = 0$$

$$\mathbf{T}_{PK1} = \det(\mathbf{F})\mathbf{F}^{-1} \cdot \boldsymbol{\sigma}, \tag{2.20}$$

or conversely,

$$\boldsymbol{\sigma} = \frac{1}{\det(\mathbf{F})}\mathbf{F} \cdot \mathbf{T}_{PK1}. \tag{2.21}$$

Although the concept of symmetry is not applicable to $\mathbf{T}_{PK1}$ because it is a two-point tensor, we can assess the symmetry of the second Piola–Kirchhoff stress. Equating equations (2.20) and (2.17) gives

$$\det(\mathbf{F})\mathbf{F}^{-1} \cdot \boldsymbol{\sigma} = \mathbf{T}_{PK2} \cdot \mathbf{F}^T \tag{2.22}$$

or

$$\mathbf{T}_{PK2} = \det(\mathbf{F})\mathbf{F}^{-1} \cdot \boldsymbol{\sigma} \cdot (\mathbf{F}^T)^{-1}. \tag{2.23}$$

Based on the symmetry of $\boldsymbol{\sigma}$, by considering the components of $\mathbf{F}^{-1} \cdot \boldsymbol{\sigma} \cdot (\mathbf{F}^T)^{-1}$ it can be shown that this tensor product is symmetric, and thus $\mathbf{T}_{PK2}$ is symmetric.

We can now revisit the conservation of linear momentum in the reference configuration using $\mathbf{T}_{PK1}$. Let the reference state of a body be defined by its mass density, surface, and volume, ρ_o, S_o and V_o, respectively. The vector sum of the forces (tractions and body force) acting on the body equals the rate of change of the momentum,

$$\int_{S_o} \mathbf{N} \cdot \mathbf{T}_{PK1} dS_o + \int_{V_o} \rho_o \mathbf{b}_o dV_o = \int_{V_o} \rho_o \dot{\mathbf{v}} dV_o, \tag{2.24}$$

where $\mathbf{b}_o$ is the body force and the dot denotes the material time derivative. The divergence theorem can be used to write the surface integral as a volume integral

$$\int_{V_o} \boldsymbol{\nabla}_X \cdot \mathbf{T}_{PK1} dV_o + \int_{V_o} \rho_o \mathbf{b}_o dV_o = \int_{V_o} \rho_o \dot{\mathbf{v}} dV_o. \tag{2.25}$$

Combining the integrals into one and knowing that V_o is arbitrary gives the equation of motion,

$$\boldsymbol{\nabla}_X \cdot \mathbf{T}_{PK1} + \rho_o \mathbf{b}_o = \rho_o \dot{\mathbf{v}}. \tag{2.26}$$

2.2.4 Constitutive relations

We accept the basic assumption of thermodynamics that the local strain energy per unit mass is determined by thermodynamic equilibrium, which is defined by a set of state variables that are in turn paired to conjugate variables through the strain energy to satisfy the equation of state. If the Lagrangian strain is taken to be the state variable, then the second Piola–Kirchhoff stress is the conjugate variable. Likewise, if the deformation gradient is the state variable, then the first Piola–Kirchhoff stress is the conjugate variable. Therefore, with W being the strain energy density per unit volume in the reference configuration,

$$\mathbf{T}_{PK2} = \frac{\partial W}{\partial \mathbf{E}} \tag{2.27}$$

$$\mathbf{T}_{PK1} = \frac{\partial W}{\partial \mathbf{F}}. \tag{2.28}$$

In the latter case, to maintain objectivity requirements, W is an implicit function of $\mathbf{F}$ through the right stretch tensor $\mathbf{U}$, which is related to the right Cauchy–Green strain tensor ($\mathbf{C} = \mathbf{U}^2$).

Hyperelastic material models follow Green's notion that there exists a strain energy function as above. Brugger [8] applied a Taylor series expansion of the strain energy in terms of the Lagrangian strain tensor, which can be written completely as

$$W = W_o + C_{ij}E_{ij} + \frac{1}{2!}C_{ijkl}E_{ij}E_{kl} + \frac{1}{3!}C_{ijklmn}E_{ij}E_{kl}E_{mn}$$
$$+ \frac{1}{4!}C_{ijklmnpq}E_{ij}E_{kl}E_{mn}E_{pq} + \cdots, \tag{2.29}$$

where W_o and C_{ij} represent the strain energy and (residual) stress, respectively, in the reference state. The Brugger elastic coefficients are

$$C_{ijkl} = \frac{\partial W}{\partial E_{ij} \partial E_{kl}}, \; C_{ijklmn} = \frac{\partial W}{\partial E_{ij} \partial E_{kl} \partial E_{mn}}, \; \cdots \tag{2.30}$$

in the reference state for any number of terms. While instead, Huang coefficients are derived from an expansion about the displacement gradient **H**, i.e.

$$W = W_o + A_{ij}H_{ij} + \frac{1}{2!}A_{ijkl}H_{ij}H_{kl} + \frac{1}{3!}A_{ijklmn}H_{ij}H_{kl}H_{mn}$$
$$+ \frac{1}{4!}A_{ijklmnpq}H_{ij}H_{kl}H_{mn}H_{pq} + \cdots. \tag{2.31}$$

A relationship between the Huang and Brugger coefficients is required to constrain the Huang coefficients to satisfy objectivity. It is found by equating terms in equations (2.29) and (2.31) that have the displacement gradient to the same power, where equation (2.7) is used in equation (2.29). The results are

$$A_{ij} = C_{ij} \tag{2.32}$$

$$A_{ijkl} = C_{ijkl} + C_{jl}\delta_{ik} \tag{2.33}$$

$$A_{ijklmn} = C_{ijklmn} + C_{jlmn}\delta_{ik} + C_{ijln}\delta_{km} + C_{jnkl}\delta_{im} \tag{2.34}$$

$$A_{ijklmnpq} = \frac{1}{2}C_{ijklmnpq} + C_{qjklmn}\delta_{ip} + C_{ijlnpq}\delta_{km} + \frac{1}{2}C_{ijklnq}\delta_{mp}$$
$$+ \frac{1}{2}C_{jlmnpq}\delta_{ik} + \frac{1}{2}C_{jlnq}\delta_{ik}\delta_{mp} + C_{jnlq}\delta_{ip}\delta_{km}, \tag{2.35}$$

where equation (2.35) has been taken from Thurston's equation (38.5) [9]. The reader should recognize C_{ij} as the residual stress from equation (2.29). Additional details on the relationships between the Brugger and Huang coefficients are provided by Cantrell's [10] section 6.1.8. Using equations (2.27) and (2.29) and the major symmetries of the stiffness tensors

$$C_{klij} = C_{ijkl}$$

$$C_{klijmn} = C_{mnklij} = C_{ijmnkl} = C_{ijklmn}$$

$$C_{klijmnpq} = C_{mnklijpq} = C_{pqklmnij} = C_{ijmnklpq} = C_{ijpqmnkl} = C_{ijklpqmn} = C_{ijklmnpq}, \tag{2.36}$$

we can write

$$T_{PK2\_ij} = \underbrace{C_{ij} + C_{ijkl}E_{kl}}_{\text{Linear}} + \underbrace{\frac{1}{2!}C_{ijklmn}E_{kl}E_{mn} + \frac{1}{3!}C_{ijklmnpq}E_{kl}E_{mn}E_{pq} + \cdots}_{\text{Nonlinear}} \tag{2.37}$$

from which the residual stress as well as linear and nonlinear stress terms are apparent. Later, we will decompose the stress tensor into linear and nonlinear parts for isotropic materials.

2.2.4.1 Linear elastic

For a linear elastic material, we can replace the Lagrangian strain with the infinitesimal strain, replace the second Piola–Kirchhoff stress with the Cauchy stress, and truncate the strain energy expansion at second order. Assuming that there is no residual stress, the constitutive relation known familiarly as Hooke's law results from equations (2.27) and (2.29):

$$\sigma_{ij} = C_{ijkl}\varepsilon_{kl}. \tag{2.38}$$

The tensor C_{ijkl} is known as the elasticity tensor (rank four). Presuming the material to be homogeneous, these $3^4 = 81$ coefficients are spatially constant. Because the C_{ijkl} are from the quadratic term in the elastic strain energy function the symmetry $C_{ijkl} = C_{klij}$ exists. The symmetry of the strain tensor gives the so-called minor symmetries $C_{ijkl} = C_{ijlk}$ and $C_{ijkl} = C_{jikl}$. Therefore, the number of independent material coefficients is reduced to 21. In addition, the material structure may have symmetries, e.g.:

- Orthotropic—three planes of symmetry give nine independent coefficients.
- Transversely isotropic—two planes of symmetry and one plane of isotropy give five independent coefficients.
- Isotropic—the same properties in all directions give just two independent coefficients.

Image credit: Jessie Lissenden

The symmetries enable the elasticity tensor C_{ijkl} to be represented by a 6×6 symmetric matrix replacing indices: $11 \rightarrow 1$, $22 \rightarrow 2$, $33 \rightarrow 3$, $23 \rightarrow 4$, $31 \rightarrow 5$, $12 \rightarrow 6$. Thus, Hooke's law in contracted form is

$$
\begin{Bmatrix} \sigma_1 = \sigma_{11} \\ \sigma_2 = \sigma_{22} \\ \sigma_3 = \sigma_{33} \\ \sigma_4 = \sigma_{23} \\ \sigma_5 = \sigma_{31} \\ \sigma_6 = \sigma_{12} \end{Bmatrix} = \begin{bmatrix} C_{11} & C_{12} & C_{13} & C_{14} & C_{15} & C_{16} \\ C_{21} & C_{22} & C_{23} & C_{24} & C_{25} & C_{26} \\ C_{31} & C_{32} & C_{33} & C_{34} & C_{35} & C_{36} \\ C_{41} & C_{42} & C_{43} & C_{44} & C_{45} & C_{46} \\ C_{51} & C_{52} & C_{53} & C_{54} & C_{55} & C_{56} \\ C_{61} & C_{62} & C_{63} & C_{64} & C_{65} & C_{66} \end{bmatrix} \begin{Bmatrix} \varepsilon_1 = \varepsilon_{11} \\ \varepsilon_2 = \varepsilon_{22} \\ \varepsilon_3 = \varepsilon_{33} \\ \varepsilon_4 = 2\varepsilon_{23} \\ \varepsilon_5 = 2\varepsilon_{31} \\ \varepsilon_6 = 2\varepsilon_{12} \end{Bmatrix}.
$$

The inverse to equation (2.38) can be written as

$$
\varepsilon_{ij} = S_{ijkl}\sigma_{kl}, \tag{2.39}
$$

where $S_{ijkl} = C_{ijkl}^{-1}$ is the compliance tensor, and it can be contracted in the same way as the elasticity tensor.

Often, the engineering properties known as Young's modulus (E), Poisson's ratio (ν), shear modulus (G), bulk modulus (K), are used to characterize the linear elastic behavior. The compliance matrix can be written in terms of the engineering properties through a superposition of uniaxial stress states to obtain for the special case of an isotropic material

$$
\mathbf{S}_{(6\times6)} = \begin{bmatrix} \dfrac{1}{E} & \dfrac{-\nu}{E} & \dfrac{-\nu}{E} & 0 & 0 & 0 \\ \dfrac{-\nu}{E} & \dfrac{1}{E} & \dfrac{-\nu}{E} & 0 & 0 & 0 \\ \dfrac{-\nu}{E} & \dfrac{-\nu}{E} & \dfrac{1}{E} & 0 & 0 & 0 \\ 0 & 0 & 0 & \dfrac{1}{G} & 0 & 0 \\ 0 & 0 & 0 & 0 & \dfrac{1}{G} & 0 \\ 0 & 0 & 0 & 0 & 0 & \dfrac{1}{G} \end{bmatrix}. \tag{2.40}
$$

Returning to equations (2.27) and (2.29), we truncate the strain energy density function at second order for an isotropic material. The most general isotropic tensor of rank 4 is

$$
G_{ijkl} = \lambda\delta_{ij}\delta_{kl} + \mu\left(\delta_{ik}\delta_{jl} + \delta_{il}\delta_{jk}\right) + \kappa\left(\delta_{ik}\delta_{jl} - \delta_{il}\delta_{jk}\right), \tag{2.41}
$$

where λ, μ and κ are constants. However, when we utilize G_{ijkl} as the elasticity tensor C_{ijkl} the symmetry of the strain tensor eliminates the third term:

$$
C_{ijkl} = \lambda\delta_{ij}\delta_{kl} + \mu\left(\delta_{ik}\delta_{jl} + \delta_{il}\delta_{jk}\right), \tag{2.42}
$$

where λ and μ are recognized as Lame's constants and can also be referred to as the second order elastic constants (SOECs) because they come from the quadratic term in the strain energy function equation (2.29). The contracted form of the elasticity tensor is

$$
\mathbf{C}_{(6\times6)} = \begin{bmatrix} \lambda+\mu & \lambda & \lambda & 0 & 0 & 0 \\ \lambda & \lambda+\mu & \lambda & 0 & 0 & 0 \\ \lambda & \lambda & \lambda+\mu & 0 & 0 & 0 \\ 0 & 0 & 0 & \mu & 0 & 0 \\ 0 & 0 & 0 & 0 & \mu & 0 \\ 0 & 0 & 0 & 0 & 0 & \mu \end{bmatrix}. \tag{2.43}
$$

Finally, for an isotropic material the linear elastic stress–strain law can be written succinctly as

$$\sigma_{ij} = \lambda \varepsilon_{kk} \delta_{ij} + 2\mu \varepsilon_{ij}. \tag{2.44}$$

2.2.4.2 Nonlinear elastic

The quadratic term in the strain energy function equation (2.29) can now be written using equation (2.42) as

$$\frac{1}{2!} C_{ijkl} E_{ij} E_{kl} = \frac{1}{2} \left(\lambda \delta_{ij} \delta_{kl} + \mu \left(\delta_{ik} \delta_{jl} + \delta_{il} \delta_{jk} \right) \right) E_{ij} E_{kl},$$

$$= \frac{\lambda}{2} E_{ii} E_{jj} + \mu E_{ij} E_{ij}$$

$$= \frac{\lambda}{2} (tr(\mathbf{E}))^2 + \mu \, tr(\mathbf{E} \cdot \mathbf{E}) \tag{2.45}$$

for an isotropic material—this is the linear elastic term with regard to the stress–strain relation, but it includes the geometric nonlinearity and is known as the Saint Venant–Kirchhoff hyperelastic model. The strain invariants ($I_1 = tr(\mathbf{E})$, $I_2 = tr(\mathbf{E} \cdot \mathbf{E})$, and $I_3 = tr(\mathbf{E} \cdot \mathbf{E} \cdot \mathbf{E})$), could also be used here. For an isotropic material the cubic term containing C_{ijklmn} can be written in terms of fifteen linearly independent combinations of Kronecker deltas (see Kearsley and Fong [11]). Landau and Lifshitz [12] chose to write it as

$$C_{ijklmn} = 2\mathscr{C} \delta_{ij} \delta_{kl} \delta_{mn} + 2\mathscr{B} \left(\delta_{ij} \delta_{lm} \delta_{kn} + \delta_{im} \delta_{jn} \delta_{kl} + \delta_{ik} \delta_{jl} \delta_{mn} \right)$$
$$+ \mathscr{A} \left(\delta_{il} \delta_{jm} \delta_{kn} + \delta_{im} \delta_{jk} \delta_{ln} \right), \tag{2.46}$$

where $\mathscr{A}$, $\mathscr{B}$, and $\mathscr{C}$ are known as the Landau–Lifshitz third order elastic constants (TOECs). The cubic term in the strain energy function equation (2.29) can now be written as

$$\frac{1}{3!} C_{ijklmn} E_{ij} E_{kl} E_{mn} = \frac{2\mathscr{C}}{6} E_{ii} E_{kk} E_{mm}$$
$$+ \frac{2\mathscr{B}}{6} \left(E_{ii} E_{km} E_{mk} + E_{kk} E_{mj} E_{jm} + E_{mm} E_{il} E_{li} \right)$$
$$+ \frac{\mathscr{A}}{6} \left(E_{ij} E_{jn} E_{nl} + E_{mj} E_{jl} E_{lm} \right) \tag{2.47}$$
$$= \frac{\mathscr{C}}{3} (tr(\mathbf{E}))^3 + \mathscr{B} tr(\mathbf{E}) tr(\mathbf{E} \cdot \mathbf{E}) + \frac{\mathscr{A}}{3} tr(\mathbf{E} \cdot \mathbf{E} \cdot \mathbf{E}).$$

Norris [13] tabulates forms of the strain energy function proposed by a number of researchers, including the Murnaghan constants l, m, and n:

$$l = \mathscr{B} + \mathscr{C}, \, m = \frac{1}{2}\mathscr{A} + \mathscr{B}, \, n = \mathscr{A}. \tag{2.48}$$

Extending the strain energy function to fourth order for an isotropic material follows Bland [14], Hamilton *et al* [15], and Liu *et al* [16]:

$$\frac{1}{4!}C_{ijklmnpq}E_{ij}E_{kl}E_{mn}E_{pq} = \mathscr{E}\mathrm{tr}(\mathbf{E})\mathrm{tr}(\mathbf{E} \cdot \mathbf{E} \cdot \mathbf{E}) + \mathscr{F}(\mathrm{tr}(\mathbf{E}))^2\mathrm{tr}(\mathbf{E} \cdot \mathbf{E})$$
$$+ \mathscr{G}(\mathrm{tr}(\mathbf{E} \cdot \mathbf{E}))^2 + \mathscr{H}(\mathrm{tr}(\mathbf{E}))^4, \tag{2.49}$$

where $\mathscr{E}$, $\mathscr{F}$, $\mathscr{G}$, and $\mathscr{H}$ are the fourth order elastic constants (FOECs). Thus, from equations (2.29), (2.45), (2.47), and (2.49) the strain energy density function for an isotropic solid without residual stress can be written as

$$W = \frac{\lambda}{2}(\mathrm{tr}(\mathbf{E}))^2 + \mu\mathrm{tr}(\mathbf{E} \cdot \mathbf{E})$$
$$+ \frac{\mathscr{A}}{3}\mathrm{tr}(\mathbf{E} \cdot \mathbf{E} \cdot \mathbf{E}) + \mathscr{B}\mathrm{tr}(\mathbf{E})\mathrm{tr}(\mathbf{E} \cdot \mathbf{E}) + \frac{\mathscr{C}}{3}(\mathrm{tr}(\mathbf{E}))^3 \tag{2.50}$$
$$+ \mathscr{E}\mathrm{tr}(\mathbf{E})\mathrm{tr}(\mathbf{E} \cdot \mathbf{E} \cdot \mathbf{E}) + \mathscr{F}(\mathrm{tr}(\mathbf{E}))^2\mathrm{tr}(\mathbf{E} \cdot \mathbf{E}) + \mathscr{G}(\mathrm{tr}(\mathbf{E} \cdot \mathbf{E}))^2 + \mathscr{H}(\mathrm{tr}(\mathbf{E}))^4$$
$$+ \mathcal{O}((\mathbf{E})^5).$$

The components of the second Piola–Kirchhoff stress tensor are then given by (2.27):

$$\begin{aligned}\mathbf{T}_{PK2} = {}&\lambda\mathrm{tr}(\mathbf{E})\mathbf{I} + 2\mu\mathbf{E}\\ &+ \mathscr{A}\mathbf{E} \cdot \mathbf{E} + \mathscr{B}\mathrm{tr}(\mathbf{E} \cdot \mathbf{E})\mathbf{I} + 2\mathscr{B}\mathrm{tr}(\mathbf{E})\mathbf{E} + \mathscr{C}(\mathrm{tr}(\mathbf{E}))^2\mathbf{I}\\ &+ \mathscr{E}\mathrm{tr}(\mathbf{E} \cdot \mathbf{E} \cdot \mathbf{E})\mathbf{I} + 3\mathscr{E}\mathrm{tr}(\mathbf{E})\mathbf{E} \cdot \mathbf{E} + 2\mathscr{F}\mathrm{tr}(\mathbf{E})\mathrm{tr}(\mathbf{E} \cdot \mathbf{E})\mathbf{I}\\ &+ 2\mathscr{F}(\mathrm{tr}(\mathbf{E}))^2\mathbf{E} + 4\mathscr{G}\mathrm{tr}(\mathbf{E} \cdot \mathbf{E})\mathbf{E} + 4\mathscr{H}(\mathrm{tr}(\mathbf{E}))^3\mathbf{I} + \mathcal{O}((\mathbf{E})^4),\end{aligned} \tag{2.51}$$

where on the right-hand side the terms in the first row are linear in strain and those in the second, third, and fourth rows are nonlinear in strain. Terms in the second row are quadratic, while those in the third and fourth rows are cubic. The nonlinear terms are written in terms of the displacement gradient in example 2.4. We conclude this section with example 2.5 presenting the strain energy function for a transversely isotropic material.

Example 2.4. Stress components decomposed into linear and nonlinear parts.

We will write out expressions for stress in terms of the displacement gradient. Motivated by a desire to separate linear and nonlinear terms when working with the equation of motion, we will decompose the stress into linear and nonlinear terms. Furthermore, as we use the first Piola–Kirchhoff stress in the equation of motion equation (2.26) we will convert equation (2.51) for the second Piola–Kirchhoff stress $\mathbf{T}_{PK2}$ into the first Piola–Kirchhoff stress $\mathbf{T}_{PK1}$:

$$\mathbf{T}_{PK2} = \mathbf{T}_{PK2}^L + \mathbf{T}_{PK2}^Q + \mathbf{T}_{PK2}^C + \mathcal{O}(\mathbf{E}^4),$$

where the superscripts L, Q, and C denote linear, quadratic, and cubic terms in strain. Use equations (2.17) and (2.2) to convert to

$$\mathbf{T}_{PK1} = (\mathbf{I} + \mathbf{H})\left[\mathbf{T}^L_{PK2} + \mathbf{T}^Q_{PK2} + \mathbf{T}^C_{PK2} + \mathcal{O}(\mathbf{E}^4)\right]$$

which decomposes to

$$\mathbf{T}^L_{PK1} = \mathbf{T}^L_{PK2}$$

$$\mathbf{T}^Q_{PK1} = \mathbf{H}\mathbf{T}^L_{PK2} + \mathbf{T}^Q_{PK2}$$

$$\mathbf{T}^C_{PK1} = \mathbf{H}\mathbf{T}^Q_{PK2} + \mathbf{T}^C_{PK2}.$$

As noted with respect to equation (2.2), relationships between $\mathbf{T}_{PK1}$ and $\mathbf{T}_{PK2}$ are valid within the Euclidean context used here. It would be equivalent to use equation (2.28) directly, knowing that $\mathbf{H} = \mathbf{F} - \mathbf{I}$.

Substituting equation (2.7) into equation (2.51) gives

$$\mathbf{T}^L_{PK1} = \frac{\lambda}{2}\mathrm{tr}\left(\mathbf{H} + \mathbf{H}^T\right)\mathbf{I} + \mu\left(\mathbf{H} + \mathbf{H}^T\right) = \lambda\mathrm{tr}(\mathbf{H})\mathbf{I} + \mu\left(\mathbf{H} + \mathbf{H}^T\right)$$

$$\mathbf{T}^Q_{PK1} = \frac{\lambda}{2}\mathrm{tr}\left(\mathbf{H} + \mathbf{H}^T\right)\mathbf{H} + \frac{\lambda}{2}\mathrm{tr}\left(\mathbf{H}^T\mathbf{H}\right)\mathbf{I} + \mu\left(\mathbf{H}\mathbf{H} + \mathbf{H}\mathbf{H}^T + \mathbf{H}^T\mathbf{H}\right)$$
$$+ \frac{\mathscr{A}}{4}\left(\mathbf{H}\mathbf{H} + \mathbf{H}\mathbf{H}^T + \mathbf{H}^T\mathbf{H} + \mathbf{H}^T\mathbf{H}^T\right)$$
$$+ \frac{\mathscr{B}}{4}\mathrm{tr}\left(\mathbf{H}\mathbf{H} + \mathbf{H}\mathbf{H}^T + \mathbf{H}^T\mathbf{H} + \mathbf{H}^T\mathbf{H}^T\right)\mathbf{I} + \mathscr{B}\mathrm{tr}(\mathbf{H})\left(\mathbf{H} + \mathbf{H}^T\right)$$
$$+ \mathscr{C}\left(\mathrm{tr}(\mathbf{H})\right)^2\mathbf{I}$$

$$\mathbf{T}^C_{PK1} = \frac{\lambda}{2}\mathrm{tr}(\mathbf{H}^T\mathbf{H})\mathbf{H} + \mu\mathbf{H}\mathbf{H}^T\mathbf{H}$$
$$+ \frac{\mathscr{A}}{4}\mathbf{H}\left(\mathbf{H}\mathbf{H} + \mathbf{H}\mathbf{H}^T + \mathbf{H}^T\mathbf{H} + \mathbf{H}^T\mathbf{H}^T\right)$$
$$+ \frac{\mathscr{A}}{4}\left(\mathbf{H}\mathbf{H}^T\mathbf{H} + \mathbf{H}^T\mathbf{H}^T\mathbf{H} + \mathbf{H}^T\mathbf{H}\mathbf{H} + \mathbf{H}^T\mathbf{H}\mathbf{H}^T\right)$$
$$+ \frac{\mathscr{B}}{4}\mathrm{tr}\left(\mathbf{H}\mathbf{H} + \mathbf{H}\mathbf{H}^T + \mathbf{H}^T\mathbf{H} + \mathbf{H}^T\mathbf{H}^T\right)\mathbf{H} + \mathscr{B}\mathrm{tr}(\mathbf{H})\left(\mathbf{H}\mathbf{H} + \mathbf{H}\mathbf{H}^T + \mathbf{H}^T\mathbf{H}\right)$$
$$+ \frac{\mathscr{B}}{2}\mathrm{tr}\left(\mathbf{H}^T\mathbf{H}\right)\left(\mathbf{H} + \mathbf{H}^T\right)$$
$$+ \frac{\mathscr{B}}{4}\mathrm{tr}\left(\mathbf{H}\mathbf{H}^T\mathbf{H} + \mathbf{H}^T\mathbf{H}\mathbf{H} + \mathbf{H}^T\mathbf{H}^T\mathbf{H} + \mathbf{H}^T\mathbf{H}\mathbf{H}^T\right)\mathbf{I}$$
$$+ \mathscr{C}\left(\mathrm{tr}(\mathbf{H})\right)^2\mathbf{H} + \mathscr{C}\mathrm{tr}(\mathbf{H})\mathrm{tr}\left(\mathbf{H}^T\mathbf{H}\right)\mathbf{I}$$
$$+ \frac{1}{8}\mathscr{E}\mathrm{tr}\begin{pmatrix} \mathbf{H}\mathbf{H}\mathbf{H} + \mathbf{H}\mathbf{H}\mathbf{H}^T + \mathbf{H}\mathbf{H}^T\mathbf{H} + \mathbf{H}\mathbf{H}^T\mathbf{H}^T \\ + \mathbf{H}^T\mathbf{H}\mathbf{H} + \mathbf{H}^T\mathbf{H}\mathbf{H}^T + \mathbf{H}^T\mathbf{H}^T\mathbf{H} + \mathbf{H}^T\mathbf{H}^T\mathbf{H}^T \end{pmatrix}\mathbf{I}$$
$$+ \frac{3}{4}\mathscr{E}\mathrm{tr}(\mathbf{H})\left(\mathbf{H}\mathbf{H} + \mathbf{H}\mathbf{H}^T + \mathbf{H}^T\mathbf{H} + \mathbf{H}^T\mathbf{H}^T\right)$$
$$+ \frac{1}{2}\mathscr{F}\mathrm{tr}(\mathbf{H})\mathrm{tr}\left(\mathbf{H}\mathbf{H} + \mathbf{H}\mathbf{H}^T + \mathbf{H}^T\mathbf{H} + \mathbf{H}^T\mathbf{H}^T\right)\mathbf{I} + \mathscr{F}\left(\mathrm{tr}(\mathbf{H})\right)^2\left(\mathbf{H} + \mathbf{H}^T\right)$$
$$+ \frac{1}{2}\mathscr{G}\mathrm{tr}\left(\mathbf{H}\mathbf{H} + \mathbf{H}\mathbf{H}^T + \mathbf{H}^T\mathbf{H} + \mathbf{H}^T\mathbf{H}^T\right)\left(\mathbf{H} + \mathbf{H}^T\right) + 4\mathscr{H}\left(\mathrm{tr}(\mathbf{H})\right)^3\mathbf{I},$$

where now the superscripts L, Q, and C denote linear, quadratic, and cubic terms in the displacement gradient. These are essentially equations (2.8)–(2.10) in Liu *et al* [16] with a couple of corrections. The equation for the quadratic nonlinearity in $\mathbf{T}^{Q}_{PK1}$ is structured such that the first three terms (in orange) provide the geometric nonlinearity from the quadratic strain–displacement relation. There are no TOECs ($\mathscr{A}$, $\mathscr{B}$, $\mathscr{C}$) in these terms; they only appear in the remaining four terms (in green), which provide the material nonlinearity. The cubic nonlinearity term also contains both geometric and material non-linearities (terms with $\mathscr{A}, \mathscr{B}, \mathscr{C}, \mathscr{E}, \mathscr{F}, \mathscr{G}, \mathscr{H}$). However, all higher orders of nonlinearity (e.g. fourth) are exclusively due to material nonlinearity.

Example 2.5. Strain energy function for transversely isotropic material.

Unidirectionally reinforced composites are often modeled as isotropic within the plane transverse to the fiber direction. In this case, whether absolutely true or an idealization, the material is transversely isotropic. Let us define the stress–strain relations for a transversely isotropic material by selecting a strain energy density function following Spencer [17]. Define the fiber direction by the unit vector **a**. For an isotropic material the strain energy density function depends only on the three strain invariants, $\text{tr}(\mathbf{E})$, $\text{tr}(\mathbf{E} \cdot \mathbf{E})$, $\text{tr}(\mathbf{E} \cdot \mathbf{E} \cdot \mathbf{E})$, as denoted in equations (2.45) and (2.47). All anisotropic materials must also exhibit dependence on direction. Thus, for a transversely isotropic material, we can write

$$W = W(\mathbf{E}, \mathbf{a}) = W(\text{tr}(\mathbf{E}), \text{tr}(\mathbf{E}^2), \text{tr}(\mathbf{E}^3), \mathbf{a} \cdot \mathbf{Ea}, \mathbf{a} \cdot \mathbf{E}^2\mathbf{a}).$$

Expanding W to third order,

$$
\begin{aligned}
W = {} & \alpha_1[\text{tr}(\mathbf{E})]^2 + \alpha_2[\text{tr}(\mathbf{E})][\mathbf{a} \cdot \mathbf{Ea}] + \alpha_3\text{tr}(\mathbf{E}^2) + \alpha_4[\mathbf{a} \cdot \mathbf{Ea}]^2 \\
& + \alpha_5[\mathbf{a} \cdot \mathbf{E}^2\mathbf{a}] + \beta_1[\text{tr}(\mathbf{E})]^3 + \beta_2\text{tr}(\mathbf{E})\text{tr}(\mathbf{E}^2) + \beta_3\text{tr}(\mathbf{E})[\mathbf{a} \cdot \mathbf{Ea}]^2 \\
& + \beta_4\text{tr}(\mathbf{E})[\mathbf{a} \cdot \mathbf{E}^2\mathbf{a}] + \beta_5[\text{tr}(\mathbf{E})]^2[\mathbf{a} \cdot \mathbf{Ea}] + \beta_6\text{tr}(\mathbf{E}^2)[\mathbf{a} \cdot \mathbf{Ea}] + \beta_7\text{tr}(\mathbf{E}^3) \\
& + \beta_8[\mathbf{a} \cdot \mathbf{Ea}]^3 + \beta_9[\mathbf{a} \cdot \mathbf{Ea}][\mathbf{a} \cdot \mathbf{E}^2\mathbf{a}] + \mathcal{O}(\mathbf{E}^4),
\end{aligned}
$$

where α_i $(i = 1, 2, \ldots, 5)$ are the coefficients of the quadratic terms and correspond to linear elastic material behavior and β_i $(i = 1, 2, \ldots, 9)$ are the coefficients of the cubic terms that correspond to quadratic nonlinearity in the stress–strain response. The strain energy density function for an isotropic material can be recovered by choosing the coefficients to be

$$\alpha_1 = \frac{\lambda}{2}, \alpha_2 = 0, \alpha_3 = \mu, \alpha_4 = 0, \alpha_5 = 0,$$

$$\beta_1 = \frac{\mathscr{C}}{3}, \beta_2 = \mathscr{B}, \beta_7 = \frac{\mathscr{A}}{3},$$

$$\beta_3 = \beta_4 = \beta_5 = \beta_6 = \beta_8 = \beta_9 = 0.$$

Applying equation (2.27) enables writing the stress–strain relation as

$$\begin{aligned}
\mathbf{T}_{PK2} = {} & 2\alpha_1(\mathrm{tr}\mathbf{E})\mathbf{I} + \alpha_2(\mathbf{a}\cdot\mathbf{Ea})\mathbf{I} + \alpha_2(\mathrm{tr}\mathbf{E})(\mathbf{a}\otimes\mathbf{a}) + 2\alpha_3\mathbf{E} \\
& + 2\alpha_4(\mathbf{a}\cdot\mathbf{Ea})(\mathbf{a}\otimes\mathbf{a}) + \alpha_5(\mathbf{a}\otimes\mathbf{Ea} + \mathbf{aE}\otimes\mathbf{a}) \\
& + 3\beta_1(\mathrm{tr}\mathbf{E})^2\mathbf{I} + \beta_2\mathrm{tr}(\mathbf{E}^2)\mathbf{I} + 2\beta_2(\mathrm{tr}\mathbf{E})\mathbf{E} + \beta_3(\mathbf{a}\cdot\mathbf{Ea})^2\mathbf{I} \\
& + 2\beta_3(\mathrm{tr}\mathbf{E})(\mathbf{a}\cdot\mathbf{Ea})(\mathbf{a}\otimes\mathbf{a}) + \beta_4(\mathbf{a}\cdot\mathbf{E}^2\mathbf{a})\mathbf{I} + \beta_4(\mathrm{tr}\mathbf{E})(\mathbf{a}\otimes\mathbf{Ea} + \mathbf{aE}\otimes\mathbf{a}) \\
& + 2\beta_5(\mathrm{tr}\mathbf{E})(\mathbf{a}\cdot\mathbf{Ea})\mathbf{I} + \beta_5(\mathrm{tr}\mathbf{E})^2(\mathbf{a}\otimes\mathbf{a}) \\
& + 2\beta_6\mathbf{E}(\mathbf{a}\cdot\mathbf{Ea}) + \beta_6(\mathrm{tr}\mathbf{E}^2)(\mathbf{a}\otimes\mathbf{a}) + 3\beta_7\mathbf{E}^2 + 3\beta_8(\mathbf{a}\cdot\mathbf{Ea})^2(\mathbf{a}\otimes\mathbf{a}) \\
& + \beta_9(\mathbf{a}\cdot\mathbf{E}^2\mathbf{a})(\mathbf{a}\otimes\mathbf{a}) + \beta_9(\mathbf{a}\cdot\mathbf{Ea})(\mathbf{a}\otimes\mathbf{Ea} + \mathbf{aE}\otimes\mathbf{a}),
\end{aligned}$$

where the operation $\mathbf{a}\otimes\mathbf{a}$ is the dyadic product whose components are a_ia_j. The second Piola–Kirchhoff stress can be transformed to the first Piola–Kirchhoff stress using equation (2.17). The strain–displacement relation, equation (2.7), can be used to write the stress in terms of the displacement gradient and then decomposed into linear and nonlinear parts. The rather lengthy result is given by Zhao *et al* [18] in their appendix A.

2.3 Elastodynamics

Above, the equation of motion (equation (2.26)) in the reference state was derived from the conservation of linear momentum in terms of the first Piola–Kirchhoff stress. Below, the equation of motion is applied to wave propagation and then attenuation is described.

2.3.1 Wave equation

Since the material's constitutive behavior has not been considered in deriving the equations of motion (equations (2.9) and (2.26)), they include both displacement and stress terms. We will now impose Hooke's law for a linear elastic material to rewrite equation (2.9) to have displacement as the primary variable. We will write the linear wave equation for an infinite domain as a precursor to solving the nonlinear wave equation in an infinite domain in chapter 3, followed by linear guided waves in chapter 4, and then nonlinear guided waves (later in chapter 4). Guided waves require us to impose boundary conditions, which is obviously not necessary in an infinite domain.

Begin with the equation of motion, linear stress–strain, and infinitesimal strain–displacement relations, which are recalled to be

$$\nabla\cdot\boldsymbol{\sigma} + \rho\mathbf{b} = \rho\dot{\mathbf{v}} \quad \text{or} \quad \sigma_{ij,j} + \rho b_i = \rho\dot{v}_i$$

$$\boldsymbol{\sigma} = \mathbf{C}\boldsymbol{\varepsilon} \quad \text{or} \quad \sigma_{ij} = C_{ijkl}\varepsilon_{kl}$$

$$\boldsymbol{\varepsilon} = \frac{1}{2}[\nabla\mathbf{u} + (\nabla\mathbf{u})^T] \quad \text{or} \quad \varepsilon_{ij} = \frac{1}{2}\left[u_{i,j} + u_{j,i}\right],$$

where we also recall that ρ is the mass density, $\mathbf{u}$ is the displacement vector, and $\boldsymbol{\sigma}$, $\boldsymbol{\varepsilon}$, $\mathbf{C}$, and $\mathbf{b}$ are the Cauchy stress, infinitesimal strain, elasticity, and body force, respectively. Substituting equation (2.8) into equation (2.38) and then into equation (2.9) and invoking the symmetry properties of the elasticity tensor, i.e. $C_{ijkl} = C_{klij} = C_{jikl} = C_{ijlk}$, we obtain the equation of motion,

$$\nabla \cdot \mathbf{C} : \nabla \mathbf{u} + \rho \mathbf{b} = \rho \dot{\mathbf{v}} \quad \text{or} \quad C_{ijkl} u_{l,\,kj} + \rho b_i = \rho \dot{v}_i \tag{2.52}$$

provided the material is homogeneous. An isotropic material has the elasticity tensor given by equation (2.42). Substituting equation (2.42) into equation (2.52) leads to Navier's equation of motion:

$$\mu \nabla^2 \mathbf{u} + [\lambda + \mu] \nabla [\nabla \cdot \mathbf{u}] + \rho \mathbf{b} = \rho \dot{\mathbf{v}} \text{ or } \mu u_{i,\,jj} + [\lambda + \mu] u_{j,\,ij} + \rho b_i = \rho \dot{v}_i, \tag{2.53}$$

which is a hyperbolic partial differential equation and must be satisfied for any motion.

Consider harmonic wave propagation of the form

$$u_i(\mathbf{x}, t) = A p_i \mathrm{e}^{\mathrm{i}[k_j x_j - \omega t]} \tag{2.54}$$

for wavevector k_j and angular frequency ω, where A is the amplitude, p_i is the polarization unit vector, and $\mathrm{i} = \sqrt{-1}$. The spatial and temporal derivatives that we need to substitute into the equation of motion (equation (2.52)) are

$$\begin{aligned} \ddot{u}_i &= -\omega^2 u_i \\ u_{l,\,kj} &= -k_j k_k u_l. \end{aligned} \tag{2.55}$$

Neglecting the body force and substituting equation (2.55) into equation (2.52) gives the well-known Christoffel equation

$$\left[C_{ijkl} k_j k_k - \rho \omega^2 \delta_{il} \right] p_l = 0. \tag{2.56}$$

Defining the direction of the wavevector $\mathbf{k}$ to be $\mathbf{d}$ (i.e. $\mathbf{k} = k\mathbf{d}$), and using the fundamental relationship for wavenumber, $k = \frac{\omega}{c}$, where c is the wave speed, we rewrite equation (2.56) as

$$[\Gamma_{il} - \rho c^2 \delta_{il}] p_l = 0, \tag{2.57}$$

where $\Gamma_{il} = C_{ijkl} d_j d_k$ is known as the acoustic tensor. This is an eigen-problem. The eigenvalues (ρc^2) of equation (2.57) are determined from the characteristic equation and then the eigenvectors (polarization direction p_l) can be found for each wave speed.

Solving the Christoffel equation (2.56) for an isotropic material given by equation (2.42) indicates that the longitudinal (dilatational) and shear (transverse) bulk waves are decoupled as one of the three roots is for the longitudinal waves and the other two (repeated) roots are for the shear waves. The next section uses the Helmholtz decomposition to arrive at the same result for isotropic materials.

2.3.2 Wave equation for isotropic materials

The Helmholtz decomposition provides a convenient tool for separating and solving the wave equations in isotropic materials using potential functions. Letting

$$\mathbf{u} = \nabla \phi + \nabla \times \boldsymbol{\psi}, \tag{2.58}$$

where ϕ is a scalar function and ψ is a vector function, decomposes the displacement field into dilatational and distortional components. Substituting into the equation of motion (equation (2.53)) and neglecting the body force can be shown to give (you can work it out on your own)

$$\nabla[[\lambda + 2\mu]\nabla^2\phi - \rho\ddot{\phi}] + \nabla \times [\mu\nabla^2\psi - \rho\ddot{\psi}] = 0 \qquad (2.59)$$

and since ϕ and ψ are independent we must have

$$c_L^2\nabla^2\phi = \ddot{\phi}, c_T^2\nabla^2\psi = \ddot{\psi}, \qquad (2.60)$$

where the longitudinal (dilatational) and shear (distortional) wave speeds have been defined by

$$c_L^2 = \frac{\lambda + 2\mu}{\rho}, c_T^2 = \frac{\mu}{\rho}. \qquad (2.61)$$

Alternatively, as in Thurston [19], Navier's equation of motion, (2.53), neglecting the body force,

$$\mu\nabla^2\mathbf{u} + [\lambda + \mu]\nabla[\nabla \cdot \mathbf{u}] = \rho\ddot{\mathbf{u}} \text{ or } \mu u_{i,jj} + [\lambda + \mu]u_{j,ij} = \rho\ddot{u}_i,$$

can be directly decoupled into equations for longitudinal waves and equations for shear waves. The identities

$$\nabla \times \nabla \times \mathbf{u} = \nabla[\nabla \cdot \mathbf{u}] - \nabla^2\mathbf{u}$$
$$\nabla \cdot \nabla \times \mathbf{u} = 0 \qquad (2.62)$$

are used to simplify the equations. Substitute the first identity into Navier's equation,

$$[\lambda + 2\mu]\nabla[\nabla \cdot \mathbf{u}] - \mu\nabla \times \nabla \times \mathbf{u} = \rho\ddot{\mathbf{u}} \qquad (2.63)$$

then take the divergence, and let $\Delta \equiv \nabla \cdot \mathbf{u}$ to obtain the longitudinal wave equation

$$[\lambda + 2\mu]\nabla^2\Delta = \rho\ddot{\Delta} \qquad (2.64)$$

$$c_L^2\nabla^2\Delta = \ddot{\Delta}, \qquad (2.65)$$

which is the same as the first of equation (2.60). Now take the curl of equation (2.63)

$$[\lambda + 2\mu]\nabla \times [\nabla[\nabla \cdot \mathbf{u}]] - \mu\nabla \times [\nabla \times \nabla \times \mathbf{u}] = \rho\nabla \times \ddot{\mathbf{u}} \qquad (2.66)$$

and let $\omega \equiv \nabla \times \mathbf{u}$ to obtain the shear wave equation (after some simplification)

$$\mu\nabla^2\omega = \rho\ddot{\omega} \qquad (2.67)$$

$$c_T^2\nabla^2\omega = \ddot{\omega}, \qquad (2.68)$$

which is the same as the second of equation (2.60). In conclusion, the longitudinal and shear wave speeds have been found in terms of linear elastic properties and the mass density.

2.3.3 Attenuation

Only in lossless materials do elastic waves propagate without amplitude changes, with solitons being the exceptional case of no amplitude change in a lossy material. A decreasing amplitude characteristic of attenuation can be caused by absorption (internal damping), scattering from microstructural features (e.g. grain boundaries and inclusions), or diffraction (beam spreading). For the simple case of exponential decay, we can rewrite equation (2.54) as

$$u_i(\mathbf{x}, t) = A(\mathbf{x})p_i \mathrm{e}^{\mathrm{i}[k_j x_j - \omega t]} \tag{2.69}$$

and for propagation in the x_1 direction

$$A(\mathbf{x}) = A_0 \mathrm{e}^{-\alpha(\omega)x_1}, \tag{2.70}$$

where A_0 is the initial amplitude at $x_1 = 0$ and $\alpha(\omega)$ is a frequency-dependent attenuation coefficient. Or we could write

$$u_i(\mathbf{x}, t) = A_0 p_i \mathrm{e}^{\mathrm{i}[(k_1 + \alpha \mathrm{i})x_1 - \omega t]}, \tag{2.71}$$

enabling us to think of the effective wavenumber as being complex-valued, $k_1 + \alpha \mathrm{i}$. Attenuation coefficient units are nepers/m, although they are often reported in terms of decibels/m (1 Np = 8.69 dB).

Interested readers are referred to Achenbach's [20] section 10.5 for viscoelastic materials having internal damping. The complex modulus is used in terms of the creep function to relate the out-of-phase stresses and strains. Destrade *et al* [21] extend application to nonlinear waves. Attenuation from wave scattering is discussed in detail by Ensminger and Bond [22] in their section 2.10.

2.4 Closure

The background necessary for modeling nonlinear ultrasonic waves in hyperelastic materials has been summarized. The defined stress, strain, and displacement gradient tensors will be used extensively throughout the remainder of this book.

References

[1] Malvern L E 1977 *Introduction to the Mechanics of a Continuous Medium* 1st edn (Englewood Cliffs, NJ: Pearson)

[2] Fung Y C and Tong P 2001 *Classical and Computational Solid Mechanics* vol 1 (Advanced Series in Engineering Science) (Singapore: World Scientific)

[3] Liu Y, Khajeh E, Lissenden C J and Rose J L 2013 Interaction of torsional and longitudinal guided waves in weakly nonlinear circular cylinders *J. Acoust. Soc. Am.* **133** 2541–53

[4] Liu Y, Lissenden C J and Rose J L 2014 Higher order interaction of elastic waves in weakly nonlinear hollow circular cylinders. I. Analytical foundation *J. Appl. Phys.* **115** 214901

[5] Liu Y, Khajeh E, Lissenden C J and Rose J L 2014 Higher order interaction of elastic waves in weakly nonlinear hollow circular cylinders. II. Physical interpretation and numerical results *J. Appl. Phys.* **115** 214902

[6] Spencer A J M 2012 *Continuum Mechanics* (Chelmsford, MA: Courier Corporation)

[7] Gurtin M 1981 *An Introduction to Continuum Mechanics* (New York: Academic)

[8] Brugger K 1964 Thermodynamic definition of higher order elastic coefficients *Phys. Rev.* A **133** 1611–2

[9] Thurston R N 1984 Waves in solids *Mechanics of Solids* (Berlin: Springer) pp 109–308

[10] Cantrell J H 2004 Fundamentals and applications of nonlinear ultrasonic nondestructive evaluation *Ultrasonic Nondestructive Evaluation: Engineering and Biological Material Characterization* ed T Kundu (Boca Raton, FL: CRC Press) pp 363–433

[11] Kearsley E A and Fong J T 1975 Linearly independent sets of isotropic Cartesian tensors of ranks up to eight *J. Res. Natl Bur. Stand.* B **79** 49

[12] Landau L D and Lifshitz E M 1959 *Theory of Elasticity* (Oxford: Pergamon)

[13] Norris A 1998 Finite amplitude waves in solids *Nonlinear Acoustics* (New York: Academic) pp 263–77

[14] Bland D R 1969 *Nonlinear Dynamic Elasticity* (Waltham, MA: Blaisdell)

[15] Hamilton M F, Ilinskii Y A and Zabolotskaya E A 2004 Separation of compressibility and shear deformation in the elastic energy density (L) *J. Acoust. Soc. Am.* **116** 41–4

[16] Liu Y, Chillara V K, Lissenden C J and Rose J L 2013 Third harmonic shear horizontal and Rayleigh Lamb waves in weakly nonlinear plates *J. Appl. Phys.* **114** 114908

[17] Spencer A J M 1972 *Deformations of Fibre-Reinforced Materials* (Oxford: Clarendon)

[18] Zhao J, Chillara V K, Ren B, Cho H, Qiu J and Lissenden C J 2016 Second harmonic generation in composites: theoretical and numerical analyses *J. Appl. Phys.* **119** 064902

[19] Thurston R N 1992 Elastic waves in rods and optical fibers *J. Sound Vib.* **159** 441–67

[20] Achenbach J D 1975 *Wave Propagation in Elastic Solids* (Amsterdam: North-Holland)

[21] Destrade M, Saccomandi G and Vianello M 2013 Proper formulation of viscous dissipation for nonlinear waves in solids *J. Acoust. Soc. Am.* **133** 1255–9

[22] Ensminger D and Bond L J 2012 *Ultrasonics Fundamentals, Technologies, and Applications* 3rd edn (Boca Raton, FL: CRC Press)

IOP Publishing

Nonlinear Ultrasonic Guided Waves

Cliff J Lissenden

Chapter 3

Nonlinear elastic waves

The equation of motion containing both material and geometric nonlinearity is solved for planar waves with longitudinal or shear polarizations in an infinite spatial domain. These solutions—with no boundary conditions—provide a solid foundation for analysing waveguides with traction free lateral boundaries. The effect of attenuation on second harmonic generation is modeled. Then we bridge the gap from modeling to measurements. Six different types of measurements that rely upon elastic nonlinearity are highlighted.

Image credit: Jessie Lissenden

We start by formulating the nonlinear wave equation in an infinite spatial domain and then examine solutions for planar problems having one-dimensional displacement fields. The equation of motion for a homogeneous elastic body, equation (2.26), repeated here for convenience, provides the starting point, except that now we ignore the body force term,

$$\nabla_X \cdot \mathbf{T}_{PK1} = \rho_o \dot{\mathbf{v}}. \tag{2.26}$$

In component form we have

$$\frac{\partial}{\partial X_j}\left(T_{ji}^{PK1}\right) = \rho_o \dot{v}_i. \tag{3.1}$$

Equation (2.28) provides the constitutive response

$$\mathbf{T}_{PK1} = \frac{\partial W}{\partial \mathbf{F}} = \frac{\partial W}{\partial \mathbf{H}}, \tag{2.28}$$

where the strain energy density is given in terms of the Huang coefficients by equation (2.29),

$$W = W_o + A_{ij}H_{ij} + \frac{1}{2!}A_{ijkl}H_{ij}H_{kl} + \frac{1}{3!}A_{ijklmn}H_{ij}H_{kl}H_{mn}$$
$$+ \frac{1}{4!}A_{ijklmnpq}H_{ij}H_{kl}H_{mn}H_{pq} + \cdots, \tag{2.29}$$

is used to obtain

$$T_{ij}^{PK1} = A_{ij} + \frac{1}{2!}\left[A_{ijrs}H_{rs} + A_{pqij}H_{pq}\right]$$
$$+ \frac{1}{3!}\left[A_{ijrsuv}H_{rs}H_{uv} + A_{pqijuv}H_{pq}H_{uv} + A_{pqrsij}H_{pq}H_{rs}\right]$$
$$+ \frac{1}{4!}\left[A_{ijrsuvkl}H_{rs}H_{uv}H_{kl} + A_{pqijuvkl}H_{pq}H_{uv}H_{kl} +\right.$$
$$\left. A_{pqrsijkl}H_{pq}H_{rs}H_{kl} + A_{pqrsuvij}H_{pq}H_{rs}H_{uv}\right] + \cdots. \tag{3.2}$$

Notice that the stress could be decomposed into linear and nonlinear parts in terms of the displacement gradient as in example 2.4. The nonlinear part is a driving force akin to a body force. Together, equations (3.1) and (3.2) represent finite amplitude elastic wave propagation in a hyperelastic material. We will solve them for longitudinal and shear bulk waves.

3.1 Bulk longitudinal waves

Consider now the one-dimensional problem of planar longitudinal waves constrained to propagate solely in the X_1 direction such that

$$\mathbf{u} = \begin{Bmatrix} u \\ 0 \\ 0 \end{Bmatrix}, \mathbf{H} = \begin{bmatrix} u' & 0 & 0 \\ 0 & 0 & 0 \\ 0 & 0 & 0 \end{bmatrix}, \mathbf{E} = \begin{bmatrix} u' + \frac{1}{2}(u')^2 & 0 & 0 \\ 0 & 0 & 0 \\ 0 & 0 & 0 \end{bmatrix}, \tag{3.3}$$

where $u' = \partial u / \partial X_1$. As there is no spatial dependence on X_2 and X_3, the only nontrivial equation in equation (3.1) occurs for $j = 1$ when $i = 1$:

$$T_{11,1}^{PK1} = \rho_o \ddot{u}. \tag{3.4}$$

Dropping the indices on the stress in equation (3.4) because only the 11 component is analysed (although the 22 and 33 components are nonzero) simplifies equation (3.2):

$$T^{PK1} = A_{11} + \frac{1}{2!}[A_{1111}H_{11} + A_{1111}H_{11}]$$

$$+ \frac{1}{3!}[A_{111111}H_{11}H_{11} + A_{111111}H_{11}H_{11} + A_{111111}H_{11}H_{11}]$$

$$+ \frac{1}{4!}[A_{11111111}H_{11}H_{11}H_{11} + A_{11111111}H_{11}H_{11}H_{11} +$$

$$+ A_{11111111}H_{11}H_{11}H_{11} + A_{11111111}H_{11}H_{11}H_{11}] + \ldots$$

$$T^{PK1} = A_{11} + A_{1111}(u') + \frac{1}{2}A_{111111}(u')^2 + \frac{1}{6}A_{11111111}(u')^3 + \ldots \quad (3.5)$$

We wish to solve a problem having quadratic nonlinearity. Therefore, the third order term (from the fourth order term in the strain energy function) and higher order terms will be neglected. Equation (3.4) becomes

$$\frac{\partial}{\partial X_1}\left[A_{11} + A_{1111}(u') + \frac{1}{2}A_{111111}(u')^2\right] = \rho_o \ddot{u} \quad (3.6)$$

and, given that the material is homogeneous,

$$A_{1111}(u'') + A_{111111}(u')(u'') = \rho_o \ddot{u}. \quad (3.7)$$

Most of the literature uses the contracted form of the elasticity coefficients, $A_{1111} \rightarrow A_{11}$ and $A_{111111} \rightarrow A_{111}$. Identifying $A_{11}/\rho_o \equiv c^2$ gives

$$c^2 u'' + \frac{A_{111}}{A_{11}}c^2(u')(u'') = \ddot{u}. \quad (3.8)$$

The so-called acoustic nonlinearity coefficient (or parameter[1]) is conveniently and naturally defined as

$$\beta \equiv -\frac{A_{111}}{A_{11}} \quad (3.9)$$

leading to

$$c^2 u'' - \beta c^2(u')(u'') = \ddot{u}$$

$$\ddot{u} = c^2[1 - \beta u']u''$$

$$\ddot{u} - c^2 u'' = -\beta c^2 u'u''. \quad (3.10)$$

The form used in equation (3.10) emphasizes the nonlinearity by placing the nonlinear term as a driving force on the right-hand side of the equation. We can impose a perturbation solution as demonstrated in examples 3.1–3.3, which leads to the final solution form

---

[1] In nonlinear acoustics of fluids β is known as the acoustic nonlinearity coefficient and the acoustic nonlinearity parameter is B/A (as in Blackstock and Hamilton [1]), but in the nonlinear ultrasonics of solids community β istypically referred to as the acoustic nonlinearity parameter.

$$u = A_1 \cos(kX_1 - \omega t) - \frac{1}{8}\beta k^2 A_1^2 X_1 \cos 2(kX_1 - \omega t). \tag{3.11}$$

This solution contains a wealth of information upon which we will build. As shown in example 3.1, the first term is the solution to the linear problem, usually referred to as the fundamental or primary waves, while the second term embodies the non-linearity through secondary waves referred to as second harmonics because their frequency is twice that of the primary waves. The second harmonics are also phase matched, or synchronized, meaning that their phase velocity, $c_p = \frac{2\omega}{2k}$, is the same as that of the primary waves, $c_p = \frac{\omega}{k}$. We can identify the amplitude of the second harmonic waves to be

$$A_2 = \frac{1}{8}\beta k^2 A_1^2 X_1, \tag{3.12}$$

displaying the interesting characteristic that the second harmonic wave amplitude increases linearly with propagation distance. Thus, synchronized second harmonics are called cumulative. It is important to keep in mind that the perturbation solution employed to arrive at equation (3.11) requires $A_2 \ll A_1$. Thus, it is not terrible that A_2 increases and A_1 remains constant in this approximate model, although in reality energy is not created, but rather comes from the primary waves (whose amplitude must decrease a small amount). Furthermore, equation (3.12) shows that A_2 increases with the square of the wavenumber, which in turn is linearly proportional to the frequency ($k = \omega/c$). Thus, $A_2 \propto \omega^2$. If we rewrite equation (3.12) as

$$\beta = \frac{8A_2}{A_1^2 k^2 X_1} \tag{3.13}$$

it appears that second harmonic generation experiments can be used to measure β. While this is true, it is not a trivial matter, as we will see in section 3.4.2 and then in chapters 11 and 12 for guided waves. We should emphasize that equation (3.13) is not a definition and should be applied heeding the context of the idealized (bulk longitudinal wave) problem from which the solution, equation (3.11), was derived. Example 3.4 walks through computing β in terms of the third order elastic constants. In practice, the relative acoustic nonlinearity coefficient (or parameter if you prefer) defined by

$$\beta' \equiv \frac{A_2}{A_1^2} \tag{3.14}$$

is often determined to assess changes in the material state because it is difficult (albeit not impossible) to calibrate the measured voltages with wave amplitudes.

If sufficiently driven, which is usually not the case, the waveform of progressive waves evolves until a discontinuity (or shock) occurs. The propagation distance necessary for the formation of a discontinuity is a subject first investigated in the classical era of acoustics in fluids. The discontinuity occurs when the waveform steepens to the extent that the particle velocity approaches infinity. As shown in

Hamilton and Blackstock's text [1] in their chapter 4 section 2, Poisson's implicit solution to the reduced wave equation can be used to derive the discontinuity distance as

$$\bar{x} = \frac{1}{|\beta| \epsilon_0 k}, \tag{3.15}$$

where $\epsilon_0 = V_0/c_0$ is the acoustic Mach number ($\epsilon_0 \ll 1$), with V_0 being the amplitude of the initial sinusoidal velocity waveform and c_0 being the small amplitude wave speed. Of course we are interested in waves in solids, so we turn to Thurston's [2] section 38 on finite amplitude waves in elastic media. Thurston derives the discontinuity distance to be

$$L = \frac{-2W_0^2 M_2}{\omega v_0 M_3} \tag{3.16}$$

in his notation, which is that W is the natural speed in the reference coordinates, M_2 and M_3 are linear combinations of the second and third order elastic constants, respectively, ω is the angular frequency, and v_0 is the initial particle velocity. There are several alternative ways to write equation (3.16), e.g.

$$L = \frac{2W_0^2}{\omega v_0 (2 + B/A)}, \tag{3.17}$$

where B/A is the conventional nonlinearity parameter. Breazeale and Ford [3] provide another form. Each of these equations for the discontinuity distance indicate dependence on frequency, amplitude, and material parameters, thus the discontinuity distance has a useful application in assessing the material nonlinearity. However, in second harmonic generation testing the nonlinearity is weak and the waveforms are not noticeably progressive.

Example 3.1. Longitudinal wave nonlinearity.

Given the nonlinear wave equation

$$\ddot{u} - c^2 u'' = -\beta c^2 u' u'' \tag{3.10}$$

with quadratic nonlinearity, formulate the solution

$$u = A_1 \cos(kX_1 - \omega t) - \frac{1}{8}\beta k^2 A_1^2 X_1 \cos 2(kX_1 - \omega t) \tag{3.11}$$

by identifying equation (3.10) as a nonhomogeneous second order partial differential equation. The solution has two components, the primary solution and the secondary solution, which is much smaller than the primary solution, i.e. $u = u_{(1)} + u_{(2)}$, where $u_{(2)} \ll u_{(1)}$. We take $u_{(1)}$ to be the linear solution, yielding the familiar primary harmonic wave solution,

$$u_{(1)} = A_1 \cos(kX_1 - \omega t).$$

We can think of the nonlinear term on the right-hand side of equation (3.10) as the nonlinear driving force obtained from the primary solution (making this a perturbation solution), i.e.

$$-c^2\beta u'u'' = -c^2\beta[-kA_1\sin(kX_1 - \omega t)][-k^2A_1\cos(kX_1 - \omega t)]$$

$$-c^2\beta u'u'' = -c^2\beta k^3 A_1^2 \sin(kX_1 - \omega t)\cos(kX_1 - \omega t)$$

$$-c^2\beta u'u'' = -c^2\beta k^3 A_1^2 \frac{1}{2}\sin 2(kX_1 - \omega t).$$

In so doing, we are implicitly imposing the solution criteria that $u_{(2)} \ll u_{(1)}$ by replacing u with $u_{(1)}$. Substituting into equation (3.10) gives

$$\ddot{u}_{(2)} - c^2 u''_{(2)} = -\frac{1}{2}c^2\beta k^3 A_1^2 \sin 2(kX_1 - \omega t),$$

which admits solutions of the form

$$u_{(2)} = D_1(X_1)\sin 2(kX_1 - \omega t) + D_2(X_1)\cos 2(kX_1 - \omega t)$$

such that the functions to be determined, $D_1(X_1)$ and $D_2(X_1)$, satisfy the known conditions at the source point, which is the origin, $D_1(X_1 = 0) = D_2(X_1 = 0) = 0$. We will now take two time derivatives and two spatial derivatives of $u_{(2)}$ and substitute them back into the differential equation for $u_{(2)}$:

$$\ddot{u}_{(2)} = -D_1(-2\omega)^2 \sin 2(kX_1 - \omega t) - D_2(-2\omega)^2 \cos 2(kX_1 - \omega t)$$

$$u''_{(2)} = (D_1'' - 4k^2 D_1 - 4kD_2')\sin 2(kX_1 - \omega t)$$
$$+ (D_2'' - 4k^2 D_2 + 4kD_1')\cos 2(kX_1 - \omega t)$$

$$-D_1(-2\omega)^2 \sin 2(kX_1 - \omega t) - D_2(-2\omega)^2 \cos 2(kX_1 - \omega t)$$
$$- c^2[(D_1'' - 4k^2 D_1 - 4kD_2')\sin 2(kX_1 - \omega t) + (D_2'' - 4k^2 D_2 + 4kD_1')\cos 2(kX_1 - \omega t)]$$
$$= -\frac{1}{2}c^2\beta k^3 A_1^2 \sin 2(kX_1 - \omega t).$$

Dividing by ω^2 and realizing that $\left(\frac{c}{\omega}\right)^2 = \frac{1}{k^2}$ gives

$$\left(D_1'' - 4kD_2'\right)\sin 2(kX_1 - \omega t) + \left(D_2'' + 4kD_1'\right)\cos 2(kX_1 - \omega t) = \frac{1}{2}\beta k^3 A_1^2 \sin 2(kX_1 - \omega t).$$

Solving for the sine and cosine terms separately:

$$D_1'' - 4kD_2' = \frac{1}{2}\beta k^3 A_1^2$$

$$D_2'' + 4kD_1' = 0.$$

If we take $D_2'' = 0$ in the latter equation, then $D_1' = 0$ as well. Thus, $D_1'' = 0$, and $D_1 = $ constant, which must be zero due to the condition that $D_1(X_1 = 0) = 0$. Using the former equation, we find that

$$D_2' = -\frac{1}{8}\beta k^2 A_1^2,$$

which is constant in terms of X_1, as required for $D_2'' = 0$. Taking the anti-derivative of D_2'

$$D_2 = \alpha_0 - \frac{1}{8}\beta k^2 A_1^2 X_1,$$

where we find that $\alpha_0 = 0$ from the condition $D_2(X_1 = 0) = 0$. Thus, we have

$$D_1(X_1) = 0$$

$$D_2(X_1) = -\frac{1}{8}\beta k^2 A_1^2 X_1,$$

and finally

$$u_{(1)} = A_1 \cos(kX_1 - \omega t)$$

$$u_{(2)} = -\frac{1}{8}\beta k^2 A_1^2 X_1 \cos 2(kX_1 - \omega t)$$

$$u = A_1 \cos(kX_1 - \omega t) - \frac{1}{8}\beta k^2 A_1^2 X_1 \cos 2(kX_1 - \omega t),$$

which is the solution we sought.

Visualization of these results helps us see more clearly what is going on. Consider bulk longitudinal waves in an aluminum alloy having the following properties:

$$\rho = 2700 \text{ kg m}^{-3},$$

$$E = 69 \text{ GPa}, \ \nu = 0.33,$$

$$\mathscr{A} = -350 \text{ GPa}, \ \mathscr{B} = -155 \text{ GPa}, \ \mathscr{C} = -95 \text{ GPa}.$$

The nonlinearity coefficient β can be computed from these properties as shown in example 3.4; $\beta = 14.8$ (dimensionless). In contrast to the definition in Guyer and Johnson [4, 5], β is positive here. Although modeling typically considers continuous waves, in nondestructive evaluations it is common to use a toneburst excitation. Therefore, let's use a 20-cycle toneburst having a center frequency of 5 MHz and a constant amplitude of 100 nm. The longitudinal and shear wave speeds computed from equation (2.61) are 6153 and 3100 m s^{-1}, respectively, and the wavenumber is 5106 rad m^{-1}. Readers with less experience in the calculation of wave propagation characteristics are encouraged to confirm these values for practice. In imaging technology, a trace of the wave amplitude with time at a point is known as an A-scan. A-scans of the fundamental (or primary) waves and secondary waves are shown in figure 3.1 for propagation distances of $X_1 = 0, 5, 10$ mm. Although it is more common to use the term fundamental waves or fundamental frequency in acoustics, in guided waves there are fundamental and higher order wave modes (as we will see for Lamb waves), which is an entirely different thing. Therefore, this book will use the term primary waves, which propagate at the primary frequency to avoid ambiguity. In this example, the amplitudes of the primary and secondary waves are

$$A_1 = 100 \text{ nm}$$

$$A_2 = -\frac{1}{8}\beta k^2 A_1^2 X_1 = \begin{cases} 0 & \text{for } X_1 = 0 \\ 2.41 \text{ nm} & \text{for } X_1 = 5 \text{ mm} \\ 4.82 \text{ nm} & \text{for } X_1 = 10 \text{ mm.} \end{cases}$$

The $u_{(1)}$ and $u_{(2)}$ signals are shown separately in figures 3.1(a)–(c), where the $u_{(2)}$ amplitude increases linearly with propagation distance. We should keep in mind that this is for planar waves in a lossless homogeneous solid. Frequency spectra obtained from the fast Fourier transform (FFT) are shown in figures 3.1(d)–(h). The 20 toneburst cycles result in the primary waves having quite a narrow frequency bandwidth. The amplitude of the second harmonic waves (at 10 MHz) is quite small relative to the primary waves, small enough to be difficult to pick out of the side lobes around the primary frequency. We can minimize the side lobes using a Hanning window, but in so doing the spectral amplitudes differ from A_1 and A_2. Since we know β for this material, it is instructive to compute the relative nonlinearity coefficient β' values from the spectrum for $X_1 = 10$ mm:

$$\text{No window: } \beta' = \frac{4.822\,94 \times 10^{-9}}{(100 \times 10^{-9})^2} = 0.482 \times 10^6.$$

$$\text{Hanning window: } \beta' = \frac{2.424\,14 \times 10^{-9}}{(50.2499 \times 10^{-9})^2} = 0.960 \times 10^6.$$

The β' values differ by a factor of 2 based on the windowing function used. Clearly, signal processing is going to be important for experimental data. As such, it will be discussed in chapter 11. Figures 3.1(i) and (j) show the A-scan for the combined signal $u_{(1)} + u_{(2)}$. The distortion of the sinusoidal signal is small, but it is evident that the signal has a tension/compression asymmetry, having larger valleys than peaks. Harmonic generation from different waveform shapes is discussed by Chillara and Lissenden [6].

Readers familiar with nonlinear ultrasonics may be wondering why the solution in equation (3.11) does not contain the quasi-static pulse as investigated in e.g. [7, 8]. The answer is simply that the chosen trial solution does not provide it. Had we used the trial function

$$u_{(2)} = -A_2 X_1 \sin^2(k X_1 - \omega t),$$

and followed the same procedure used in this example, and then imposed the trigonometric identity,

$$\cos 2\theta = 1 - 2\sin^2\theta,$$

the solution found is

$$u_{(2)} = \frac{1}{8}\beta k^2 A_1^2 X_1 [1 - \cos 2(k X_1 - \omega t)].$$

The quasi-static pulse term has the same cumulative amplitude as does the second harmonic.

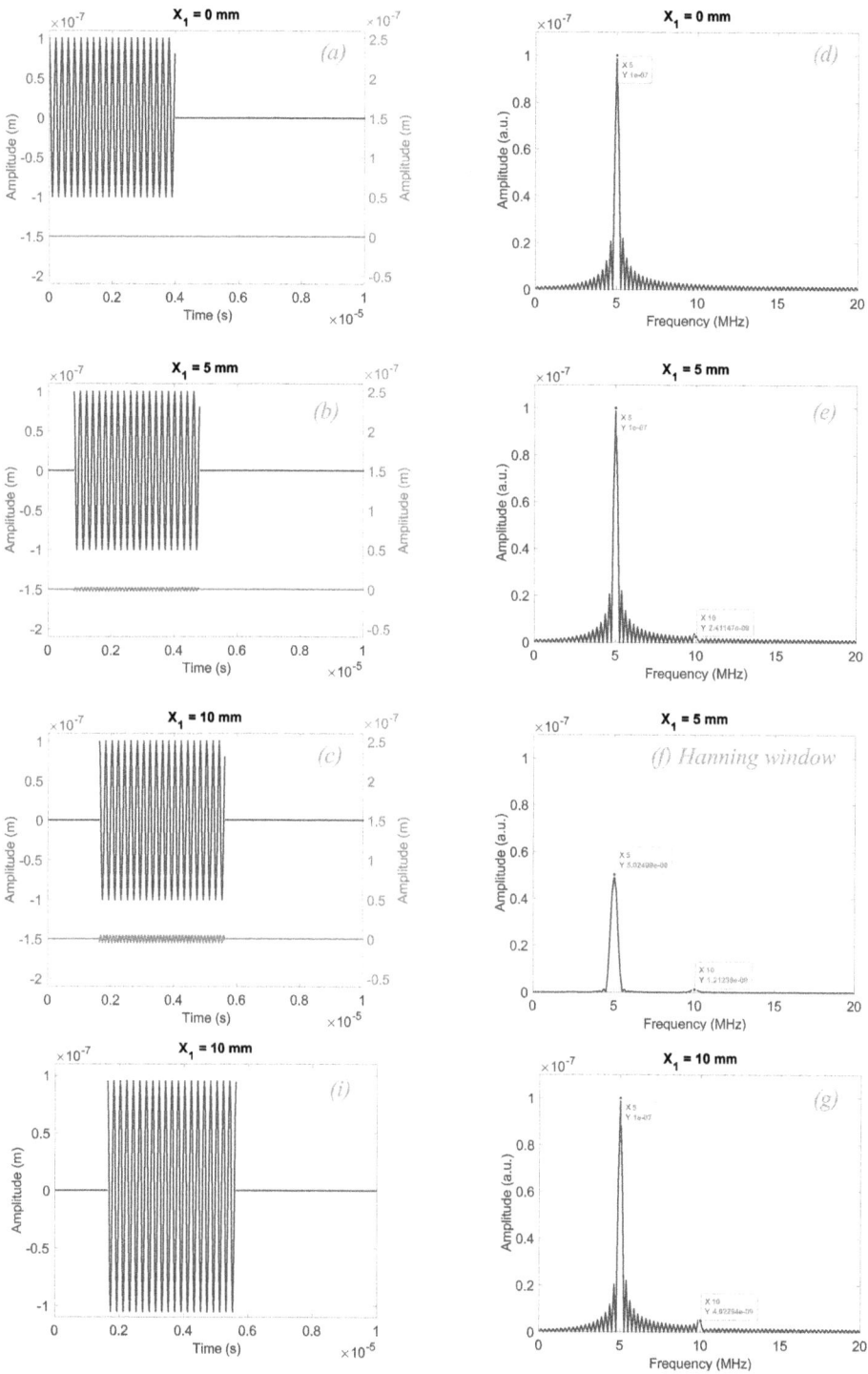

Figure 3.1. Nonlinear longitudinal waves at $X_1 = 0, 5, 10$ mm: (a)–(c) A-scan signals of primary and secondary waves, (d)–(h) frequency spectra, and (i)–(j) A-scans of combined waves. Continued over.

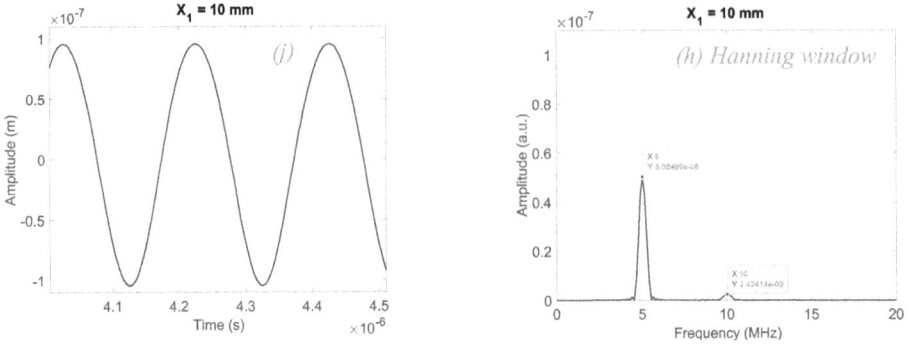

Figure 3.1. (Continued.)

Example 3.2. Regular perturbation approach to the nonlinear longitudinal wave problem.

As an alternative to the solution in example 3.1, we can solve the problem employing a more formal (regular) perturbation approach with a power series expansion as shown below. The general idea is to convert the nonlinear problem into a series of linear problems. Interested readers will find more detailed information about perturbation methods for nonlinear wave propagation and vibration in Hamilton and Blackstock [1] and Nayfeh and Mook [9]. Starting the problem over, we define the boundary value problem (BVP) to be given by the governing differential equation

$$\ddot{u} - c^2 u'' = -\beta c^2 u' u'' \tag{3.10}$$

and the initial conditions

$$u(X_1 = 0, t = 0) = A_1$$
$$\dot{u}(X_1 = 0, t = 0) = 0.$$

These initial conditions give the primary solution as $A_1 \cos(kX_1 - \omega t)$. Let's be wise and nondimensionalize the variables

$$y = kX_1 = \frac{\omega}{c} X_1$$

$$\tau = \omega t$$

$$v = ku$$

and write the acoustic nonlinearity coefficient β in terms of the acoustic Mach number ϵ defined after equation (3.15) because ϵ is sufficiently small to use as the parameter in the expansion of the displacement field

$$\beta = \alpha \epsilon$$

where it can be shown that

$$\alpha = \frac{3C_{11} + C_{111}}{\rho c V_0}.$$

The governing differential equation (GDE) and initial conditions (ICs) become

$$\frac{\partial^2 v}{\partial \tau^2} = \left(1 - \alpha\epsilon \frac{\partial v}{\partial y}\right)\frac{\partial^2 v}{\partial y^2}$$

$$v(y = 0, \tau = 0) = A_1 k = B_1$$

$$\frac{\partial v}{\partial \tau}(y = 0, \tau = 0) = 0.$$

Try substituting a first order expansion of the displacement field in terms of the acoustic Mach number ϵ

$$v(y, \tau, \epsilon) = v_0(y, \tau) + \epsilon v_1(y, \tau) + O(\epsilon^2)$$

into the GDE:

$$\frac{\partial^2}{\partial \tau^2}[v_0(y, \tau) + \epsilon v_1(y, \tau) + \cdots]$$

$$= \left(1 - \alpha\epsilon \frac{\partial}{\partial y}[v_0(y, \tau) + \epsilon v_1(y, \tau) + \cdots]\right)\frac{\partial^2}{\partial y^2}[v_0(y, \tau) + \epsilon v_1(y, \tau) + \cdots]$$

$$\frac{\partial^2 v_0}{\partial \tau^2} + \epsilon\frac{\partial^2 v_1}{\partial \tau^2} + \cdots = \left(1 - \alpha\epsilon\frac{\partial v_0}{\partial y} - \alpha\epsilon^2\frac{\partial v_1}{\partial y} - \cdots\right)\left[\frac{\partial^2 v_0}{\partial y^2} + \epsilon\frac{\partial^2 v_1}{\partial y^2} + \cdots\right].$$

Neglecting all terms having ϵ raised to a power greater than one

$$\frac{\partial^2 v_0}{\partial \tau^2} + \epsilon\frac{\partial^2 v_1}{\partial \tau^2} = \frac{\partial^2 v_0}{\partial y^2} + \epsilon\frac{\partial^2 v_1}{\partial y^2} - \alpha\epsilon\frac{\partial v_0}{\partial y}\frac{\partial^2 v_0}{\partial y^2} + O(\epsilon^2)$$

and since ϵ is small we can separate into equations for ϵ^0 and ϵ^1

$$O(\epsilon^0): \frac{\partial^2 v_0}{\partial \tau^2} = \frac{\partial^2 v_0}{\partial y^2}$$

$$O(\epsilon^1): \frac{\partial^2 v_1}{\partial \tau^2} = \frac{\partial^2 v_1}{\partial y^2} - \alpha\frac{\partial v_0}{\partial y}\frac{\partial^2 v_0}{\partial y^2}.$$

Also substitute the expansion into the ICs,

$$v_0(y = 0, \tau = 0) + \epsilon v_1(y = 0, \tau = 0) = B_1$$

$$\frac{\partial v_0}{\partial \tau}(y = 0, \tau = 0) + \epsilon\frac{\partial v_1}{\partial \tau}(y = 0, \tau = 0) = 0,$$

and by again separating terms, we obtain that the ICs for the v_0 and v_1 problems are

$$v_0(y = 0, \tau = 0) = v_o;\ v_1(y = 0, \tau = 0) = 0$$

$$\frac{\partial v_0}{\partial \tau}(y = 0, \tau = 0) = 0;\ \frac{\partial v_1}{\partial \tau}(y = 0, \tau = 0) = 0.$$

Thus, the ϵ^0 BVP is

$$\frac{\partial^2 v_0}{\partial \tau^2} - \frac{\partial^2 v_0}{\partial y^2} = 0$$

$$v_0(y = 0, \tau = 0) = B_1$$

$$\frac{\partial v_0}{\partial \tau}(y = 0, \tau = 0) = 0$$

whose solution can be verified to be

$$v_0(y, \tau) = B_1 \cos(y - \tau).$$

More interestingly, the ϵ^1 BVP is

$$\frac{\partial^2 v_1}{\partial \tau^2} - \frac{\partial^2 v_1}{\partial y^2} = -\alpha \frac{\partial v_0}{\partial y} \frac{\partial^2 v_0}{\partial y^2}$$

$$v_1(y = 0, \tau = 0) = 0$$

$$\frac{\partial v_1}{\partial \tau}(y = 0, \tau = 0) = 0.$$

Notice that the nonhomogeneity in the ϵ^1 GDE comes only from the ϵ^0 BVP, whose solution is now known:

$$-\alpha \frac{\partial v_0}{\partial y} \frac{\partial^2 v_0}{\partial y^2} = -\alpha[-B_1 \sin(y - \tau)][-B_1 \cos(y - \tau)]$$

$$-\alpha \frac{\partial v_0}{\partial y} \frac{\partial^2 v_0}{\partial y^2} = -\alpha B_1^2 \sin(y - \tau) \cos(y - \tau)$$

$$-\frac{\partial u_0}{\partial y} \frac{\partial^2 u_0}{\partial y^2} = -\alpha B_1^2 \frac{1}{2} \sin 2(y - \tau).$$

The ϵ^1 GDE is now

$$\frac{\partial^2 v_1}{\partial \tau^2} - \frac{\partial^2 v_1}{\partial y^2} = -\alpha B_1^2 \frac{1}{2} \sin 2(y - \tau),$$

which admits solutions of the form

$$v_1(y, \tau) = D_1(y) \sin 2(y - \tau) + D_2(y) \cos 2(y - \tau).$$

Determine $D_1(y)$ and $D_2(y)$ by substituting into the GDE and then satisfying the ICs. The GDE gives

$$- D_1''(y)\sin 2(y - \tau) - 4D_1' \cos 2(y - \tau) - D_2''(y)\cos 2(y - \tau)$$

$$+ 4D_2'(y) \sin 2(y - \tau) = -\frac{1}{2}\alpha v_o^2 \sin 2(y - \tau)$$

and separating the $\sin 2(y - \tau)$ and $\cos 2(y - \tau)$ terms gives the two equations

$$\sin 2(y - \tau) : - D_1''(y) + 4D_2'(y) = -\frac{1}{2}\alpha B_1^2$$

$$\cos 2(y - \tau): - 4D_1' - D_2''(y) = 0.$$

The $\cos 2(y - \tau)$ equation dictates that $D_2''(y) = -4D_1' = $ constant, and therefore $D_1''(y) = 0$. Now, the $\sin 2(y - \tau)$ equation requires $D_2'(y) = -\frac{1}{8}\alpha B_1^2$. Hence, we have learned from the GDE that

$$D_1(y) = C_1 y + C_2$$

$$D_2(y) = -\frac{1}{8}\alpha B_1^2 y + C_3.$$

Satisfying the ICs requires that

$$v_1(0, 0) = D_2(y = 0) = 0 = C_3$$

$$\frac{\partial v_1}{\partial \tau}(0, 0) = -2D_1(0)\cos(0) + 2D_2(0)\sin(0) = 2D_1(0)1 = 0$$

$$\therefore C_2 = 0.$$

Thus,

$$D_1(y) = C_1 y$$

$$D_2(y) = -\frac{1}{8}\alpha B_1^2 y$$

$$v_1(y, \tau) = -\frac{1}{8}\alpha B_1^2 y \cos 2(y - \tau),$$

where, since C_1 is arbitrary, we have taken the liberty to set it to zero. We see that since the $v_1(y, \tau)$ solution grows with y, the solution is nonuniformly accurate, having a singularity as $y \longrightarrow \infty$. If desired, this can be improved upon by using a more sophisticated perturbation method. Nonetheless, our final solution from

$$v(y, \tau, \epsilon) = v_0(y, \tau) + \epsilon v_1(y, \tau) + \cdots$$

is reasonably good within the limitations on y and can now be written as

$$v(y, \tau, \epsilon) = B_1 \cos(y - \tau) - \frac{1}{8}\epsilon \alpha B_1^2 y \cos 2(y - \tau) + \cdots,$$

where

$$y = kX_1, \quad \tau = \omega t, \quad \beta = \alpha \epsilon, \quad B_1 = kA_1.$$

Transforming back to the original problem with physical variables and units we have:

$$u(X_1, t) = A_1 \cos(kX_1 - \omega t) - \frac{1}{8}\beta A_1^2 k^2 X_1 \cos 2(kX_1 - \omega t) + \cdots. \qquad \text{(E1)}$$

Example 3.3. Nonlinear longitudinal wave solution using the method of multiple scales.

The perturbation solutions applied thus far to the nonlinear longitudinal wave problem have the limitations that: (i) the secondary displacement field must be small relative to the primary displacement field and (ii) although the energy driving the secondary waves to be cumulative comes from the primary waves, the amplitude of the primary waves remains constant. These limitations can be overcome by implementing a more advanced perturbation technique. We will apply a version of the method of multiple scales known as two-timing [9]. In their section 8.1, Nayfeh and Mook [9] use the method of multiple scales to solve the problem of nonlinear longitudinal waves in a bar. The basis for applying this approach is that nonlinear wave propagation has two distinct length scales—a short one based on the wavelength of the primary waves and a long one associated with the discontinuity distance.

For clarity, we once again start from the beginning and define the boundary value problem in terms of the governing differential equation (GDE)

$$\ddot{u} - c^2 u'' = -\beta c^2 u' u'' \tag{3.10}$$

and the initial conditions (ICs)

$$u(X_1 = 0, t = 0) = u_0$$

$$\dot{u}(X_1 = 0, t = 0) = 0.$$

Let's be wise and nondimensionalize the variables

$$y = kX_1 = \frac{\omega}{c} X_1$$

$$\tau = \omega t$$

$$v = ku$$

and write the acoustic nonlinearity coefficient β in terms of the acoustic Mach number ϵ defined after equation (3.15) (which is sufficiently small to use as the parameter in the expansion),

$$\beta = \alpha \epsilon$$

where it can be shown that

$$\alpha = \frac{3C_{11} + C_{111}}{\rho c V_0}$$

giving the governing differential equation (GDE) and initial conditions (ICs)

$$\frac{\partial^2 v}{\partial \tau^2} = \left(1 - \alpha \epsilon \frac{\partial v}{\partial y}\right) \frac{\partial^2 v}{\partial y^2}$$

$$v(y = 0, \tau = 0) = u_0 k$$

$$\frac{\partial v}{\partial \tau}(y = 0, \tau = 0) = 0.$$

The nonuniform accuracy of the regular perturbation solution suggests that there are two length scales in the problem. Thus, we will define two length scales,

$$y_0 = y$$

$$y_1 = \epsilon y,$$

which enables us to write

$$v(y, \tau, \epsilon) = v(y_0, y_1, \tau, \epsilon)$$

except we will treat y_1 as a dependent variable. However, Nayfeh and Mook [9] suggest that the solution is more straightforward if the characteristics are used,

$$s_1 = \tau - y$$

$$s_2 = \tau + y$$

$$y_1 = \epsilon y,$$

which enables us to write

$$v(y, \tau, \epsilon) = v(s_1, s_2, y_1, \tau, \epsilon).$$

Thus, the derivatives we need can be written as

$$\frac{\partial}{\partial \tau} = \frac{\partial}{\partial s_2} + \frac{\partial}{\partial s_1}$$

$$\frac{\partial^2}{\partial \tau^2} = \left(\frac{\partial}{\partial s_2} + \frac{\partial}{\partial s_1} \right)^2$$

$$\frac{\partial}{\partial y} = \frac{\partial}{\partial s_2} - \frac{\partial}{\partial s_1} + \frac{\partial}{\partial y_1} \frac{\partial y_1}{\partial y} = \frac{\partial}{\partial s_2} - \frac{\partial}{\partial s_1} + \epsilon \frac{\partial}{\partial y_1}$$

$$\frac{\partial^2}{\partial y^2} = \left(\frac{\partial}{\partial s_2} - \frac{\partial}{\partial s_1} \right)^2 + 2\epsilon \frac{\partial}{\partial y_1} \left(\frac{\partial}{\partial s_2} - \frac{\partial}{\partial s_1} \right) + \epsilon^2 \frac{\partial^2}{\partial y_1^2}.$$

Trying a first order expansion of the displacement field in terms of the acoustic Mach number ϵ,

$$v(y, \tau, \epsilon) = v_0(s_2, s_1, y_1, \tau) + \epsilon v_1(s_2, s_1, y_1, \tau) + O(\epsilon^2),$$

the GDE becomes

$$\left(\frac{\partial}{\partial s_2} + \frac{\partial}{\partial s_1} \right)^2 [v_0(s_2, s_1, y_1, \tau) + \epsilon v_1(s_2, s_1, y_1, \tau) + O(\epsilon^2)]$$

$$= \left[1 - \alpha\epsilon \left(\frac{\partial}{\partial s_2} - \frac{\partial}{\partial s_1} + \epsilon \frac{\partial}{\partial y_1} \right) \right]$$

$$[v_0(s_2, s_1, y_1, \tau) + \epsilon v_1(s_2, s_1, y_1, \tau) + O(\epsilon^2)]$$

$$\times \left[\left(\frac{\partial}{\partial s_2} - \frac{\partial}{\partial s_1} \right)^2 + 2\epsilon \frac{\partial}{\partial y_1} \left(\frac{\partial}{\partial s_2} - \frac{\partial}{\partial s_1} \right) + \epsilon^2 \frac{\partial^2}{\partial y_1^2} \right]$$

$$[v_0(s_2, s_1, y_1, \tau) + \epsilon v_1(s_2, s_1, y_1, \tau) + O(\epsilon^2)].$$

Separating the GDE into $O(\epsilon^0)$ and $O(\epsilon^1)$ terms and ignoring the $O(\epsilon^2)$ terms gives

$$O(\epsilon^0): \left(\frac{\partial}{\partial s_2} + \frac{\partial}{\partial s_1}\right)^2 [v_0(s_2, s_1, y_1, \tau)] = \left(\frac{\partial}{\partial s_2} - \frac{\partial}{\partial s_1}\right)^2 v_0(s_2, s_1, y_1, \tau)$$

$$O(\epsilon^1): \left(\frac{\partial}{\partial s_2} + \frac{\partial}{\partial s_1}\right)^2 v_1(s_2, s_1, y_1, \tau)$$

$$= 2\frac{\partial}{\partial y_1}\left(\frac{\partial}{\partial s_2} - \frac{\partial}{\partial s_1}\right)v_0(s_2, s_1, y_1, \tau)$$

$$- \alpha\left(\frac{\partial}{\partial s_2} - \frac{\partial}{\partial s_1}\right)v_0(s_2, s_1, y_1, \tau)\left(\frac{\partial}{\partial s_2} - \frac{\partial}{\partial s_1}\right)^2 v_0(s_2, s_1, y_1, \tau)$$

$$+ \left(\frac{\partial}{\partial s_2} - \frac{\partial}{\partial s_1}\right)^2 v_1(s_2, s_1, y_1, \tau),$$

which reduces to

$$O(\epsilon^0): 4\frac{\partial^2 v_0}{\partial s_2 \partial s_1} = 0$$

$$O(\epsilon^1): 4\frac{\partial^2 v_1}{\partial s_2 \partial s_1} = 2\frac{\partial}{\partial y_1}\left(\frac{\partial v_0}{\partial s_2} - \frac{\partial v_0}{\partial s_1}\right) - \alpha\left(\frac{\partial v_0}{\partial s_2} - \frac{\partial v_0}{\partial s_1}\right)\left(\frac{\partial}{\partial s_2} - \frac{\partial}{\partial s_1}\right)^2 v_0.$$

We take the solution of the $O(\epsilon^0)$ equation to be left-to-right propagating waves, and thus we can write

$$v_0 = f(s_1, y_1).$$

The $O(\epsilon^1)$ equation is now

$$4\frac{\partial^2 v_1}{\partial s_2 \partial s_1} = -2\frac{\partial^2 f}{\partial y_1 \partial s_1} - \alpha\frac{\partial f}{\partial s_1}\frac{\partial^2 f}{\partial s_1^2}.$$

For the expansion to be uniformly valid, v_1/v_0 is bounded. In other words, the terms on the right-hand side are known as secular terms and should be set to zero, which constitutes the so-called amplitude equation (or solvability condition). By satisfying the amplitude equation, v_1 and v_0 have the same form, $f(s_1, y_1)$. Solving the amplitude equation we have

$$\frac{\partial^2 f}{\partial y_1 \partial s_1} + \frac{\alpha}{2}\frac{\partial f}{\partial s_1}\frac{\partial^2 f}{\partial s_1^2} = 0,$$

for which Nayfeh and Mook [9] give the general solution to be

$$f(s_1, y_1) = F(\xi) + \frac{\alpha}{4}y_1[F'(\xi)]^2,$$

where ξ is defined by

$$s_1 = \xi + \frac{\alpha}{2}y_1 F'(\xi).$$

Proof that this is indeed the solution is given at the end of this example. So far, we have that the approximation of the nondimensional displacement is

$$v = F(\xi) + \frac{\alpha}{4}\epsilon y [F'(\xi)]^2$$

$$s_1 = \xi + \frac{\alpha}{2}\epsilon y F'(\xi) = \tau - y.$$

Let's now consider the boundary conditions at the origin at any time to be

$$v(y = 0, \tau) = \phi(\tau)$$

$$\frac{\partial v}{\partial \tau}(y = 0, \tau) = \dot{\phi}(\tau)$$

and once again apply the expansion

$$v_0(\tau, \tau, 0, \tau) + \epsilon v_1(\tau, \tau, 0, \tau) + O(\epsilon^2) = \phi(\tau).$$

The zeroth and first order terms for the displacement BC are

$$O(\epsilon^0): v_0(\tau, \tau, 0, \tau) = F(\xi)\,|_{(\tau, \tau, 0, \tau)} = \phi(\tau)$$

$$O(\epsilon^1): v_1(\tau, \tau, 0, \tau) = \frac{\alpha}{4}y\Big[F'(\xi)\,|_{(\tau, \tau, 0, \tau)}\Big]^2 = 0.$$

The zeroth and first order terms for the velocity BC are

$$O(\epsilon^0): \frac{\partial v_0}{\partial \tau}(\tau, \tau, 0, \tau) = F'(\xi)\frac{\partial \xi}{\partial \tau}\Big|_{(\tau, \tau, 0, \tau)} = \dot{\phi}(\tau)$$

$$O(\epsilon^1): \frac{\partial v_1}{\partial \tau}(\tau, \tau, 0, \tau) = \frac{\alpha}{4}y\left[F''(\xi)\frac{\partial \xi}{\partial \tau}\Big|_{(\tau, \tau, 0, \tau)}\right]^2 = 0.$$

Now we can specify a harmonic (sinusoidal) boundary condition, i.e.

$$\phi(\tau) = B_1 \cos \tau,$$

where B_1 has been nondimensionalized, $B_1 = kA_1$. The displacement BC yields

$$F(\tau) = \phi(\tau) = B_1 \cos \tau.$$

Therefore,

$$v(y, \tau) = B_1 \cos \xi + \frac{\alpha}{4}\epsilon y B_1^2 \sin^2 \xi$$

$$\tau - y = \xi - \frac{\alpha}{2}\epsilon y B_1 \sin \xi.$$

When transformed back to the dimensional variables, the solution is

$$u(X_1, t) = A_1 \cos \xi + \frac{\beta}{4}k^2 A_1^2 X_1 \sin^2 \xi$$

$$\omega t - kX_1 = \xi - \frac{\beta}{2} k^2 A_1 X_1 \sin \xi.$$

To compare this solution to the solution from the regular perturbation approach in example 3.1 we again use the trigonometric identity $\cos 2\theta = 1 - 2\sin^2 \theta$:

$$u(X_1, t) = A_1 \cos \xi + \frac{\beta}{8} k^2 A_1^2 X_1 [1 - \cos 2\xi]. \tag{E2}$$

The only difference between the two solutions (equations (E1) and (E2)) is the argument of the cosine functions (and that equation (E1) lacks the quasi-static pulse term, which can easily be added). For the regular perturbation approach, the argument is $(kX_1 - \omega t)$ while for the method of multiple scales it is $\xi = \left(\omega t - kX_1 + \frac{\beta}{2} k^2 A_1 X_1 \sin \xi\right)$. Let's rewrite this as $\xi = -\eta + \psi$, where $\eta = kX_1 - \omega t$ and $\psi = \frac{\beta}{2} k^2 A_1 X_1 \sin \xi$ and use the trigonometric identity $\cos(\psi - \eta) = \cos \psi \cos \eta + \sin \psi \sin \eta$. Hence, we have

$$A_1 \cos \xi = A_1 [\cos \psi \cos \eta + \sin \psi \sin \eta]$$

$$A_1 \cos \xi = A_1 \cos \eta [\cos \psi + \sin \psi \tan \eta].$$

The term $[\cos \psi + \sin \psi \tan \eta]$ can be identified as the correction that the method of multiple scales applies to the primary wave amplitude from the regular perturbation method. When $t = X_1/c_L$ the phase angle $\eta = 0$ and $\tan \eta = 0$, therefore the correction is simply $\cos \psi$.

The discontinuity distance is $\bar{x} = \frac{1}{|\beta| \epsilon k}$, which for this example is 25.9 mm. The corrections are tabulated below.

$X = X_1/\bar{x}$	Amp(ψ)	$\cos (\text{Amp}(\psi))$
0	0	1
0.2	0.10	0.9950
0.4	0.20	0.9801
0.6	0.30	0.9553
0.8	0.40	0.9211
1.0	0.50	0.8776
1.5	0.75	0.7317
2.0	1.00	0.5403

The progressive waveform evolution is depicted below as $X = X_1/\bar{x}$ progresses from 0.1 to 0.5 to 1.0.

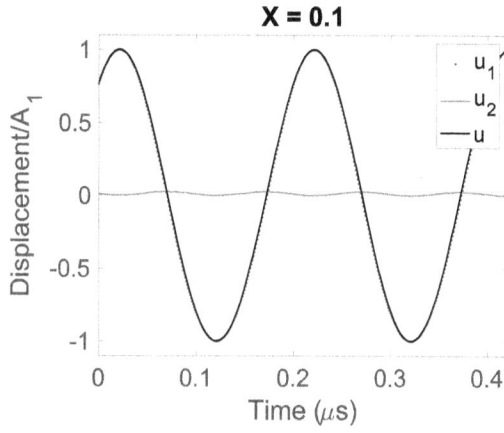

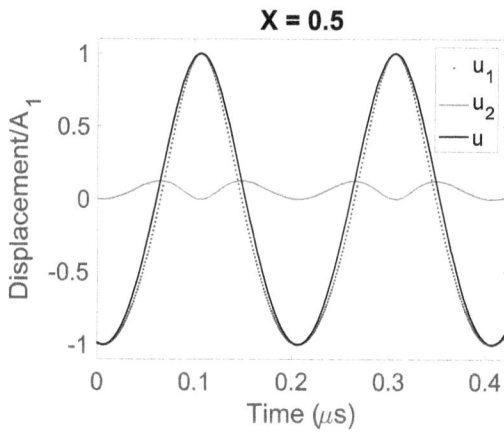

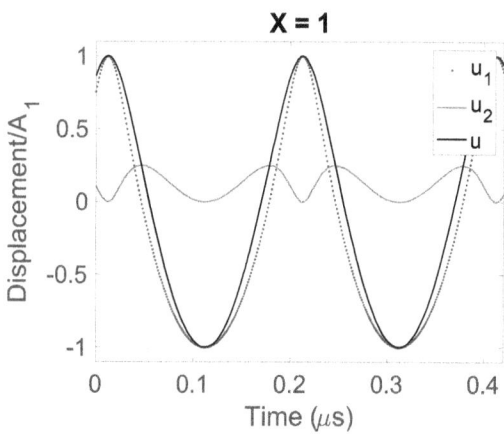

Appendix: Prove that $f(s_1, y_1) = F(\xi) + \frac{\alpha}{2}y_1[F'(\xi)]^2$ is the solution to the amplitude equation $\frac{\partial^2 f}{\partial y_1 \partial s_1} + \frac{\alpha}{2}\frac{\partial f}{\partial s_1}\frac{\partial^2 f}{\partial s_1^2} = 0$.

Start by differentiating $s_1 = \xi + \frac{\alpha}{2}y_1 F'(\xi)$ with respect to s_1 and y_1:

$$1 = \frac{\partial \xi}{\partial s_1} + \frac{\alpha}{2}y_1 F''(\xi)\frac{\partial \xi}{\partial s_1} \rightarrow \frac{\partial \xi}{\partial s_1} = \frac{1}{1 + \frac{\alpha}{2}y_1 F''(\xi)}$$

$$0 = \frac{\partial \xi}{\partial y_1} + \frac{\alpha}{2}F'(\xi) + \frac{\alpha}{2}y_1 F''(\xi)\frac{\partial \xi}{\partial y_1} \rightarrow \frac{\partial \xi}{\partial y_1} = -\frac{\frac{\alpha}{2}F'(\xi)}{1 + \frac{\alpha}{2}y_1 F''(\xi)}.$$

Also differentiate f with respect to s_1,

$$\frac{\partial f}{\partial s_1} = F'(\xi)\frac{\partial \xi}{\partial s_1} + \frac{1}{2}\frac{\alpha}{2}y_1 2F'(\xi)F''(\xi)\frac{\partial \xi}{\partial s_1} = F'(\xi)\left[1 + \frac{\alpha}{2}y_1 F''(\xi)\right]\frac{\partial \xi}{\partial s_1} = F'(\xi).$$

Now we have

$$\frac{\partial^2 f}{\partial s_1^2} = F''(\xi)\frac{\partial \xi}{\partial s_1}$$

$$\frac{\partial^2 f}{\partial s_1 \partial y_1} = F''(\xi)\frac{\partial \xi}{\partial y_1}$$

and the amplitude equation becomes

$$\frac{\partial^2 f}{\partial y_1 \partial s_1} + \frac{\alpha}{2}\frac{\partial f}{\partial s_1}\frac{\partial^2 f}{\partial s_1^2} = F''(\xi)\frac{\partial \xi}{\partial y_1} + \frac{\alpha}{2}F'(\xi)F''(\xi)\frac{\partial \xi}{\partial s_1} = F''(\xi)\left[\frac{\partial \xi}{\partial y_1} + \frac{\alpha}{2}F'(\xi)\frac{\partial \xi}{\partial s_1}\right]$$

$$\frac{\partial^2 f}{\partial y_1 \partial s_1} + \frac{\alpha}{2}\frac{\partial f}{\partial s_1}\frac{\partial^2 f}{\partial s_1^2} = F''(\xi)\left[-\frac{\frac{\alpha}{2}F'(\xi)}{1 + \frac{\alpha}{2}y_1 F''(\xi)} + \frac{\alpha}{2}F'(\xi)\frac{1}{1 + \frac{\alpha}{2}y_1 F''(\xi)}\right] = 0.$$

The end.

Example 3.4. Determine β in terms of Landau–Lifshitz TOECs for an isotropic material.

The acoustic nonlinearity coefficient β is defined for longitudinal bulk waves in terms of the contracted Huang coefficients in equation (3.9):

$$\beta \equiv -\frac{A_{111}}{A_{11}}.$$

Derive β in terms of the Brugger coefficients and then write it in terms of the Landau–Lifshitz third order elastic constants (TOECs). The relationships between the Huang and Brugger coefficients are given in tensor notation in equations (2.32)–(2.35). We will compute A_{111111} and A_{1111} and then contract them to obtain A_{111} and A_{11}:

$$A_{ijklmn} = C_{ijklmn} + C_{jlmn}\delta_{ik} + C_{ijln}\delta_{km} + C_{jnkl}\delta_{im} \qquad (2.34)$$

$$A_{111111} = C_{111111} + C_{1111}\delta_{11} + C_{1111}\delta_{11} + C_{1111}\delta_{11}$$
$$A_{111111} = C_{111111} + 3C_{1111},$$

likewise

$$A_{ijkl} = C_{ijkl} + C_{jl}\delta_{ik} \qquad (2.33)$$
$$A_{1111} = C_{1111} + C_{11}\delta_{11} = C_{1111} + C_{11}.$$

In contracted notation we have

$$A_{111} = C_{111} + 3C_{11}$$
$$A_{11} = C_{11} + C_{1},$$

where C_1 is the residual stress in the X_1 direction. Therefore,

$$\beta \equiv -\frac{A_{111}}{A_{11}} = -\frac{3C_{11} + C_{111}}{\sigma_1 + C_{11}}.$$

For an isotropic material

$$C_{11} = \lambda + 2\mu$$
$$C_{111} = 2\mathscr{C} + 6\mathscr{B} + 2\mathscr{A}$$

from equations (2.42) and (2.46), respectively. Finally, we have

$$\beta = -\frac{3(\lambda + 2\mu) + 2\mathscr{C} + 6\mathscr{B} + 2\mathscr{A}}{\sigma_1 + (\lambda + 2\mu)}.$$

This definition is used by many, including for example Cantrell [10] and Kundu [11]. However, other definitions are used for β, notably

$$\beta \equiv -\left(\frac{3}{2} + \frac{C_{111}}{2\rho c^2}\right)$$

used by Norris [12] and de Lima and Hamilton [13], which is simply one-half of equation (3.9).

3.2 Bulk shear waves

Consider now planar shear waves constrained to propagate solely in the X_1 direction such that

$$\mathbf{u} = \begin{Bmatrix} 0 \\ v \\ 0 \end{Bmatrix}, \mathbf{H} = \begin{bmatrix} 0 & 0 & 0 \\ v' & 0 & 0 \\ 0 & 0 & 0 \end{bmatrix}, \mathbf{E} = \frac{1}{2}\begin{bmatrix} (v')^2 & v' & 0 \\ v' & 0 & 0 \\ 0 & 0 & 0 \end{bmatrix}, \qquad (3.18)$$

where $v' = \partial v/\partial X_1$. As there is no spatial dependence on X_2 and X_3, the only nontrivial equation in equation (3.1) occurs for $j = 1$ when $i = 2$:

$$T_{12,\,1}^{PK1} = \rho_o \ddot{v}. \qquad (3.19)$$

We will write out the 12 and 21 stress components from equation (3.2), which is helped by there being only one nonzero component of **H**. In so doing we must be careful with the indices:

$$T_{ij}^{PK1} = A_{ij} + \tfrac{1}{2!}\left[A_{ijrs}H_{rs} + A_{pqij}H_{pq}\right]$$
$$+ \tfrac{1}{3!}\left[A_{ijrsuv}H_{rs}H_{uv} + A_{pqijuv}H_{pq}H_{uv} + A_{pqrsij}H_{pq}H_{rs}\right]$$
$$+ \tfrac{1}{4!}\left[A_{ijrsuvkl}H_{rs}H_{uv}H_{kl} + A_{pqijuvkl}H_{pq}H_{uv}H_{kl} + A_{pqrsijkl}H_{pq}H_{rs}H_{kl} + A_{pqrsuvij}H_{pq}H_{rs}H_{uv}\right] + \dots$$

$$T_{12}^{PK1} = A_{12} + \tfrac{1}{2!}[A_{1221}H_{21} + A_{2112}H_{21}]$$
$$+ \tfrac{1}{3!}[A_{122121}H_{21}H_{21} + A_{211221}H_{21}H_{21} + A_{212112}H_{21}H_{21}]$$
$$+ \tfrac{1}{4!}[A_{12212121}H_{21}H_{21}H_{21} + A_{21122121}H_{21}H_{21}H_{21} + A_{21211221}H_{21}H_{21}H_{21} + A_{21212112}H_{21}H_{21}H_{21}] + \dots$$

$$T_{12}^{PK1} = A_{12} + A_{1221}(v') + \frac{1}{2}A_{122121}(v')^2 + \frac{1}{6}A_{12212121}(v')^3 + \dots \tag{3.20}$$

$$T_{21}^{PK1} = A_{21} + \tfrac{1}{2!}[A_{2121}H_{21} + A_{2121}H_{21}]$$
$$+ \tfrac{1}{3!}[A_{212121}H_{21}H_{21} + A_{212121}H_{21}H_{21} + A_{212121}H_{21}H_{21}]$$
$$+ \tfrac{1}{4!}[A_{21212121}H_{21}H_{21}H_{21} + A_{21212121}H_{21}H_{21}H_{21} + A_{21212121}H_{21}H_{21}H_{21} + A_{21212121}H_{21}H_{21}H_{21}] + \dots$$

$$T_{21}^{PK1} = A_{21} + A_{2121}(v') + \frac{1}{2}A_{212121}(v')^2 + \frac{1}{6}A_{21212121}(v')^3 + \dots \tag{3.21}$$

We can now compare the shear stress terms in equations (3.20) and (3.21) by converting the Huang coefficients into Brugger coefficients (whose symmetries are known because they relate strain to a symmetric stress tensor) by using equations (2.32)–(2.35). This is straightforward, but tedious, so the analysis will not be presented here. We find that

$$A_{2121} = A_{1221} + C_{11}$$

$$A_{212121} = A_{122121} + 2C_{1211}$$

$$A_{21212121} \neq A_{12212121}$$

and therefore $T_{21}^{PK1} \neq T_{12}^{PK1}$.

We wish to solve a problem having quadratic nonlinearity. Therefore, the third order term (from the fourth order term in the strain energy function) and higher order terms will be neglected. Equation (3.19) becomes

$$\frac{\partial}{\partial X_1}\left[A_{12} + A_{1221}(v') + \frac{1}{2}A_{122121}(v')^2\right] = \rho_o \ddot{v} \tag{3.22}$$

and given that the material is homogeneous,

$$A_{1221}(v'') + A_{122121}(v')(v'') = \rho_o \ddot{v}. \tag{3.23}$$

Before proceeding further, we should evaluate A_{122121} from

$$A_{ijklmn} = C_{ijklmn} + C_{jlmn}\delta_{ik} + C_{ijln}\delta_{km} + C_{jnkl}\delta_{im} \tag{2.34}$$

$$A_{122121} = C_{122121} + C_{2121}\delta_{12} + C_{1211}\delta_{22} + C_{2121}\delta_{12}$$

$$A_{122121} = C_{122121} + C_{1211}.$$

If we limit consideration to an isotropic material, then we can evaluate C_{122121} from

$$\begin{aligned} C_{ijklmn} &= 2\mathscr{C}\delta_{ij}\delta_{kl}\delta_{mn} + 2\mathscr{B}\big(\delta_{ij}\delta_{lm}\delta_{kn} + \delta_{im}\delta_{jn}\delta_{kl} + \delta_{ik}\delta_{jl}\delta_{mn}\big) \\ &\quad + \mathscr{A}\big(\delta_{il}\delta_{jm}\delta_{kn} + \delta_{im}\delta_{jk}\delta_{ln}\big) \end{aligned} \tag{2.46}$$

$$\begin{aligned} C_{122121} &= 2\mathscr{C}\delta_{12}\delta_{21}\delta_{21} + 2\mathscr{B}(\delta_{12}\delta_{12}\delta_{21} + \delta_{12}\delta_{21}\delta_{21} + \delta_{12}\delta_{21}\delta_{21}) \\ &\quad + \mathscr{A}(\delta_{11}\delta_{22}\delta_{21} + \delta_{12}\delta_{22}\delta_{11}) \end{aligned}$$

$$C_{122121} = 0.$$

Also,

$$C_{ijkl} = \lambda\delta_{ij}\delta_{kl} + \mu\big(\delta_{ik}\delta_{jl} + \delta_{il}\delta_{jk}\big) \tag{2.42}$$

$$C_{1211} = \lambda\delta_{12}\delta_{11} + \mu(\delta_{11}\delta_{21} + \delta_{11}\delta_{21})$$

$$C_{1211} = 0,$$

therefore,

$$A_{122121} = 0. \tag{3.24}$$

Now we see that for an isotropic material with quadratic nonlinearity equation (3.19) reduces to

$$A_{1221}(v'') = \rho_o\ddot{v}, \tag{3.25}$$

which is the linear wave equation for transverse waves as A_{1221} is given by

$$A_{ijkl} = C_{ijkl} + C_{jl}\delta_{ik} \tag{2.37}$$

$$A_{1221} = C_{1221} + C_{21}\delta_{12}$$

$$A_{1221} = C_{1221} + 0$$

$$C_{ijkl} = \lambda\delta_{ij}\delta_{kl} + \mu\big(\delta_{ik}\delta_{jl} + \delta_{il}\delta_{jk}\big) \tag{2.42}$$

$$C_{1221} = \lambda\delta_{12}\delta_{21} + \mu(\delta_{12}\delta_{21} + \delta_{11}\delta_{22}) = \mu$$

$$A_{1221} = \mu$$

and the wave equation becomes

$$\frac{\mu}{\rho_o}(v'') = \ddot{v}. \tag{3.26}$$

Therefore, finite amplitude shear waves propagating in isotropic materials do not generate second harmonics. Green [14] and Norris [15] arrive at this conclusion based directly on the symmetry of the third order elastic constants. However, example 3.5 demonstrates that shear waves do generate third harmonics.

The example problems in this chapter illustrate that the ultrasonic nonlinearity depends on the TOECs and FOECs. These elasticity parameters can be determined from acoustoelasticity measurements or second and third harmonic generation measurements. TOECs and FOECs are not nearly as available as the second order elastic constants (SOECs) used for linear elasticity and readily available from ASM International or in online databases such as matweb.com. However, the Landolt–Börnstein database (Group III condensed matter 29A) provides TOECs and some FOECs. Lurie [18] provides tables with TOECs for steels (his table 1), aluminum and copper alloys (his table 2), and miscellaneous materials (his table 3). A subset of these data, provided in table 3.1, have been converted to the Landau–Lifshitz constants $\mathcal{A}$, $\mathcal{B}$, $\mathcal{C}$ using the conversions given by Norris [12] for the third order Lamé constants [19]. These data are given as representative examples, as the sources

Table 3.1. Representative third order elastic constants from Lurie [18].

Material	ρ kg m^{-3}	λ GPa	μ GPa	$\mathcal{A}$ GPa	$\mathcal{B}$ GPa	$\mathcal{C}$ GPa
Steel Hecla 37 (0.4%C, 0.3%Si, 0.8%Mn)	7823	111 ± 1	82.1 ± 0.5	−708 ± 32	−282 ± 30	−179 ± 35
Steel Hecla 17 (0.6%C, 0.2%Si, 0.8%Mn)	7825	110 ± 1	82.0 ± 0.5	−668 ± 24	−261 ± 20	−67 ± 10
Steel Hecla 138A (0.4%C, 0.4%Cr, 0.5%Mo, 2.5%Ni)	7843	109 ± 1	81.9 ± 0.5	−708 ± 40	−265 ± 30	−161.5 ± 25
Steel Hecla ATV (3.6%Ni, 10%Cr, 1%Mn)	8065	87 ± 2	71.6 ± 3	−400	−552 ± 80	17 ± 10
Aluminum B53S (2.8%Mg, 0.8% Mn, 0.1%Cr)	2677	58.0 ± 0.2	26.0 ± 0.1	−240 ± 20	−99 ± 10	−124.5 ± 30
Aluminum D54S (4.5%Mg, 0.8% Mn, 1%Cr)	2719	49.1 ± 0.1	26.0 ± 0.1	−320 ± 12	−198 ± 9	−189.5 ± 1.2
Copper	—	107	47.7	−1592 ± 20	172 ± 14	−280 ± 70
Magnesium	1716	25.9 ± 0.2	16.6 ± 3	−168.4 ± 16	−57.4 ± 1	−32.7 ± 0.5
Acrylic	1160	39	18.6	18.8	−7	−3.9
Fused quartz	2203	15.872	31.261	−44 ±12	93 ± 8	27 ± 6.5
Glass (pyrex)	—	13.53 ± 0.03	27.5 ± 0.3	420 ± 352	−168 ± 138	130 ± 185

Example 3.5. Shear wave third harmonic generation.

Solve the wave equation with cubic nonlinearity for bulk shear waves. Equation (3.20) is

$$T_{12}^{PK1} = A_{12} + A_{1221}(v') + \frac{1}{2}A_{122121}(v')^2 + \frac{1}{6}A_{12212121}(v')^3 + \dots.$$

We know that $A_{1221} = \mu$ and $A_{122121} = 0$ and need to find $A_{12212121}$ from

$$A_{ijklmnpq} = \frac{1}{2}C_{ijklmnpq} + C_{qjklmn}\delta_{ip} + C_{ijlnpq}\delta_{km} + \frac{1}{2}C_{ijklnq}\delta_{mp}$$
$$+ \frac{1}{2}C_{jlmnpq}\delta_{ik} + \frac{1}{2}C_{jlnq}\delta_{ik}\delta_{mp} + C_{jnlq}\delta_{ip}\delta_{km} \tag{2.35}$$

$$A_{12212121} = \frac{1}{2}C_{12212121} + C_{122121}\delta_{12} + C_{121121}\delta_{22} + \frac{1}{2}C_{122111}\delta_{22} + \frac{1}{2}C_{212121}\delta_{12}$$
$$+ \frac{1}{2}C_{2111}\delta_{12}\delta_{22} + C_{2111}\delta_{12}\delta_{22}$$

$$A_{12212121} = \frac{1}{2}C_{12212121} + 0 + C_{121121} + \frac{1}{2}C_{122111} + 0 + 0 + 0$$

$$C_{ijklmn} = 2\mathscr{C}\delta_{ij}\delta_{kl}\delta_{mn} + 2\mathscr{B}\big(\delta_{ij}\delta_{lm}\delta_{kn} + \delta_{im}\delta_{jn}\delta_{kl} + \delta_{ik}\delta_{jl}\delta_{mn}\big)$$
$$+ \mathscr{A}\big(\delta_{il}\delta_{jm}\delta_{kn} + \delta_{im}\delta_{jk}\delta_{ln}\big) \tag{2.46}$$

$$C_{121121} = 0 + 2\mathscr{B}(0 + 0 + 0) + \mathscr{A}(1 + 0) = \mathscr{A}$$

$$C_{122111} = 0 + 2\mathscr{B}(0 + 0 + 0) + \mathscr{A}(0 + 1) = \mathscr{A}.$$

We do not have an expression for $C_{ijklmnpq}$ in terms of the fourth order elastic constants, but we know that

$$\frac{1}{4!}C_{ijklmnpq}E_{ij}E_{kl}E_{mn}E_{pq}$$
$$= \mathscr{E}\mathrm{tr}(\mathbf{E})\mathrm{tr}(\mathbf{E}\cdot\mathbf{E}\cdot\mathbf{E}) + \mathscr{F}(\mathrm{tr}(\mathbf{E}))^2\mathrm{tr}(\mathbf{E}\cdot\mathbf{E}) + \mathscr{G}(\mathrm{tr}(\mathbf{E}\cdot\mathbf{E}))^2 \tag{2.49}$$
$$+ \mathscr{H}(\mathrm{tr}(\mathbf{E}))^4$$

and given the strain tensor in equation (3.18), after some work we arrive at

$$\frac{1}{4!}C_{ijklmnpq}E_{ij}E_{kl}E_{mn}E_{pq}$$
$$= \mathscr{E}E_{11}(E_{11}^3 + 3E_{11}E_{12}E_{21}) + \mathscr{F}E_{11}^2(E_{11}^2 + 2E_{12}E_{21})$$
$$+ \mathscr{G}(E_{11}^2 + 2E_{12}E_{21})^2 + \mathscr{H}E_{11}^4,$$

which enables us to write

$$\frac{1}{4!}C_{12212121}E_{12}E_{21}E_{21}E_{21} = \mathscr{G}(2E_{12}E_{21})^2$$

$$C_{12212121} = 4 \cdot 4!\mathscr{G}.$$

Therefore,

$$A_{12212121} = \frac{1}{2} \cdot 4 \cdot 4!\mathscr{G} + \mathscr{A} + \frac{1}{2}\mathscr{A} = \frac{3}{2}\mathscr{A} + 48\mathscr{G}.$$

From equation (3.19), the shear wave equation for cubic nonlinearity can now be written

$$\frac{\partial}{\partial X_1}\left[A_{12} + \mu(v') + \frac{1}{6}A_{12212121}(v')^3\right] = \rho_o \ddot{v}$$

$$\mu v'' + \frac{1}{2}A_{12212121}(v')^2 v '' = \rho_o \ddot{v}.$$

Defining

$$c_T^2 \equiv \frac{\mu}{\rho_o}$$

$$\gamma \equiv -\frac{A_{12212121}}{2\mu}$$

we obtain the nonlinear wave equation

$$\ddot{v} - c_T^2 v '' = -\gamma c_T^2(v')^2 v'',$$

which we will solve using the approach of example 3.1. Start with the primary wave solution,

$$v_{(1)} = A_1 \cos(kX_1 - \omega t).$$

Now substitute the primary wave solution into the nonlinear driving force

$$-\gamma c_T^2(v')^2 v'' = -\gamma c_T^2[-kA_1 \sin(kX_1 - \omega t)]^2 [-k^2 A_1 \cos(kX_1 - \omega t)]$$

$$-\gamma c_T^2(v')^2 v'' = \gamma c_T^2 k^4 A_1^3 \sin^2(kX_1 - \omega t)\cos(kX_1 - \omega t).$$

Using the trigonometric identities

$$\cos(3\theta) = 4\cos^3(\theta) - 3\cos(\theta)$$

$$\cos^2(\theta) + \cos^2(\theta) = 1$$

we can write

$$\sin^2(kX_1 - \omega t)\cos(kX_1 - \omega t) = \frac{1}{4}[\cos(kX_1 - \omega t) - \cos 3(kX_1 - \omega t)]$$

and give the driving force to be

$$-\gamma c_T^2(v')^2 v '' = -\gamma c_T^2 k^4 A_1^3 \frac{1}{4}[\cos 3(kX_1 - \omega t) - \cos(kX_1 - \omega t)].$$

The partial differential equation for the secondary wave solution becomes

$$\ddot{v}_{(2)} - c_T^2 v''_{(2)} = -\gamma c_T^2 k^4 A_1^3 \frac{1}{4}[\cos 3(kX_1 - \omega t) - \cos(kX_1 - \omega t)].$$

Let's pick the trial solution

$$v_{(2)} = D_3(X_1)\sin 3(kX_1 - \omega t) + D_4(X_1)\cos 3(kX_1 - \omega t)$$
$$+ D_5(X_1)\sin(kX_1 - \omega t) + D_6(X_1)\cos(kX_1 - \omega t)$$

that enables us to determine the unknown functions $D_3(X_1) - D_6(X_1)$ by requiring that $v_{(2)}(X_1 = 0) = 0$. Grinding through the substitution of $v_{(2)}$ into the partial differential equation we set $D_3'' = 0$, which gives $D_4' = 0$, $D_4'' = 0$, $D_4 = 0$. Then we set $D_5'' = 0$, giving $D_6' = 0$, $D_6'' = 0$, $D_6 = 0$. The nonzero results are

$$D_3 = \frac{1}{24}\gamma k^3 A_1^3 X_1$$

$$D_5 = -\frac{1}{8}\gamma k^3 A_1^3 X_1.$$

The primary and secondary solutions are now added together:

$$v = v_{(1)} + v_{(2)} = A_1 \cos(kX_1 - \omega t) + \frac{1}{24}\gamma k^3 A_1^3 X_1[\sin 3(kX_1 - \omega t) - 3\sin(kX_1 - \omega t)]$$

and we can identify the amplitude of the third harmonic waves to be

$$A_3 = \frac{1}{24}\gamma k^3 A_1^3 X_1,$$

where

$$\gamma = -\frac{3\mathscr{A} + 24\mathscr{G}}{4\mu}.$$

The amplitude of the third harmonic A_3 depends on the cube of the primary wave amplitude, the cube of the wavenumber, and grows linearly with propagation distance X_1. It is worth noting that this solution agrees with (12.26) in Truell *et al* [16]. Interested readers are referred to Wang and Achenbach's [17] analysis (and its erratum) of the cubic nonlinearity in torsional waves in a pipe.

of the data cited by Lurie [18] have not been vetted by the author. Readers are cautioned to use these data at their own risk.

3.3 Attenuation

The acoustic nonlinearity coefficient β was defined in equation (3.13), based on equation (3.11), in terms of the amplitudes of planar elastic waves in a lossless solid. Example 3.4 is even more specific in that it relates β to the elastic material parameters for an isotropic material. However, real materials suffer attenuation from absorption, scattering, diffraction, etc. The longer the propagation distance, the more likely the attenuation is to affect the measurement of β. Thus, it behooves us to be able to correct measurements for the effects of attenuation. We use the measurement of β as an illustrative example following Hikata and Elbaum [20] and Cantrell [10].

Consider a lossy nonlinear material. The primary and second harmonic amplitudes depend on propagation distance X_1,

$$A_1(X_1) = A_{1\_0}e^{-\alpha_1 X_1} \tag{3.27}$$

$$A_2(X_1) = A_{2\_0}e^{-\alpha_2 X_1} + \frac{1}{8}\beta A_1^2 k^2 X_1, \tag{3.28}$$

where any interaction between attenuation and second harmonic generation is ignored. The change in A_2 with propagation distance X_1 is

$$\frac{dA_2}{dX_1} = -\alpha_2 A_{2\_0} e^{-\alpha_2 X_1} + \frac{1}{8} \beta A_1^2 k^2, \tag{3.29}$$

which we recognize to be a first order ordinary differential equation

$$\frac{dA_2}{dX_1} + \alpha_2 A_2 = \frac{1}{8} \beta A_1^2 k^2 \tag{3.30}$$

that can be solved subject to the initial condition $A_2(X_1 = 0) = 0$. This is a nonhomogeneous equation having the general solution

$$A_{2\_gen} = C_1 e^{-\alpha_2 X_1} \tag{3.31}$$

and a particular solution

$$A_{2\_part} = \frac{\beta k^2 A_{1\_0}^2}{8(\alpha_2 - 2\alpha_1)} e^{-2\alpha_1 X_1}. \tag{3.32}$$

Therefore,

$$A_2 = A_{2\_gen} + A_{2\_part} = C_1 e^{-\alpha_2 X_1} + \frac{\beta k^2 A_{1\_0}^2}{8(\alpha_2 - 2\alpha_1)} e^{-2\alpha_1 X_1}. \tag{3.33}$$

The remaining constant C_1 is determined from the initial condition to be

$$C_1 = -\frac{\beta k^2 A_{1_0}^2}{8(\alpha_2 - 2\alpha_1)}, \tag{3.34}$$

giving the simplified solution

$$A_2(X_1) = \frac{\beta k^2 A_{1\_0}^2}{8(\alpha_2 - 2\alpha_1)} [e^{-2\alpha_1 X_1} - e^{-\alpha_2 X_1}]. \tag{3.35}$$

Now the solution can be applied to the measurement of the acoustic nonlinearity coefficient in a lossy material. Compute β_{meas} from the measurement of wave amplitudes using equation (3.13)

$$\beta_{\text{meas}} = \frac{8A_2}{k^2 A_1^2 X_1}. \tag{3.36}$$

It is noteworthy that wave amplitude measurements are commonly done in units of volts, thus a calibration between volts and distance units (e.g. nm) is needed. Dace et al [21] provide a procedure for piezoelectric transducers. Alternatively, equation (3.14) is used to compute the relative nonlinearity coefficient β'. Moving forward, we now substitute equations (3.27) and (3.35) into equation (3.36):

$$\beta_{\text{meas}} = \frac{8\left\{\frac{\beta k^2 A_{1\_0}^2}{8(\alpha_2 - 2\alpha_1)}[e^{-2\alpha_1 X_1} - e^{-\alpha_2 X_1}]\right\}}{k^2[A_{1\_0}e^{-\alpha_1 X_1}]^2 X_1}$$

$$\vdots$$

$$\beta_{\text{meas}} = \frac{\beta}{(\alpha_2 - 2\alpha_1)X_1}[1 - e^{-(\alpha_2 - 2\alpha_1)X_1}]. \tag{3.37}$$

Finally, we solve for β

$$\beta = \beta_{\text{meas}}\frac{(\alpha_2 - 2\alpha_1)X_1}{1 - e^{-(\alpha_2 - 2\alpha_1)X_1}}. \tag{3.38}$$

If the attenuation is small, i.e. $(\alpha_2 - 2\alpha_1)X_1 \ll 1$, then $\beta \approx \beta_{\text{meas}}$ by using L'Hopital's rule, as expected. Norris [12] suggests an alternative approach to including material damping in the problem formulation through a viscous stress term. We close this section by mentioning that Hurley and Fortunko [22] provide a diffraction correction for when the waves are not planar.

3.4 Measurements of nonlinearity

As discussed in chapter 1, an application for nonlinear ultrasonics is to identify material degradation at an early stage. Thus, measurements of ultrasonic nonlinearity require a source of nonlinearity indicative of the degradation. The aim is to measure material nonlinearity associated with the effects of microscopic defects or discontinuities such as cracks. The interest is not in the gross plastic deformation that occurs as metals permanently deform (for example in forging operations), but rather subtle deviations in the elasticity. While engineers commonly treat engineering materials as being linear elastic, this is not entirely accurate. Interatomic elastic potentials that are non-quadratic give rise to anharmonicity due to the lattice structure. This anharmonicity is characteristic of the material in its pristine condition and only valuable as a reference state for nondestructive evaluation. It is the effect that lattice defects (such as voids, interstitial atoms, dislocations, and grain boundaries), microcracks, precipitates, and inclusions have on the material nonlinearity that is sought, as well as the contact nonlinearity from breathing cracks, disbonds, and delaminations. Measurements reliant upon the nonlinear wave equation are introduced in this section, beginning with acoustoelastic determination of stress.

3.4.1 Acoustoelasticity

Acoustoelasticity is a nonlinear phenomenon whereby the elastic wave speed depends on the current stress state. Although a higher order effect, acoustoelasticity can be very useful for residual stress determination and characterizing the third order elastic constants (TOECs). Acoustoelastic measurements typically rely upon linear ultrasound characteristics (i.e. wave speed) to quantify a nonlinear phenomenon. Hughes and Kelly [23] developed an early set of acoustoelasticity equations relating wave speeds to stresses and used them to determine the third order elastic constants in Murnaghan's model [24] for glass, iron, and polystyrene. Thurston and Brugger [25] developed expressions for the acoustic tensors and tabulated results for anisotropic crystals and isotropic materials. Green [26] indicates that TOECs are typically 3–10 times the magnitude of the second order elastic constants, and they can be positive or negative. Pao *et al* [27] provide a comprehensive review of acoustoelasticity and ultrasonic measurements of residual stress. Rocks have an unusually high degree of nonlinear elasticity, and some TOECs are measured and tabulated by Winkler and Liu [28].

A sampling of some interesting applications of acoustoelasticity with guided waves follows. Gandhi *et al* [29] developed a theory for acoustoelastic Lamb wave propagation in the presence of a uniform biaxial stress field. The dispersion curves show that both stress state and propagation direction play a role in determining the wave speed, which is supported by experiments. Based on these results, Shi *et al* [30] developed a numerical approach to predict a uniform biaxial stress state from wave speed measurements in different directions. Fatigue cracks were found to cause wave scattering that complicates accurate measurements. Nucera and Lanza di Scalea [31, 32] propose a modeling approach to determine the stress-free temperature (known as the neutral temperature) of a constrained solid. Rather than being based on wave speed (and acoustoelasticity), it is based on second harmonic generation, which clearly shows the connection between acoustoelasticity and second harmonic generation is the material nonlinearity. The application driving this research is thermal buckling-related failures of continuous welded rail. Related to the connection between acoustoelasticity and second harmonic generation, Pau and Lanza di Scalea [33] developed an approach for modeling Lamb wave propagation that encompasses both effects, i.e. wave distortion had not been addressed in the prior models. Motivated by the lack of stress sensitivity for bulk waves and some Lamb wave modes, Pei and Bond [34] investigated using higher order Lamb wave modes (e.g. the S1 mode) which have been shown by models to have an unusually high stress dependence. They found that the S1 mode excited near the cut-off frequency (where the group velocity is low) is roughly ten times more sensitive to stress aligned perpendicular to the wave propagation direction. Peddeti and Santhanam [35] implemented the acoustoelastic equations into a semi-analytical finite element (SAFE) analysis for application to waveguides having arbitrary cross-sections. Dubuc *et al* [36] were interested in determining the prestress in multi-wire strands and used high-frequency higher order wave modes to do so. In the most recent article mentioned here, Mohabuth *et al* [37] proposed using a combination of Rayleigh waves and bulk waves to characterize the TOECs in metals.

3.4.2 Second harmonic generation

As already demonstrated in example 3.1, waveform distortion associated with material nonlinearity is evident in the frequency spectrum through second harmonic generation. Matlack *et al* [38] and Jhang *et al* [39] provide thorough descriptions of

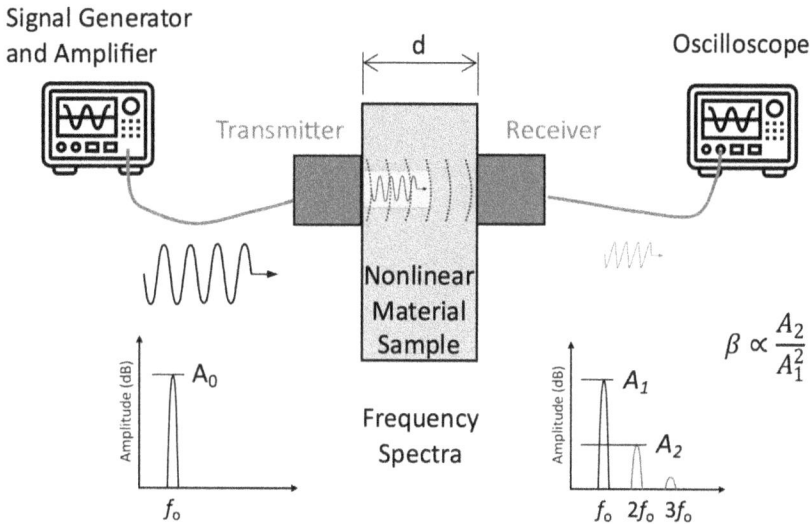

Figure 3.2. Self-interaction of waves in a nonlinear material leading to second harmonic generation.

second harmonic ultrasonic testing of materials. Almost without exception, second harmonic testing is performed in the through-transmission mode. Pulse–echo is not a viable option because the reflected second harmonic waves are not cumulative [40, 41]. The through-transmission mode imposes some accessibility limitations that can be avoided by using guided waves because single-sided access is sufficient for Rayleigh, Lamb, and SH waves. Moreover, guided waves can provide more efficient volumetric coverage of a structure because of the reduction in the scanning requirements for full coverage.

Figure 3.2 illustrates the essence of second harmonic testing. The resulting frequency spectra can be analysed to compute the relative nonlinearity β' from a range of primary wave amplitudes as depicted in figure 3.3 or to assess the cumulative nature of the second harmonic as shown in figure 3.4. Measurements to determine β' typically utilize received voltages rather than actual amplitudes, thereby relying upon the relationship between amplitude in voltage and amplitude of the ultrasonic waves in nanometers. Measurements to determine the absolute β must use the amplitude in nanometers; a capacitive receiver is described by Breazeale and Philip [42] and a calibration procedure for piezoelectric transducers is outlined by Dace *et al* [21]. Through-thickness measurements with longitudinal bulk waves in the 5–10 MHz range are common. But if these are to be nondestructive tests then measurements with increasing propagation distances are often not practical. Rayleigh waves and Lamb waves are more conducive to adjustable propagation distances and accordingly comprise a significant portion of this book.

3.4.3 Wave mixing

Waves having different frequencies, say f_a and f_b, mix together in nonlinear media to generate waves at combinational frequencies. The second order mutual interactions

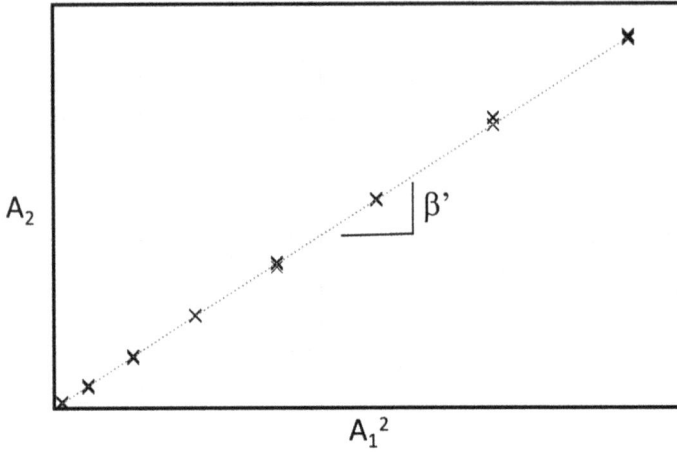

Figure 3.3. Recording many A-scans as a function of the primary wave amplitude (in volts) enables the relative nonlinearity coefficient to be fit by linear regression for fixed propagation distance.

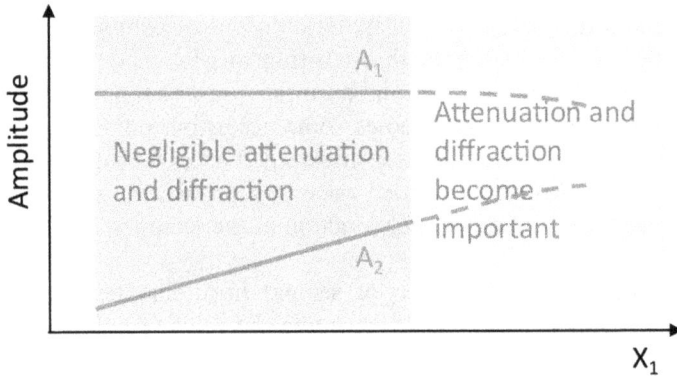

Figure 3.4. Many A-scans and their frequency spectra recorded over a range of propagation distances reveal the cumulative nature of the second harmonic.

result in the sum and difference frequencies $f_a + f_b$ and $f_a - f_b$, respectively, see for example [43, 44]. The wave vectors for the mixing waves may be collinear or non-collinear. Wave mixing is appealing because it enables selection of primary frequencies that generate secondary waves at frequencies sufficiently removed from frequencies that carry measurement system nonlinearities, and which might otherwise confound the detection of weak material nonlinearity indicative of material degradation. Additionally, non-collinear wave mixing geometries can be designed such that the secondary waves propagate in a direction that the primary waves and system nonlinearity-generated waves do not, making them easier to detect. Moreover, wave mixing enlarges the design space in which nonlinear measurements can be made in domains having limited accessibility. Figure 3.5

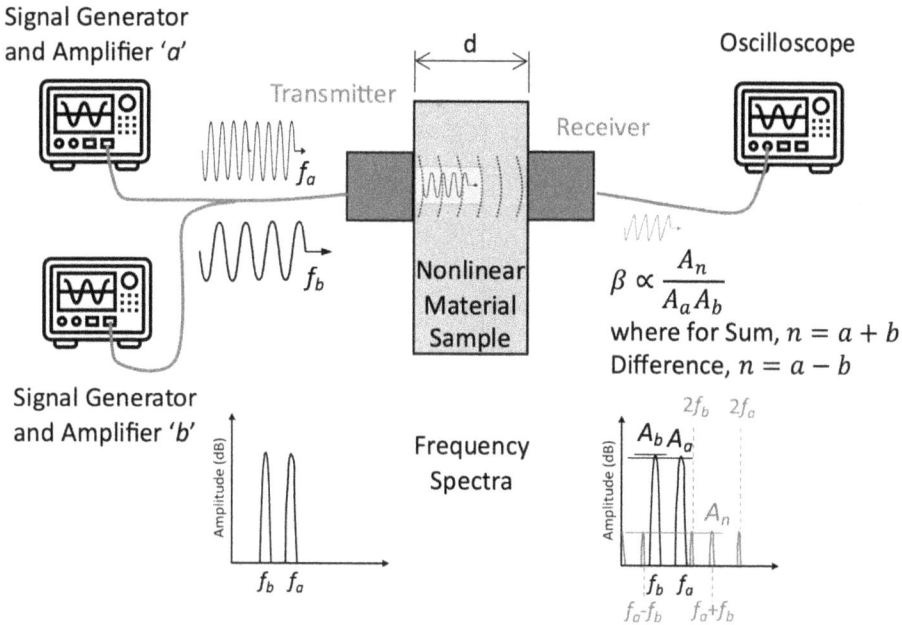

Figure 3.5. Mutual interaction from wave mixing in a nonlinear material leading to combinational harmonic generation.

illustrates the essentials of a collinear wave mixing application. We will discuss a number of guided wave mixing arrangements later in this book.

3.4.4 Nonlinear resonant ultrasound spectroscopy (NRUS)

The resonant frequencies of a solid body depend on its elasticity, mass density, and geometric configuration. A subset of the infinite number of resonant frequencies of a continuous solid body can be used via an inversion process to determine the elasticity tensor of a homogeneous body provided its mass and geometry are well-known. This is the basis of resonant ultrasound spectroscopy (RUS), where dynamic loading is applied over a range of frequencies to identify a few of the resonant frequencies such that the inverse problem can be solved for the elasticity tensor coefficients. Since material degradation is typically evident from evolution of the elasticity tensor, this is an excellent method to identify damage. A drawback is that the inversion process is very sensitive to the body's geometry and boundary conditions. If the material is nonlinear or contains discontinuities with contact nonlinearities, then the resonant frequencies are dependent on the dynamic loading amplitude. In nonlinear resonant ultrasound spectroscopy (NRUS) selected resonant frequencies are excited by a swept-frequency signal for a progressively increasing (or decreasing) amplitude. The shift of a resonant frequency with amplitude, determined from the frequency spectrum, is indicative of the nonlinearity. The dynamic loading can range from a sequence of continuous sinusoidal

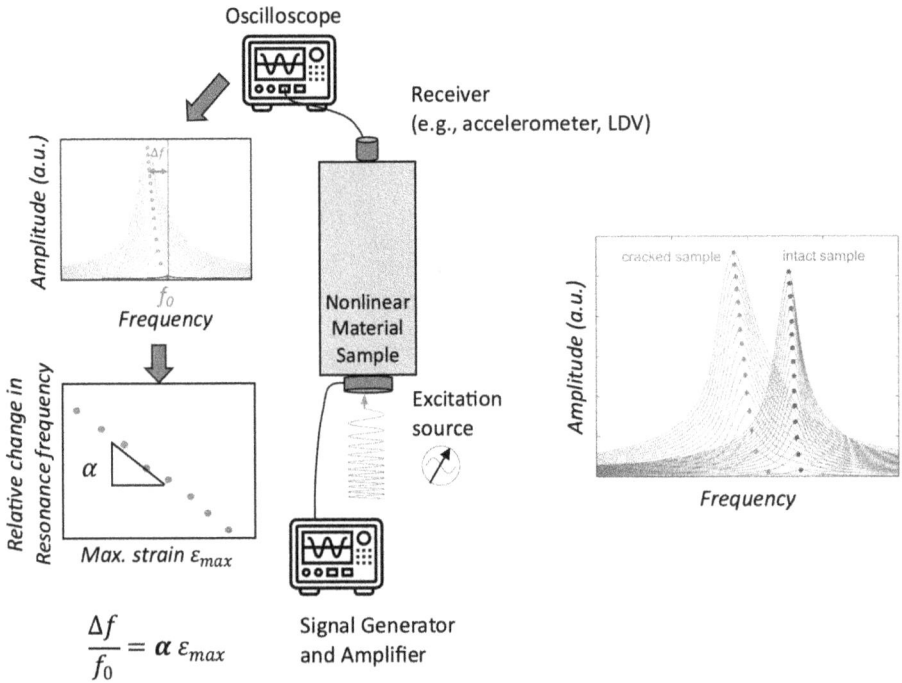

Figure 3.6. Nonlinear resonant ultrasound spectroscopy (NRUS) testing to characterize the strain amplitude sensitivity. The experimental result at right is for concrete. (Figure courtesy of Professor Parisa Shokouhi.)

waves, a chirp, or an impact. TenCate and Johnson [45] provide an excellent summary of NRUS methods and applications.

Figure 3.6 illustrates one version of NRUS testing whereby a chirp excitation is sent to a transducer that actuates waves in the nonlinear material sample, which are then received and the frequency spectrum computed. The test is repeated for increasing strain amplitudes and the relationship between the frequency shift and the maximum strain amplitude indicates the material nonlinearity through the linear regression coefficient α shown in the figure.

The resonances of localized defects can be leveraged by tuning the driving frequency of the standing waves to the local resonances. For example, Solodov and colleagues [39, 46, 47] have developed sensitive imaging techniques based on exciting these local defect resonances.

3.4.5 Vibro-acoustics

Vibro-acoustic methods are infused into nondestructive testing protocols in several ways. We refer here to pump–probe methods where a low frequency vibration, often having high amplitude, *pumps* energy into a material and at the same time high-frequency acoustic waves, often having low amplitude, *probe* for nonlinearity-induced modulation around the probe frequency due to the mutual interactions between the energies at different frequencies. An illustration of a typical set-up is shown in figure 3.7.

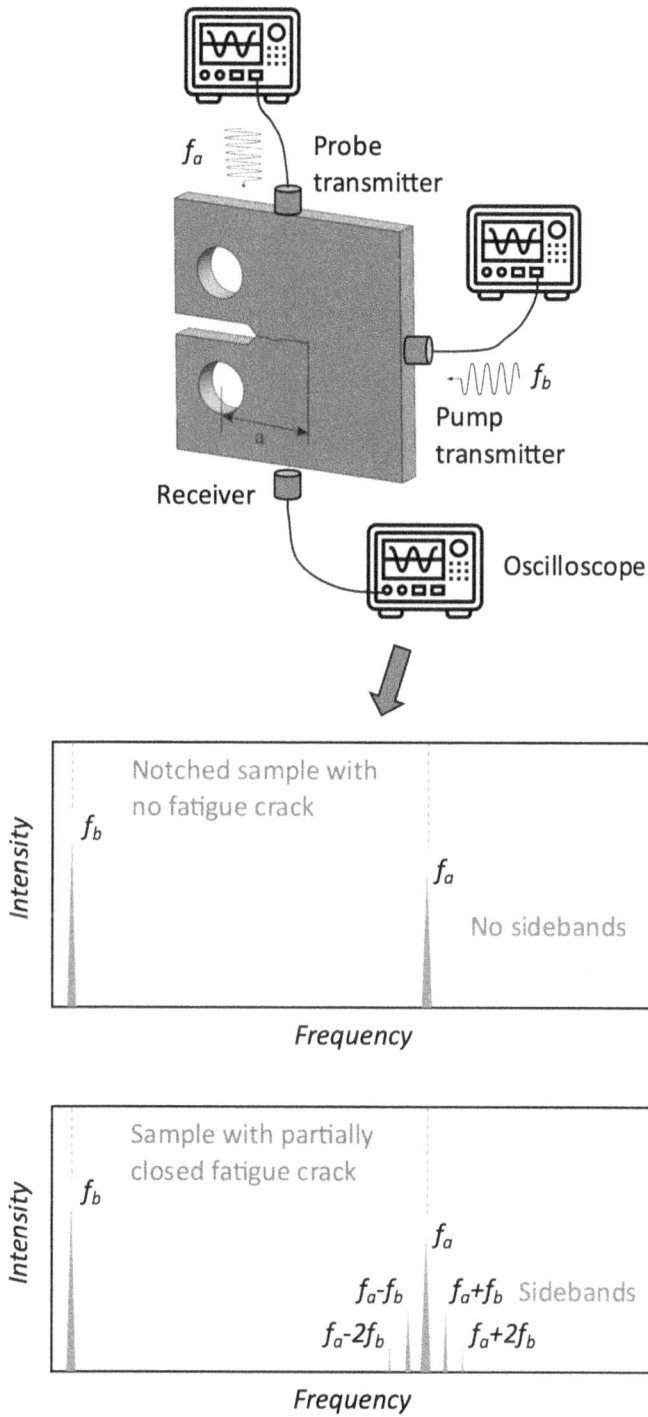

Figure 3.7. Vibro-acoustics testing with the pump and probe method based on the presence of sidebands.

The nonlinearity associated with the targeted damage causes sidebands to occur at integer multiples of the pump frequency around the probe frequency. Specific methods are often named as explicit types of spectroscopy due to the heavy reliance on the frequency spectrum. Fatigue cracks, delaminations, and interfacial debonding can all exhibit the contact acoustic nonlinearity (CAN) and are often the target of vibro-acoustic methods, see e.g. [48–51].

3.4.6 Dynamic acoustoelastic testing

While traditional acoustoelastic testing is conducted with a static stress field, dynamic acoustoelastic testing (DAET) employs a low frequency pump to generate a dynamic stress field that oscillates between tension and compression. DAET uses a short-duration probe to determine the effect of the dynamic stress on the elastic wave speed of the probe. Syncing the pump and probe waves is crucial to recording the

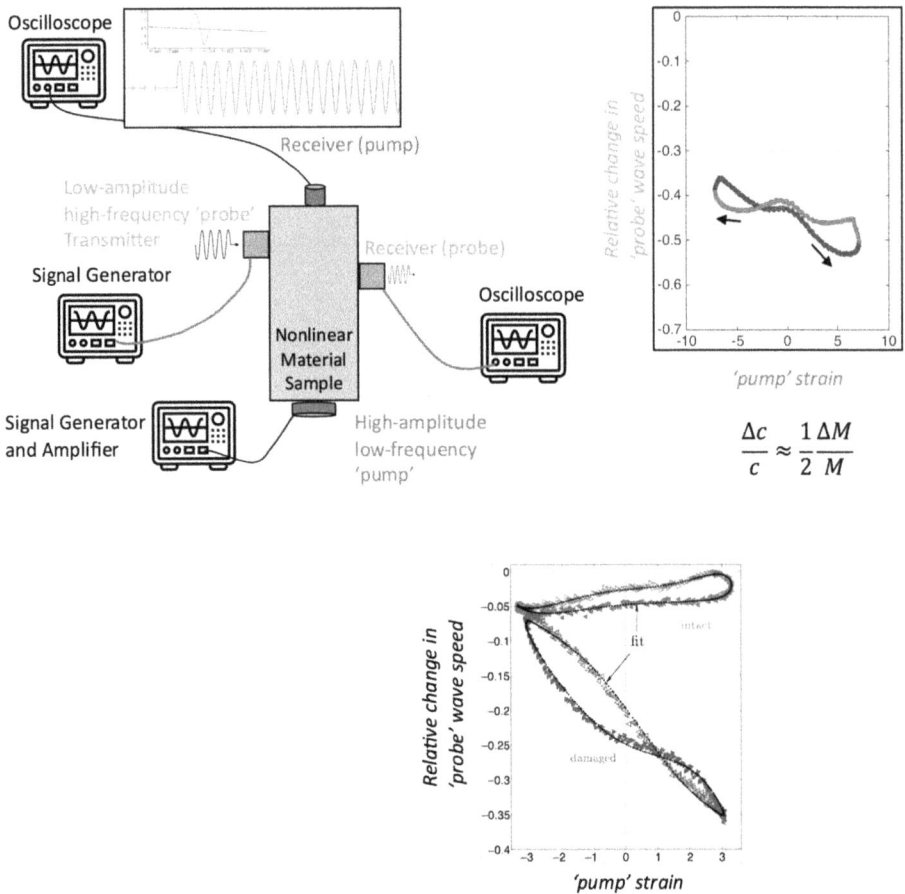

$$\frac{\Delta c}{c} \approx \frac{1}{2} \frac{\Delta M}{M}$$

Figure 3.8. Dynamic acoustoelastic testing (DAET) of a nonlinear material uses a pump and a probe to assess the change in wave speed as a function of the stress induced by the pump. The experimental data shown at the bottom are for intact and damaged concrete. (Figure courtesy of Professor Parisa Shokouhi.)

effect of the pump-induced stress on the probe wave speed. The key measurements are the strain produced by the pump and the time-of-flight of the probe. A sample set-up is depicted in figure 3.8. The figure notes that relative wave speed change is proportional to the relative elastic modulus change, see [52, 53]. Typically the pump is a standing wave and the probe is a pulse, but there is sufficient flexibility in the DAET methodology for creativity, details can be found in Haupert *et al*'s chapter 13 in [47] and Shokouhi *et al* [52].

3.5 Closure

To provide a basis for nonlinear ultrasonic guided waves, the nonlinear ultrasonics problem in an unbounded solid was solved for longitudinal and shear waves. Acoustic nonlinearity coefficients were determined in terms of the third order elastic constants for bulk longitudinal waves and third and fourth order elastic constants for bulk shear waves. In preparation for measuring nonlinear ultrasonic guided waves, nonlinear ultrasonic measurement methodologies were highlighted.

References

[1] Hamilton M F and Blackstock D T 1997 *Nonlinear Acoustics* 1st edn (San Diego, CA: Academic)

[2] Thurston R N 1984 Waves in solids *Mechanics of Solids* vol 4 (Berlin: Springer) pp 109–308

[3] Breazeale M A and Ford J 1965 Ultrasonic studies of the nonlinear behavior of solids *J. Appl. Phys.* **36** 3486–90

[4] Guyer R A and Johnson P A 2009 *Nonlinear Mesoscopic Elasticity: The Complex Behaviour of Granular Media Including Rocks and Soil* (Weinheim: Wiley)

[5] Lissenden C J 2021 Nonlinear ultrasonic guided waves—principles for nondestructive evaluation *J. Appl. Phys.* **129** 021101

[6] Chillara V K and Lissenden C J 2015 On some aspects of material behavior relating microstructure and ultrasonic higher harmonic generation *Int. J. Eng. Sci.* **94** 59–70

[7] Nagy P B, Qu J and Jacobs L J 2013 Finite-size effects on the quasistatic displacement pulse in a solid specimen with quadratic nonlinearity *J. Acoust. Soc. Am.* **134** 1760–74

[8] Jiang C, Li W, Deng M and Ng C-T 2022 Quasistatic pulse generation of ultrasonic guided waves propagation in composites *J. Sound Vib.* **524** 116764

[9] Nayfeh A H and Mook D T 1979 *Nonlinear Oscillations* (New York: Wiley)

[10] Cantrell J H 2004 Fundamentals and applications of nonlinear ultrasonic nondestructive evaluation *Ultrasonic Nondestructive Evaluation: Engineering and Biological Material Characterization* ed T Kundu (Boca Raton, FL: CRC Press) pp 363–433

[11] Kundu T, Eiras J, Li W, Liu P and Sohn H 2019 Fundamentals of nonlinear acoustical techniques and sideband peak count *Nonlinear Ultrasonic and Vibro-Acoustical Techniques for Nondestructive Evaluation* (Cham: Springer Nature) pp 1–88

[12] Norris A 1998 Finite amplitude waves in solids *Nonlinear Acoustics* (New York: Academic) pp 263–77

[13] de Lima W J N and Hamilton M F 2003 Finite-amplitude waves in isotropic elastic plates *J. Sound Vib.* **265** 819–39

[14] Green W A 1965 The growth of plane discontinuities propagating into a homogeneously deformed elastic material *Arch Ration. Mech. Anal.* **19** 20–3

[15] Norris A N 1991 Symmetry conditions for third order elastic moduli and implications in nonlinear wave theory *J. Elast.* **25** 247–57

[16] Truell R, Elbaum C and Chick B B 1969 *Ultrasonic Methods in Solid State Physics* (New York: Academic)

[17] Wang Y and Achenbach J D 2016 The effect of cubic material nonlinearity on the propagation of torsional wave modes in a pipe *J. Acoust. Soc. Am.* **140** 3874–83

[18] Lurie A I 1990 *Nonlinear Theory of Elasticity* (Amsterdam: North-Holland)

[19] Toupin R A and Bernstein B 1961 Sound waves in deformed perfectly elastic materials. Acoustoelastic effect *J. Acoust. Soc. Am.* **33** 216–25

[20] Hikata A and Elbaum C 1966 Generation of ultrasonic second and third harmonics due to dislocations. I *Phys. Rev.* **144** 469–77

[21] Dace G E, Thompson R B and Buck O 1992 Measurement of the acoustic harmonic generation for materials characterization using contact transducers *Review of Progress in Quantitative Nondestructive Evaluation* (New York: Plenum) pp 2069–76

[22] Hurley D C and Fortunko C M 1997 Determination of the nonlinear ultrasonic parameter using a Michelson interferometer *Meas. Sci. Technol.* **8** 634–42

[23] Hughes D S and Kelly J L 1953 Second-order elastic deformation of solids *Phys. Rev.* **92** 1145–9

[24] Murnaghan F D 1937 Finite deformations of an elastic solid *Am. J. Math.* **59** 235

[25] Thurston R N and Brugger K 1964 Third-order elastic constants and the velocity of small amplitude elastic waves in homogeneously stressed media *Phys. Rev.* A **133** 1604–10

[26] Green R E 1973 Nonlinear elastic waves *Ultrasonic Investigation of Mechanical Properties: Treatise on Materials Science and Technology* vol 3 (New York: Academic) pp 73–144

[27] Pao Y-H, Fukuoka H and Sachse W 1984 Acoustoelasticity and ultrasonic measurements of residual stress *Physical Acoustics* vol XVII (Orlando, FL: Academic) pp 62–144

[28] Winkler K W and Liu X 1996 Measurements of third-order elastic constants in rocks *J. Acoust. Soc. Am.* **100** 1392–8

[29] Gandhi N, Michaels J E and Lee S J 2012 Acoustoelastic Lamb wave propagation in biaxially stressed plates *J. Acoust. Soc. Am.* **132** 1284–93

[30] Shi F, Michaels J E and Lee S J 2013 *In situ* estimation of applied biaxial loads with Lamb waves *J. Acoust. Soc. Am.* **133** 677–87

[31] Nucera C and Lanza di Scalea F 2014 Nonlinear wave propagation in constrained solids subjected to thermal loads *J. Sound Vib.* **333** 541–54

[32] Nucera C and Lanza di Scalea F 2014 Nondestructive measurement of neutral temperature in continuous welded rails by nonlinear ultrasonic guided waves *J. Acoust. Soc. Am.* **136** 2561–74

[33] Pau A and Lanza di Scalea F 2015 Nonlinear guided wave propagation in prestressed plates *J. Acoust. Soc. Am.* **137** 1529–40

[34] Pei N and Bond L J 2016 Higher order acoustoelastic Lamb wave propagation in stressed plates *J. Acoust. Soc. Am.* **140** 3834–43

[35] Peddeti K and Santhanam S 2018 Dispersion curves for Lamb wave propagation in prestressed plates using a semi-analytical finite element analysis *J. Acoust. Soc. Am.* **143** 829–40

[36] Dubuc B, Ebrahimkhanlou A and Salamone S 2018 Higher order longitudinal guided wave modes in axially stressed seven-wire strands *Ultrasonics* **84** 382–91

[37] Mohabuth M, Khanna A, Hughes J, Vidler J, Kotousov A and Ng C-T 2019 On the determination of the third-order elastic constants of homogeneous isotropic materials utilising Rayleigh waves *Ultrasonics* **96** 96–103

[38] Matlack K H, Kim J-Y, Jacobs L J and Qu J 2015 Review of second harmonic generation measurement techniques for material state determination in metals *J. Nondestruct. Eval.* **34** 273

[39] Jhang K-Y, Choi S and Kim J 2020 Measurement of nonlinear ultrasonic parameters from higher harmonics *Measurement of Nonlinear Ultrasonic Characteristics* (Measurement Science and Technology) ed K-Y Jhang (Singapore: Springer Nature) pp 9–60

[40] Breazeale M A and Lester W W 1961 Demonstration of the 'least stable waveform' of finite amplitude waves *J. Acoust. Soc. Am.* **33** 1803

[41] Best S R, Croxford A J and Neild S A 2014 Pulse-echo harmonic generation measurements for non-destructive evaluation *J. Nondestruct. Eval.* **33** 205–15

[42] Breazeale M A and Philip J 1984 Determination of third order elastic constants from ultrasonic harmonic generation measurements *Physical Acoustics* vol 17 (Orlando, FL: Academic) pp 1–60

[43] Jones G L and Kobett D R 1963 Interaction of elastic waves in an isotropic solid *J. Acoust. Soc. Am.* **35** 5–10

[44] Korneev V A and Demčenko A 2014 Possible second-order nonlinear interactions of plane waves in an elastic solid *J. Acoust. Soc. Am.* **135** 591–8

[45] TenCate J A and Johnson P A 2019 Nonlinear resonant ultrasound spectroscopy: assessing global damage *Nonlinear Ultrasonic and Vibro-Acoustical Techniques for Nondestructive Evaluation* (Cham: Springer Nature) pp 89–101

[46] Solodov I, Bai J, Bekgulyan S and Busse G 2011 A local defect resonance to enhance acoustic wave-defect interaction in ultrasonic nondestructive evaluation *Appl. Phys. Lett.* **99** 211911

[47] 2019 *Nonlinear Ultrasonic and Vibro-Acoustical Techniques for Nondestructive Evaluation* ed T Kundu (Cham: Springer International)

[48] Solodov I Y, Krohn N and Busse G 2002 CAN: an example of nonclassical acoustic nonlinearity in solids *Ultrasonics* **40** 621–5

[49] Klepka A, Staszewski W, Jenal R, Szwedo M, Iwaniec J and Uhl T 2012 Nonlinear acoustics for fatigue crack detection—experimental investigations of vibro-acoustic wave modulations *Struct. Health Monit.* **11** 197–211

[50] Van Den Abeele K E-A, Johnson P A and Sutin A 2000 Nonlinear elastic wave spectroscopy (NEWS) techniques to discern material damage, part I: nonlinear wave modulation spectroscopy (NWMS) *Res. Nondestruct. Eval.* **12** 17–30

[51] Van Den Abeele K E-A, Carmeliet J, Ten Cate J A and Johnson P A 2000 Nonlinear elastic wave spectroscopy (NEWS) techniques to discern material damage, part II: single-mode nonlinear resonance acoustic spectroscopy *Res. Nondestruct. Eval.* **12** 31–42

[52] Shokouhi P, Rivière J, Lake C R, Le Bas P-Y and Ulrich T J 2017 Dynamic acousto-elastic testing of concrete with a coda-wave probe: comparison with standard linear and nonlinear ultrasonic techniques *Ultrasonics* **81** 59–65

[53] Renaud G, Rivière J, Le Bas P-Y and Johnson P A 2013 Hysteretic nonlinear elasticity of Berea sandstone at low-vibrational strain revealed by dynamic acousto-elastic testing *Geophys. Res. Lett.* **40** 715–9

IOP Publishing

Nonlinear Ultrasonic Guided Waves

Cliff J Lissenden

Chapter 4

Boundary value problem formulation

We transition from waves in unbounded solids to waveguides that channel or guide the energy transport. In a waveguide the wavefield must satisfy traction-free boundary conditions on free surfaces as well as be continuous at interfaces between different materials. Boundary value problems are formulated, with the final solutions given in upcoming chapters. Dispersion is a distinguishing feature of guided waves; at each frequency there are multiple modes whose speed depends on the frequency. There are many structural components that we would like to monitor that serve as excellent waveguides.

Image credit: Jessie Lissenden

Chapter 4 has the modest goal of formulating the boundary value problems (BVPs) whose solutions provide the characteristics of freely propagating waves in specific types of waveguides, whether in the linear or nonlinear regime. The actual solutions are the subject of chapters 5–8. That guided waves are affected by the boundary of the material domain in which they propagate indicates that the boundary conditions are essential ingredients that augment the wave equation. The equation of motion was formulated in chapter 2. Some nonlinear wave equation problems in unbounded domains were investigated in chapter 3, but waveguides were not analysed. We start by applying the linear wave equation to different types of waveguides before moving to nonlinear problems.

4.1 Linear BVPs

For familiarity, we will use the Cauchy stress and infinitesimal strain tensors for the linear analysis before switching to Piola–Kirchhoff stress and Lagrangian strain tensors for nonlinear analysis. For convenience, the wave equations (hyperbolic partial differential equations) for anisotropic and isotropic materials are repeated here:

$$\text{Anisotropic } \nabla \cdot \mathbf{C} : \nabla \mathbf{u} + \rho \mathbf{b} = \rho \ddot{\mathbf{u}} \quad \text{or} \quad C_{ijkl} u_{l,kj} + \rho b_i = \rho \ddot{u}_i \qquad (2.52)$$

$$\text{Isotropic } \mu \nabla^2 \mathbf{u} + (\lambda + \mu) \nabla [\nabla \cdot \mathbf{u}] + \rho \mathbf{b} = \rho \ddot{\mathbf{u}}$$
$$\text{or } \mu u_{i,jj} + (\lambda + \mu) u_{j,ij} + \rho b_i = \rho \ddot{u}_i. \qquad (2.53)$$

Likewise, we consider harmonic wave propagation of the form

$$u_i(\mathbf{x}, t) = A p_i e^{i(k_j x_j - \omega t)} \qquad (2.54)$$

to compute the (complex-valued) wavenumbers that satisfy the characteristic equation of the eigenproblem obtained from the boundary conditions. In all problems of interest in this book the waveguide has traction-free lateral boundaries (Neumann boundary conditions). Strictly speaking, a traction-free boundary of a solid occurs when the waveguide is immersed in a vacuum. However, it is typically sufficient to consider a solid–air interface to be traction-free. Solid–liquid boundaries are another matter, see Rose [1].

Take for example a lateral free surface defined by $x_3 = 0$ with an outward unit normal vector $\mathbf{n}$ in the x_3 direction. The boundary conditions are

$$\mathbf{t}(x_3 = 0) = \sigma \cdot \mathbf{n} = (\mathbf{C} \cdot \nabla \mathbf{u}) \cdot \mathbf{n} = 0, \qquad (4.1)$$

where Hooke's law and the strain–displacement relation have been used to replace stress with displacement since the governing differential equation is in terms of displacement. We will formulate the solution to the linear problem for waves along surfaces, in plates and pipes as well as in waveguides having arbitrary cross-sections. The general form of the BVP is the same in each case, only the geometry of the traction-free surface changes. Let's start with surface waves.

4.1.1 Free surfaces

Waves that propagate in a solid along a free surface are typically referred to as surface acoustic waves (SAWs). In many applications SAWs have high frequencies, which results in the energy being localized very near the surface. SAWs can propagate in anisotropic media and in heterogeneous granular media. If the solid can be treated as a homogeneous isotropic half-space, then the waves that travel along the surface are known as Rayleigh waves [2]. Figure 4.1 is a sketch of the

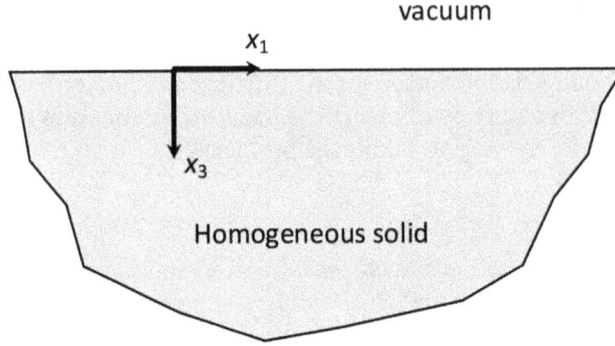

Figure 4.1. Homogeneous half-space.

waveguide in which Rayleigh waves propagate. This is a plane strain problem, with $\varepsilon_{2i} = 0$ and $\frac{\partial(\cdot)}{\partial x_2} = 0$. The traction-free boundary conditions are

$$\begin{bmatrix} \sigma_{11} & \sigma_{12} & \sigma_{13} \\ \sigma_{21} & \sigma_{22} & \sigma_{23} \\ \sigma_{31} & \sigma_{32} & \sigma_{33} \end{bmatrix} \begin{Bmatrix} 0 \\ 0 \\ -1 \end{Bmatrix} = \begin{Bmatrix} \sigma_{13} \\ \sigma_{23} \\ \sigma_{33} \end{Bmatrix} = \begin{Bmatrix} 0 \\ 0 \\ 0 \end{Bmatrix} \quad @\ x_3 = 0. \tag{4.2}$$

Using Hooke's law (equation (2.44)) and the strain–displacement relations (equation (2.8)) we can write the boundary conditions in terms of the primary variables, i.e. displacement

$$\sigma_{13} = \mu(u_{1,3} + u_{3,1}) = 0 \quad @\ x_3 = 0 \tag{4.3}$$

$$\sigma_{33} = (\lambda + 2\mu)u_{3,3} = 0 \ @\ x_3 = 0. \tag{4.4}$$

The stress component σ_{23} is identically zero because this is a plane strain problem. We analyse Rayleigh waves propagating in the x_1 direction in order to find the phase speed (and wavenumber) and the wavestructure (i.e. the displacement profile in the cross-section of the waveguide). The BVP is

$$\mu u_{i,jj} + (\lambda + \mu)u_{j,ij} + \rho b_i = \rho \ddot{u}_i \text{ for } x_3 \geqslant 0 \tag{4.5}$$

$$u_{1,3} + u_{3,1} = 0, \ u_{3,3} = 0 \ @\ x_3 = 0. \tag{4.6}$$

Neglecting the body force term, we showed in chapter 2 that the Helmholtz decomposition decouples the wave equation into

$$c_L^2 \nabla^2 \phi = \ddot{\phi}, \quad c_T^2 \nabla^2 \psi = \ddot{\psi}. \tag{2.60}$$

If we let $\psi = \{0 \ \ \psi \ \ 0\}^T$ and substitute the trial solution

$$\phi = D_1(x_3)e^{i(kx_1 - \omega t)}, \quad \psi = D_2(x_3)e^{i(kx_1 - \omega t)}, \tag{4.7}$$

into equation (4.5) we obtain

$$D_{1,33} + p^2 D_1 = 0, \quad D_{2,33} + q^2 D_2 = 0, \tag{4.8}$$

where

$$p^2 \equiv \left(\frac{\omega}{c_L}\right)^2 - k^2, \quad q^2 \equiv \left(\frac{\omega}{c_T}\right)^2 - k^2, \tag{4.9}$$

which admit the general solutions

$$D_1 = A_1 \sin(px_3) + A_2 \cos(px_3), \quad D_2 = B_1 \sin(qx_3) + B_2 \cos(qx_3). \tag{4.10}$$

However, this is an oscillating solution, whereas we seek a solution that decays with distance from the free surface. Thus, define

$$p \equiv ir, \quad q \equiv is \tag{4.11}$$

to obtain

$$D_{1,33} - r^2 D_1 = 0, \quad D_{2,33} - s^2 D_2 = 0 \tag{4.12}$$

having the general solution

$$D_1(x_3) = A_1 e^{-rx_3} + A_2 e^{rx_3}, \quad D_2(x_3) = B_1 e^{-sx_3} + B_2 e^{sx_3}. \tag{4.13}$$

Set $A_2 = B_2 = 0$ to obtain decaying solutions. Thus, the general solutions are

$$\phi = A_1 e^{-rx_3} e^{i(kx_1 - \omega t)} \tag{4.14}$$

$$\psi = B_1 e^{-sx_3} e^{i(kx_1 - \omega t)}, \tag{4.15}$$

which give the displacements

$$u_1 = \phi_{,1} - \psi_{,3} \tag{4.16}$$

$$u_3 = \phi_{,3} + \psi_{,1} \tag{4.17}$$

and the stresses involved in the boundary conditions are

$$\sigma_{13} = \mu\left(\phi_{,13} - \psi_{,33} + \phi_{,31} + \psi_{,11}\right) \tag{4.18}$$

$$\sigma_{33} = \lambda\left(\phi_{,11} - \psi_{,31}\right) + (\lambda + 2\mu)\left(\phi_{,33} + \psi_{,13}\right). \tag{4.19}$$

Substituting the nondimensionalized variables

$$\hat{r} \equiv r/k, \quad \hat{s} \equiv s/k, \quad \hat{t} \equiv 2 - (c/c_T)^2 \tag{4.20}$$

into equations (4.18) and (4.19) and then into the boundary conditions, equation (4.6), we obtain

$$\begin{bmatrix} \hat{t} & -2i\hat{s} \\ 2i\hat{r} & \hat{t} \end{bmatrix} \begin{Bmatrix} A_1 \\ B_1 \end{Bmatrix} = \begin{Bmatrix} 0 \\ 0 \end{Bmatrix}. \tag{4.21}$$

The characteristic equation comes from setting the determinant of the 2×2 matrix equal to zero,

$$\hat{t}^2 - 4\hat{r}\hat{s} = 0. \tag{4.22}$$

It is convenient to define two more nondimensional variables,

$$\eta \equiv k_T/k = c/c_T, \quad \zeta \equiv k_L/k_T = c_T/c_L$$

$$\hat{r} = 1 - (\zeta\eta)^2, \quad \hat{s} = 1 - \eta^2, \quad \hat{t} \equiv 2 - \eta^2, \tag{4.23}$$

enabling the characteristic equation to be rewritten as

$$\eta^2[\eta^6 - 8\eta^4 + 8\eta^2(3 - 2\zeta^2) + 16(\zeta^2 - 1)] = 0 \tag{4.24}$$

and solved for η to obtain the Rayleigh wave speed

$$c_R = \eta c_T, \tag{4.25}$$

which is nondispersive since it does not depend on the frequency. Viktorov's [3] approximate solution, based upon $\zeta = \frac{c_T}{c_L} = \sqrt{\frac{1 - 2\nu}{2(1 - \nu)}}$, is

$$\eta \approx \frac{0.87 + 1.12\nu}{1 + \nu}. \tag{4.26}$$

Likewise, the wavestructure is described by the relationship between amplitude coefficients obtained from equation (4.21),

$$A_1 = \left(\frac{i\hat{t}}{2\hat{r}}\right)B_1, \tag{4.27}$$

indicating that u_1 and u_3 are out of phase (since $\hat{t}$ and $\hat{r}$ have the same phase) and causing elliptical particle motion.

It can be shown that shear-horizontal (SH) surface waves do not exist in an isotropic half-space, but SH waves propagate as Love waves in a layer on a half-space. Sezawa waves propagate as Lamb waves in a layer on a half-space, while Scholte waves propagate when there is a liquid layer on a half-space. The commonality between Love waves, Sezawa waves, and Scholte waves is that each BVP has a free surface and an interface. Stoneley waves are different in that they propagate at the interface between two half-spaces.

4.1.2 Plates

A plate is defined to have a thickness that is much smaller than the dimensions in the other two coordinate directions. As shown in figure 4.2, we take the plate thickness $d = 2h$ to be in the x_3 direction. We analyse freely propagating shear-horizontal (SH) waves and Lamb waves [4] in a traction-free homogeneous isotropic plate of infinite extent whose midplane lies in the $x_1 - x_2$ plane. The traction-free boundary conditions (BCs) are applied at $x_3 = \pm h$. Readers experienced in vibrations and guided wave propagation know that the modes can be subdivided based on symmetric and antisymmetric displacement patterns and perhaps even that SH waves and Lamb waves are decoupled in isotropic plates. But it is reassuring to know that we do not need that experience to properly formulate the problem. As in

vacuum

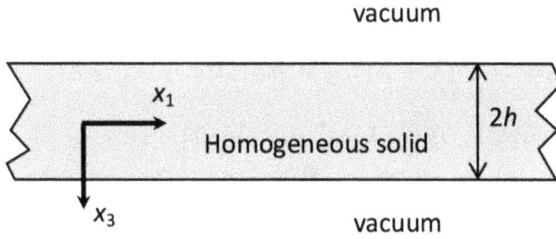

Figure 4.2. Traction-free plate.

the above Rayleigh wave problem, we analyse straight-crested wave propagation in the x_1-direction and thus select the Helmholtz potential functions to be

$$\phi = f(x_3)e^{i(kx_1-\omega t)}, \quad \psi = \psi_i(x_3)e^{i(kx_1-\omega t)}, \tag{4.28}$$

which give the displacement field,

$$\mathbf{u} = \nabla\phi + \nabla \times \psi, \tag{2.58}$$

$$\begin{aligned}
u_1 &= \varphi_{,1} + \psi_{3,2} - \psi_{2,3} \\
u_2 &= \varphi_{,2} - \psi_{3,1} + \psi_{1,3} \\
u_3 &= \varphi_{,3} + \psi_{2,1} - \psi_{1,2},
\end{aligned} \tag{4.29}$$

where ϕ and ψ are associated with volume change and distortion, respectively. To ensure there is no volume change associated with ψ, we impose the so-called gauge invariance condition,

$$\nabla \cdot \psi = 0. \tag{4.30}$$

To satisfy the traction-free BCs,

$$\mathbf{t}(x_3 = \pm h) = \sigma \cdot \mathbf{n} = \mathbf{0} \rightarrow \sigma_{3i}(x_3 = \pm h) = 0, \tag{4.1}$$

we write the stresses in terms of displacement using Hooke's law, equation (2.44), and the linearized strain–displacement relations, equation (2.8):

$$\sigma_{33} = \lambda(u_{1,1} + u_{2,2} + u_{3,3})1 + 2\mu u_{3,3}$$

$$\sigma_{33} = \lambda(u_{1,1} + u_{2,2}) + (\lambda + 2\mu)u_{3,3}$$

$$\sigma_{32} = 0 + 2\mu\frac{1}{2}(u_{2,3} + u_{3,2})$$

$$\sigma_{32} = \mu(u_{2,3} + u_{3,2})$$

$$\sigma_{31} = 0 + 2\mu\frac{1}{2}(u_{1,3} + u_{3,1})$$

$$\sigma_{31} = \mu(u_{1,3} + u_{3,1}).$$

The BVP in terms of displacements is

$$\mu u_{i,jj} + [\lambda + \mu]u_{j,ij} = \rho \ddot{u}_i \quad \text{for } |x_3| \leqslant h \tag{4.31}$$

$$\left.\begin{array}{c} \lambda(u_{1,1} + u_{2,2}) + (\lambda + 2\mu)u_{3,3} = 0 \\ u_{2,3} + u_{3,2} = 0 \\ u_{1,3} + u_{3,1} = 0 \end{array}\right\} @\ x_3 = \pm h. \tag{4.32}$$

Substituting the potential functions into the BCs, we obtain

$$\sigma_{33} \rightarrow \lambda\left(\varphi_{,11} - \psi_{2,31}\right) + (\lambda + 2\mu)\left(\varphi_{,33} + \psi_{2,13}\right) = 0$$

$$\sigma_{32} \rightarrow -\psi_{3,13} + \psi_{1,33} = 0$$

$$\sigma_{31} \rightarrow \varphi_{,13} - \psi_{2,33} + \varphi_{,31} + \psi_{2,11} = 0,$$

where we have used $\frac{\partial(\cdot)}{\partial x_2} = 0$ because the waves are straight-crested. The gauge invariance condition

$$\boldsymbol{\nabla} \cdot \boldsymbol{\psi} = 0$$

$$\psi_{1,1} + \psi_{3,3} = 0 \tag{4.33}$$

is applied at the boundary as well. The decoupled BVP is now

$$\nabla^2 \varphi = \frac{1}{c_L^2}\ddot{\varphi}, \ \nabla^2 \psi = \frac{1}{c_T^2}\ddot{\psi} \text{ for } |x_3| \leqslant h \tag{4.34}$$

$$\left.\begin{array}{c} \lambda\left(\varphi_{,11} - \psi_{2,31}\right) + (\lambda + 2\mu)\left(\varphi_{,33} + \psi_{2,13}\right) = 0 \\ -\psi_{3,13} + \psi_{1,33} = 0 \\ \varphi_{,13} - \psi_{2,33} + \varphi_{,31} + \psi_{2,11} = 0 \\ \psi_{1,1} + \psi_{3,3} = 0 \end{array}\right\} @\ x_3 = \pm h. \tag{4.35}$$

Substituting equation (4.28) into equation (4.34) eliminates the time variable, resulting in the ODEs

$$f''(x_3) + p^2 f(x_3) = 0$$

$$\psi_i''(x_3) + q^2\psi_i(x_3) = 0 \ i = 1, 2, 3, \tag{4.36}$$

whose general solutions are

$$f(x_3) = A\cos(px_3) + B\sin(px_3)$$

$$\psi_i(x_3) = C_i\cos(qx_3) + D_i\sin(qx_3) \ i = 1, 2, 3. \tag{4.37}$$

The equations that must be satisfied on the boundary become

$$\sigma_{33} = A(-c)\cos(px_3) + B(-c)\sin(px_3) + C_2(-f)\sin(qx_3) + D_2(f)\cos(qx_3)$$

$$\sigma_{32} = \mu[C_3(\ell)\sin(qx_3) - C_1 q^2\cos(qx_3) + D_3(-\ell)\cos(qx_3) - D_1 q^2\sin(qx_3)]$$

$$\sigma_{31} = \mu\Big[A(-d)\sin(px_3) + B(d)\cos(px_3) - C_2(g)\cos(qx_3) - D_2(g)\sin(qx_3)\Big]$$

$$C_1 ik \cos(qx_3) + D_1 ik \sin(qx_3) - C_3 q \sin(qx_3) + D_3 q \cos(qx_3) = 0,$$

where we have defined the intermediate variables,

$$\ell = ikq, \quad c = (\lambda + 2\mu)p^2 + \lambda k^2$$

$$d = ikp, \quad f = 2i\mu kq, \quad g = k^2 - q^2.$$

The six stress BCs and two gauge invariance conditions written in matrix form are

$$\mathbb{G}\mathbb{x} = 0, \tag{4.38}$$

where

$$\mathbb{G} = \begin{bmatrix}
-c\cos(ph) & -c\sin(ph) & 0 & 0 & -f\sin(qh) & f\cos(qh) & 0 & 0 \\
-c\cos(ph) & c\sin(ph) & 0 & 0 & f\sin(qh) & f\cos(qh) & 0 & 0 \\
0 & 0 & -q^2\cos(qh) & -q^2\sin(qh) & 0 & 0 & \ell\sin(qh) & -\ell\cos(qh) \\
0 & 0 & -q^2\cos(qh) & q^2\sin(qh) & 0 & 0 & -\ell\sin(qh) & -\ell\cos(qh) \\
-d\sin(ph) & d\cos(ph) & 0 & 0 & -g\cos(qh) & -g\sin(qh) & 0 & 0 \\
d\sin(ph) & d\cos(ph) & 0 & 0 & -g\cos(qh) & g\sin(qh) & 0 & 0 \\
0 & 0 & ik\cos(qh) & ik\sin(qh) & 0 & 0 & -q\sin(qh) & q\cos(qh) \\
0 & 0 & ik\cos(qh) & -ik\sin(qh) & 0 & 0 & q\sin(qh) & q\cos(qh)
\end{bmatrix}
\begin{matrix}
\sigma_{33}\,@ + h \\
\sigma_{33}\,@ - h \\
\sigma_{32}\,@ + h \\
\sigma_{32}\,@ - h \\
\sigma_{31}\,@ + h \\
\sigma_{31}\,@ - h \\
\nabla\cdot\psi\,@ + h \\
\nabla\cdot\psi\,@ - h
\end{matrix}$$

$$\mathbb{x} = \begin{Bmatrix} A \\ B \\ C_1 \\ D_1 \\ C_2 \\ D_2 \\ C_3 \\ D_3 \end{Bmatrix}.$$

We recognize that the coefficients of the $\mathbb{G}$ matrix contain frequency and wavenumber through the variables p and q. Given the frequency, we would like to find the wavenumbers, or vice versa. Equation (4.38) is an eigenproblem whose solution gives the wavenumbers for a prescribed frequency and the wavestructure for each wavenumber. The eigenvalues are found by requiring that det $[\mathbb{G}] = 0$. We now simplify $[\mathbb{G}]$ by adding and subtracting the pairs of equations to make use of the powerful symmetric and antisymmetric nature of the sine and cosine functions:

$$\mathbb{G} = \begin{bmatrix}
-2c\cos(ph) & 0 & 0 & 0 & 0 & 2f\cos(qh) & 0 & 0 \\
0 & 2c\sin(ph) & 0 & 0 & 2f\sin(qh) & 0 & 0 & 0 \\
0 & 0 & -2q^2\cos(qh) & 0 & 0 & 0 & 0 & -2\ell\cos(qh) \\
0 & 0 & 0 & 2q^2\sin(qh) & 0 & 0 & -2\ell\sin(qh) & 0 \\
0 & 2d\cos(ph) & 0 & 0 & -2g\cos(qh) & 0 & 0 & 0 \\
2d\sin(ph) & 0 & 0 & 0 & 0 & 2g\sin(qh) & 0 & 0 \\
0 & 0 & 2ik\cos(qh) & 0 & 0 & 0 & 0 & 2q\cos(qh) \\
0 & 0 & 0 & -2ik\sin(qh) & 0 & 0 & 2q\sin(qh) & 0
\end{bmatrix}.$$

Re-arranging the equations gives a banded diagonal matrix, comprised of four 2×2 matrices $\mathbb{G}_1 - \mathbb{G}_4$ and a re-ordered vector of undetermined coefficients:

$$\mathbb{G} = \begin{bmatrix} -2c\cos{(ph)} & 2f\cos{(qh)} & 0 & 0 & 0 & 0 & 0 & 0 \\ 2d\sin{(ph)} & 2g\sin{(qh)} & 0 & 0 & 0 & 0 & 0 & 0 \\ 0 & 0 & -2\ell\sin{(qh)} & 2q^2\sin{(qh)} & 0 & 0 & 0 & 0 \\ 0 & 0 & 2q\sin{(qh)} & -2ik\sin{(qh)} & 0 & 0 & 0 & 0 \\ 0 & 0 & 0 & 0 & 2f\sin{(qh)} & 2c\sin{(ph)} & 0 & 0 \\ 0 & 0 & 0 & 0 & -2g\cos{(qh)} & 2d\cos{(ph)} & 0 & 0 \\ 0 & 0 & 0 & 0 & 0 & 0 & -2q^2\cos{(qh)} & -2\ell\cos{(qh)} \\ 0 & 0 & 0 & 0 & 0 & 0 & 2ik\cos{(qh)} & 2q\cos{(qh)} \end{bmatrix}$$

$$\mathbb{x} = \begin{Bmatrix} A \\ D_2 \\ C_3 \\ D_1 \\ C_2 \\ B \\ C_1 \\ D_3 \end{Bmatrix}.$$

Expanding the determinant of $[\mathbb{G}]$ we obtain

$$\det{[\mathbb{G}_1]} \times \det{[\mathbb{G}_2]} \times \det{[\mathbb{G}_3]} \times \det{[\mathbb{G}_4]} = 0 \tag{4.39}$$

and the decoupled equations are

$$[\mathbb{G}_1]\begin{Bmatrix} A \\ D_2 \end{Bmatrix} = \begin{bmatrix} -c\cos{(ph)} & f\cos{(qh)} \\ d\sin{(ph)} & g\sin{(qh)} \end{bmatrix}\begin{Bmatrix} A \\ D_2 \end{Bmatrix} = \begin{Bmatrix} 0 \\ 0 \end{Bmatrix}$$

$$[\mathbb{G}_2]\begin{Bmatrix} C_3 \\ D_1 \end{Bmatrix} = \begin{bmatrix} -\ell\sin{(qh)} & q^2\sin{(qh)} \\ q\sin{(qh)} & -ik\sin{(qh)} \end{bmatrix}\begin{Bmatrix} C_3 \\ D_1 \end{Bmatrix} = \begin{Bmatrix} 0 \\ 0 \end{Bmatrix}$$

$$[\mathbb{G}_3]\begin{Bmatrix} C_2 \\ B \end{Bmatrix} = \begin{bmatrix} f\sin{(qh)} & c\sin{(ph)} \\ -g\cos{(qh)} & d\cos{(ph)} \end{bmatrix}\begin{Bmatrix} C_2 \\ B \end{Bmatrix} = \begin{Bmatrix} 0 \\ 0 \end{Bmatrix}$$

$$[\mathbb{G}_4]\begin{Bmatrix} C_1 \\ D_3 \end{Bmatrix} = \begin{bmatrix} q^2\cos{(qh)} & \ell\cos{(qh)} \\ ik\cos{(qh)} & q\cos{(qh)} \end{bmatrix}\begin{Bmatrix} C_1 \\ D_3 \end{Bmatrix} = \begin{Bmatrix} 0 \\ 0 \end{Bmatrix}$$

$$\tag{4.40}$$

which we will examine case by case.

Case 1. Demand that $B = C_1 = D_1 = C_2 = C_3 = D_3 = 0$, while $A \neq 0$, $D_2 \neq 0$ to force $\det{[\mathbb{G}_1]} = 0$. The characteristic equation is

$$cg\tan{(qh)} + df\tan{(ph)} = 0. \tag{4.41}$$

Substituting for the intermediate variables we obtain the Rayleigh–Lamb dispersion relation

$$\frac{\tan{(qh)}}{\tan{(ph)}} = -\frac{4k^2pq}{(k^2 - q^2)^2}, \tag{4.42}$$

where we manipulated c_g using

$$\rho c_L^2 = \lambda + 2\mu, \quad \rho c_L^2 - 2\rho c_T^2 = \lambda,$$

and the displacement components are

$$u_1 = \varphi_{,1} - \psi_{2,3} = [ikA\cos(px_3) - qD_2\cos(qx_3)]e^{i(kx_1-\omega t)} \tag{4.43a}$$

$$u_2 = -\psi_{3,1} + \psi_{1,3} = 0 \tag{4.43b}$$

$$u_3 = \varphi_{,3} + \psi_{2,1} = [-pA\sin(px_3) + ikD_2\sin(qx_3)]e^{i(kx_1-\omega t)}. \tag{4.43c}$$

Lamb wave modes are defined to be symmetric or antisymmetric based on the u_1 component, thus these are the symmetric modes.

Case 2. Demand that $A = C_1 = D_1 = D_2 = C_3 = D_3 = 0$, while $B \neq 0$, $C_2 \neq 0$ to force det $[\mathbb{G}_3] = 0$. The characteristic equation is

$$d_f\tan(qh) + c_g\tan(ph) = 0. \tag{4.44}$$

Substituting for the intermediate variables we obtain the other Rayleigh–Lamb dispersion relation

$$\frac{\tan(qh)}{\tan(ph)} = -\frac{(k^2 - q^2)^2}{4k^2pq}, \tag{4.45}$$

and the displacement components are

$$u_1 = \varphi_{,1} - \psi_{2,3} = [ikB\sin(px_3) + qC_2\sin(qx_3)]e^{i(kx_1-\omega t)} \tag{4.46a}$$

$$u_2 = -\psi_{3,1} + \psi_{1,3} = 0 \tag{4.46b}$$

$$u_3 = \varphi_{,3} + \psi_{2,1} = [pB\cos(px_3) + ikC_2\cos(qx_3)]e^{i(kx_1-\omega t)}. \tag{4.46c}$$

Thus, these are the antisymmetric Lamb wave modes.

Case 3. Demand that $A = B = C_1 = C_2 = D_2 = D_3 = 0$, while $D_1 \neq 0$, $C_3 \neq 0$ to force det $[\mathbb{G}_2] = 0$. The characteristic equation is

$$ik\ell\sin^2(qh) - q^3\sin^2(qh) = 0$$

$$-q(k^2 + q^2)\sin^2(qh) = 0$$

$$q\sin^2(qh) = 0, \tag{4.47}$$

for which there are two solutions. If $q = 0$, then we have bulk shear waves. Thus, we are interested in solutions to $\sin^2(qh) = 0$, $qh = n\pi$ for $n = 0, 2, 4, \ldots$. The displacement components are

$$u_1 = \varphi_{,1} - \psi_{2,3} = 0 \tag{4.48a}$$

$$u_2 = -\psi_{3,1} + \psi_{1,3} = [-ikC_3 + qD_1]\cos(qx_3)e^{i(kx_1-\omega t)} \tag{4.48b}$$

$$u_3 = \varphi_{,3} + \psi_{2,1} = 0. \tag{4.48c}$$

Thus, these are the symmetric SH wave modes.

Case 4. Demand that $A = B = D_1 = C_2 = D_2 = C_3 = 0$, while $C_1 \neq 0$, $D_3 \neq 0$ to force det $[\mathbb{G}_4] = 0$. The characteristic equation is

$$q^3 \cos^2(qh) - ik\ell \cos^2(qh) = 0$$

$$q(k^2 + q^2) \cos^2(qh) = 0$$

$$q \cos^2(qh) = 0, \tag{4.49}$$

for which there are two solutions. If $q = 0$ then we have bulk shear waves. Thus, we are interested in solutions to $\cos^2(qh) = 0$, $qh = n\pi$ for $n = 1, 3, 5, \ldots$. The displacement components are

$$u_1 = \varphi_{,1} - \psi_{2,3} = 0 \tag{4.50a}$$

$$u_2 = -\psi_{3,1} + \psi_{1,3} = -[ikD_3 + qC_1]\sin(qx_3)e^{i(kx_1 - \omega t)} \tag{4.50b}$$

$$u_3 = \varphi_{,3} + \psi_{2,1} = 0. \tag{4.50c}$$

Thus, these are the antisymmetric SH wave modes.

4.1.2.1 Shear-horizontal waves

Alternatively, we can simplify the formulation for SH waves by assuming the particle displacement is only in the x_2 direction, i.e. $u_1 = u_3 = 0$. Since u_2 is the only nonzero displacement component we can substitute the trial solution

$$u_2 = f(x_3)e^{i(kx_1 - \omega t)} \tag{4.51}$$

directly into the partial differential equation, giving

$$u_{2,11} + u_{2,33} = \frac{1}{c_T^2}\ddot{u}_2 \tag{4.52}$$

$$f'' + q^2 f = 0, \tag{4.53}$$

where q is defined in (4.9), and which admits the general solution

$$f(x_3) = A_1 \sin(qx_3) + A_2 \cos(qx_3). \tag{4.54}$$

The traction-free BCs reduce to

$$\sigma_{23} = \mu u_{2,3} = u_{2,3} = 0 \ @ \ x_3 = \pm h. \tag{4.55}$$

Consider the symmetric and antisymmetric displacement fields separately, i.e.

$$u_2^{\text{sym}} = A_2 \cos(qx_3)e^{i(kx_1 - \omega t)} \tag{4.56}$$

$$u_2^{\text{anti}} = A_1 \sin(qx_3)e^{i(kx_1 - \omega t)}. \tag{4.57}$$

Impose the BCs on the symmetric solution first

$$u_{2,3}^{\mathrm{sym}}(x_3 = \pm h) = -qA_2 \sin(qh)e^{i(kx_1 - \omega t)} = 0 \tag{4.58}$$

$$\therefore q \sin(qh) = 0. \tag{4.59}$$

The $q = 0$ solution corresponds to bulk shear waves. We are interested in the other solution, $\sin(qh) = 0$, which gives $qh = n\pi$, where $n = 0, 1, 2, \dots$ or equivalently

$$q = \frac{n\pi}{2h} \text{ for } n = 0, 2, 4, \dots. \tag{4.60}$$

The wavenumber k, and then the phase speed c, can now be computed from equations (4.9) and (4.60) for symmetric modes. Now impose the BCs on the antisymmetric solution using an analogous procedure to obtain

$$q = \frac{n\pi}{2h} \text{ for } n = 1, 3, 5, \dots \tag{4.61}$$

for antisymmetric modes. Therefore, the particle displacement field is

$$u_2 = \begin{cases} A_2 \cos(qx_3)e^{i(kx_1 - \omega t)} & \text{symmetric} \\ A_1 \sin(qx_3)e^{i(kx_1 - \omega t)} & \text{antisymmetric.} \end{cases} \tag{4.62}$$

From equations (4.9) and (4.60) the wavenumber is

$$k^2 = \left(\frac{\omega}{c_T}\right)^2 - \left(\frac{n\pi}{2h}\right)^2, \tag{4.63}$$

which can either be real or imaginary, giving propagating or evanescent waves, respectively. The phase velocity is computed using $\omega = ck$ and (4.9):

$$\left(\frac{1}{c_p}\right)^2 = \left(\frac{1}{c_T}\right)^2 - \left(\frac{n\pi}{2\omega h}\right)^2. \tag{4.64}$$

SH waves are dispersive because their phase velocity depends on frequency. The wavestructure depends on the mode n, but not the frequency. Finally, equation (4.64) indicates that $c_p > c_T$, which means that q is real. We close by acknowledging that the u_2 displacement fields in equations (4.48a) and (4.50a) appear to have different amplitudes than in equation (4.62), which can be remedied by setting $C_3 = D_3 = 0$.

4.1.2.2 Lamb waves

Lamb waves [4] are decoupled from SH waves when plane strain ($u_2 = 0$) is assumed. Wave motion is in the sagittal plane (i.e. the x_1–x_3 plane). Similar to the formulation for Rayleigh waves, the traction-free BCs are

$$\sigma_{13} = \mu\left(\phi_{,13} - \psi_{,33} + \phi_{,31} + \psi_{,11}\right) = 0 \quad @\, x_3 = \pm h \tag{4.65}$$

$$\sigma_{33} = \lambda\left(\phi_{,11} - \psi_{,31}\right) + (\lambda + 2\mu)\left(\phi_{,33} + \psi_{,13}\right) = 0 \,@\, x_3 = \pm h, \tag{4.66}$$

where equations (4.18) and (4.19) were used. The wave equation is manipulated as described for Rayleigh waves from immediately after equation (4.6) to equation (4.10). The text and equations are repeated verbatim here for convenience.

Neglecting the body force term, chapter 2 showed that the Helmholtz decomposition decouples the wave equation into

$$c_L^2 \nabla^2 \phi = \ddot{\phi}, \; c_T^2 \nabla^2 \psi = \ddot{\psi} \tag{2.60}$$

If we let $\psi = \{0 \; \psi \; 0\}^T$ and substitute the trial solution

$$\phi = D_1(x_3)e^{i(kx_1 - \omega t)}, \quad \psi = D_2(x_3)e^{i(kx_1 - \omega t)}, \tag{4.7}$$

into equation (4.5) we obtain

$$D_{1,33} + p^2 D_1 = 0, \quad D_{2,33} + q^2 D_2 = 0, \tag{4.8}$$

where

$$p^2 \equiv \left(\frac{\omega}{c_L}\right)^2 - k^2, \quad q^2 \equiv \left(\frac{\omega}{c_T}\right)^2 - k^2 \tag{4.9}$$

which admits the general solution

$$D_1 = A_1 \sin(px_3) + A_2 \cos(px_3), \quad D_2 = B_1 \sin(qx_3) + B_2 \cos(qx_3). \tag{4.10}$$

Therefore, the displacement field can be written as

$$u_1 = ik[A_1 \sin(px_3) + A_2 \cos(px_3)]e^{i(kx_1 - \omega t)}$$

$$-q[B_1 \cos(qx_3) - B_2 \sin(qx_3)]e^{i(kx_1 - \omega t)} \tag{4.67}$$

$$u_3 = p[A_1 \cos(px_3) - A_2 \sin(px_3)]e^{i(kx_1 - \omega t)}$$

$$+ik[B_1 \sin(qx_3) + B_2 \cos(qx_3)]e^{i(kx_1 - \omega t)}. \tag{4.68}$$

Notice that u_1 and u_3 are out of phase. Defining symmetry based on the u_1 component, set $A_1 = B_2 = 0$ for symmetric modes, giving

$$\phi_{sym} = A_2 \cos(px_3)e^{i(kx_1 - \omega t)} \tag{4.69}$$

$$\psi_{sym} = B_1 \sin(qx_3)e^{i(kx_1 - \omega t)}. \tag{4.70}$$

Computing the derivatives and substituting into the BCs, equations (4.65) and (4.66) give

$$\begin{bmatrix} -[\lambda(k^2 + p^2) + 2\mu p^2]\cos(ph) & 2\mu ikq\cos(qh) \\ -2ikp\sin(ph) & (q^2 - k^2)\sin(qh) \end{bmatrix} \begin{Bmatrix} A_2 \\ B_1 \end{Bmatrix} = \begin{Bmatrix} 0 \\ 0 \end{Bmatrix}. \tag{4.71}$$

Setting the determinant to zero gives the characteristic equation and then doing some manipulating yields the Rayleigh–Lamb dispersion relation for symmetric modes:

$$\frac{\tan(qh)}{\tan(ph)} = \frac{-4k^2 pq}{(q^2 - k^2)^2}. \tag{4.72}$$

In addition, the wavestructure is computed as the ratio A_2/B_1 from equation (4.71) after equation (4.72) has been solved.

Likewise, setting $A_2 = B_1 = 0$ for antisymmetric modes gives

$$\phi_{\text{anti}} = A_1 \sin{(px_3)}e^{\mathrm{i}(kx_1-\omega t)} \tag{4.73}$$

$$\psi_{\text{anti}} = B_2 \cos{(qx_3)}e^{\mathrm{i}(kx_1-\omega t)}, \tag{4.74}$$

and ultimately yields the Rayleigh–Lamb dispersion relation for antisymmetric modes (following the same procedure):

$$\frac{\tan{(qh)}}{\tan{(ph)}} = \left[\frac{-4k^2pq}{(q^2-k^2)^2}\right]^{-1}. \tag{4.75}$$

The Rayleigh–Lamb dispersion relations are transcendental equations that can be solved at a prescribed frequency ω for k, p, and q. Once the wavenumber k is known, then phase velocity can be computed from $c_p = \omega/k$.

4.1.2.3 Partial waves method

For isotropic materials, the Helmholtz decomposition is effective at decoupling longitudinal and shear waves in an infinite solid. Although traction-free surfaces couple them back together in finite waveguides, the approach works well. But it is limited to isotropic materials, and hence for anisotropic materials we need another approach. The Christoffel equation (2.57) has six roots, indicating that six unique bulk waves propagate in an infinite anisotropic solid; they are L (longitudinal), SV (shear-vertical), and SH (shear-horizontal) type waves, and each type propagates in both forward and backward directions. The interferences between these waves at boundaries sets up the guided wave propagation. Thus, we can think of guided waves as being the superposition of six bulk waves that Solie and Auld [5] refer to as partial waves. The superposed wavefield is

$$u_i = \sum_{n=1}^{6} C_n \alpha_i^n e^{\mathrm{i}(k_1^n x_1 + k_3^n x_3 - \omega t)}, \tag{4.76}$$

but according to Snell's law, $k_1^n = k_1$, enabling us to write

$$u_i = \left[\sum_{n=1}^{6} C_n \alpha_i^n e^{\mathrm{i}k(x_1 + l_3^n x_3)}\right]e^{-\mathrm{i}\omega t}, \tag{4.77}$$

where $l_3^n = k_3^n/k$ and $k = |\mathbf{k}|$. The polarization of partial wave n is given by α_i^n. The to-be-determined coefficients C_n prescribe the relative composition of the overall wavefield.

Considering a surface or interface having a unit normal vector in the x_3 direction, the partial waves may be homogeneous or nonhomogeneous depending on the wavevector component in the x_3 direction. If k_3^n is real, the partial wave is homogeneous and propagates at an oblique angle to the surface (interface). If k_3^n is imaginary, the partial wave is nonhomogeneous and decays exponentially from the surface (interface).

The method starts by solving Christoffel's equation for the polarization α_i^n and direction l_3^n of each partial wave n. The system matrix is then formed based on the traction-free boundary conditions at free surfaces and displacement and traction continuity conditions at interfaces using either the transfer matrix method or the global matrix method as described so well by Lowe [6].

For a homogeneous plate of thickness $2h$, the six traction-free boundary conditions are

$$t_i(x_3 = \pm h) = \sigma_{3i}(x_3 = \pm h) = 0. \qquad (4.78)$$

Thus, we use Hooke's law and the strain–displacement relation with equation (4.76) to write σ_{3i} and formulate the eigenproblem

$$[\mathbb{G}]\{x\} = \{0\}, \qquad (4.79)$$

where $\{x\}$ contains the six unknown C_n. Setting $\det[\mathbb{G}] = 0$ gives the wavenumbers, and for each wavenumber there is a set of partial wave coefficients C_n.

For a layered plate, the six traction-free boundary conditions still apply at the top and bottom surfaces, but we must add the displacement and traction continuity conditions between layers. If the interface between layers m and $m + 1$ is at $x_3 = b$, then the six continuity conditions are

$$u_i^{(m)}(x_3 = b^-) = u_i^{(m+1)}(x_3 = b^+) \qquad (4.80)$$

$$t_i^{(m)}(x_3 = b^-) = -t_i^{(m+1)}(x_3 = b^+)$$

or

$$\sigma_{3i}^{(m)}(x_3 = b^-) = \sigma_{3i}^{(m+1)}(x_3 = b^+). \qquad (4.81)$$

In the global matrix method, equation (4.79) is formed for each of the M layers, resulting in M sets of the unknown C_n coefficients, for a total of $6M$ unknowns. Therefore, the global matrix is $6M \times 6M$, being formed by 6 BCs and $6(M - 1)$ continuity conditions.

4.1.3 Hollow cylinders

The initial solution for wave propagation along a hollow (circular) cylinder that is homogeneous and isotropic is from Gazis [7, 8]. We will follow the problem formulation in Rose [1] using notation consistent with the remainder of this book. The cylindrical geometry calls for implementing the cylindrical coordinate system shown in figure 4.3. While we are still solving the wave equation subject to traction-free boundary conditions for harmonic waves propagating in the z-direction, the differential operators are implemented in the cylindrical coordinate system.

The BVP is given by equations (2.53) and (4.5)

$$\mu\nabla^2\mathbf{u} + [\lambda + \mu]\nabla[\nabla \cdot \mathbf{u}] + \rho\mathbf{b} = \rho\ddot{\mathbf{u}} \qquad (2.53)$$

$$\mathbf{t}(r = r_1, \theta, z, t) = \boldsymbol{\sigma}(r = r_1, \theta, z, t) \cdot \mathbf{n} = 0 \qquad (4.82)$$

$$\mathbf{t}(r = r_2, \theta, z, t) = \boldsymbol{\sigma}(r = r_2, \theta, z, t) \cdot \mathbf{n} = 0, \qquad (4.83)$$

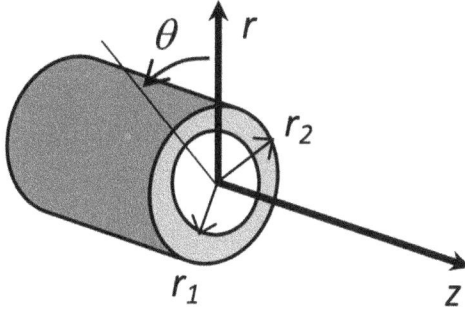

Figure 4.3. Hollow cylinder.

where **n** is in the radial direction. The Helmholtz decomposition is used to separate the dilatational (longitudinal) and distortional (shear) wave types in isotropic materials giving

$$c_L^2 \nabla^2 \phi = \ddot{\phi}, \, c_T^2 \nabla^2 \psi = \ddot{\psi}, \tag{2.60}$$

where $\psi = \psi_r \mathbf{e}_r + \psi_\theta \mathbf{e}_\theta + \psi_z \mathbf{e}_z$. The Laplacian operating on a scalar in cylindrical coordinates results in

$$\nabla^2 \phi = \frac{\partial^2 \phi}{\partial r^2} + \frac{1}{r} \frac{\partial \phi}{\partial r} + \frac{1}{r^2} \frac{\partial^2 \phi}{\partial \theta^2} + \frac{\partial^2 \phi}{\partial z^2}.$$

Abbreviated notation will be used for partial derivatives (subscripts with a comma followed by the partial derivative—note that this is not index notation used in Cartesian coordinates throughout this book where summation is implied by repeated indices):

$$\nabla^2 \phi = \phi_{,rr} + \frac{1}{r} \phi_{,r} + \frac{1}{r^2} \phi_{,\theta\theta} + \phi_{,zz}. \tag{4.84}$$

The Laplacian operating on a vector in cylindrical coordinates results in

$$\nabla^2 \psi = \left[\nabla^2 \psi_r - \frac{1}{r^2} \psi_r - \frac{2}{r^2} \psi_{\theta,\theta} \right] \mathbf{e}_r + \left[\nabla^2 \psi_\theta - \frac{1}{r^2} \psi_\theta + \frac{2}{r^2} \psi_{r,\theta} \right] \mathbf{e}_\theta + \nabla^2 \psi_z \mathbf{e}_z. \tag{4.85}$$

We will also need the divergence, gradient, and curl in cylindrical coordinates:

$$\nabla \cdot \psi = \frac{1}{r} \psi_r + \psi_{r,r} + \frac{1}{r} \psi_{\theta,\theta} + \psi_{z,z} \tag{4.86}$$

$$\nabla \phi = \phi_{,r} \mathbf{e}_r + \frac{1}{r} \phi_{,\theta} \mathbf{e}_\theta + \phi_{,z} \mathbf{e}_z \tag{4.87}$$

$$\nabla \times \psi = \left[\frac{1}{r} \psi_{z,\theta} - \psi_{\theta,z} \right] \mathbf{e}_r + \left[\psi_{r,z} - \psi_{z,r} \right] \mathbf{e}_\theta + \frac{1}{r} \left[\psi_\theta + r \psi_{\theta,r} - \psi_{r,\theta} \right] \mathbf{e}_z. \tag{4.88}$$

Choose the potential functions to be harmonic in the z-direction, periodic in the θ-direction, and undetermined in the r-direction:

$$\phi = f(r)e^{im\theta}e^{i(kz-\omega t)} \tag{4.89}$$

$$\psi = [\psi_r(r)\mathbf{e}_r + \psi_\theta(r)\mathbf{e}_\theta + \psi_z(r)\mathbf{e}_z]e^{im\theta}e^{i(kz-\omega t)}. \tag{4.90}$$

To determine $f(r)$, substitute equation (4.89) into the wave equation (2.60) to obtain

$$r^2f''(r) + rf'(r) + \left[\left(\left(\frac{\omega}{c_L}\right)^2 - k^2\right)r^2 - m^2\right]f(r) = 0. \tag{4.91}$$

Equivalent analyses of equation (4.90) lead to

$$r^2\psi_r''(r) + r\psi_r'(r) + \left[\left(\left(\frac{\omega}{c_T}\right)^2 - k^2\right)r^2 - (m^2+1)\right]\psi_r(r) - 2mi\psi_\theta(r) = 0 \tag{4.92}$$

$$r^2\psi_\theta''(r) + r\psi_\theta'(r) + \left[\left(\left(\frac{\omega}{c_T}\right)^2 - k^2\right)r^2 - (m^2+1)\right]\psi_\theta(r) + 2mi\psi_r(r) = 0 \tag{4.93}$$

$$r^2\psi_z''(r) + r\psi_z'(r) + \left[\left(\left(\frac{\omega}{c_T}\right)^2 - k^2\right)r^2 - m^2\right]\psi_z(r) = 0. \tag{4.94}$$

Using equation (4.9) for q and transforming to variables

$$\psi_1 \equiv \frac{1}{2}(i\psi_r - \psi_\theta), \quad \psi_2 \equiv \frac{1}{2}(i\psi_r + \psi_\theta), \quad \psi_3 \equiv \psi_z \tag{4.95}$$

simplifies the three equations for the components of ψ:

$$r^2\psi_1''(r) + r\psi_1'(r) + [(qr)^2 - (m+1)^2]\psi_1(r) = 0 \tag{4.96}$$

$$r^2\psi_2''(r) + r\psi_2'(r) + [(qr)^2 - (m-1)^2]\psi_2(r) = 0 \tag{4.97}$$

$$r^2\psi_3''(r) + r\psi_3'(r) + [(qr)^2 - m^2]\psi_3(r) = 0. \tag{4.98}$$

Using equation (4.9) for p gives

$$r^2f''(r) + rf'(r) + [(pr)^2 - m^2]f(r) = 0. \tag{4.99}$$

These four equations can be transformed into Bessel's equation

$$x^2g''(x) + xg'(x) + [x^2 - n^2]g(x) = 0 \tag{4.100}$$

by letting $x = pr$ or $x = qr$. The variables p and q can be either real or imaginary depending on equation (4.9). If x is real, then the solution to equation (4.100) is a linear combination of the Bessel functions of the first and second kind:

$$g(x) = A_i J_n(x) + B_i Y_n(x), \qquad (4.101)$$

where A_i and B_i are undetermined constants. If x is imaginary, then equation (4.100) becomes the modified Bessel equation and the solution is a linear combination of the modified Bessel functions of the first and second kind:

$$g(x) = A_i I_n(x) + B_i K_n(x). \qquad (4.102)$$

The four equations are embodied in Bessel's equation, as shown in table 4.1.

For a given frequency ω, we are trying to find the wavenumber k, thus we do not know whether p and q are real or imaginary, making it convenient to write the solution generically as

$$g(x) = A_i Z_n(x) + B_i W_n(x), \qquad (4.103)$$

where for the first kind

$$Z_n(x) = \begin{cases} J_n(x) \text{ if } x \text{ is real} \\ I_n(x) \text{ if } x \text{ is imaginary} \end{cases}$$

and for the second kind

$$W_n(x) = \begin{cases} Y_n(x) \text{ if } x \text{ is real} \\ K_n(x) \text{ if } x \text{ is imaginary}. \end{cases}$$

To impose the traction-free BCs, we must write the stress components σ_{rr}, $\sigma_{r\theta}$, and σ_{rz} in terms of the displacements (using Hooke's law and the strain–displacement relations),

$$\sigma_{rr} = \lambda(\varepsilon_{rr} + \varepsilon_{\theta\theta} + \varepsilon_{zz}) + 2\mu\varepsilon_{rr}$$

$$\sigma_{r\theta} = 2\mu\varepsilon_{r\theta} \qquad (4.104)$$

$$\sigma_{rz} = 2\mu\varepsilon_{rz}$$

$$\varepsilon_{rr} = u_{r,r}; \ \varepsilon_{r\theta} = \frac{1}{2}\left(\frac{1}{r}u_{r,\theta} + u_{\theta,r} - \frac{1}{r}u_\theta\right)$$

$$\varepsilon_{\theta\theta} = \frac{1}{r}(u_{\theta,\theta} + u_r); \ \varepsilon_{rz} = \frac{1}{2}(u_{r,z} + u_{z,r}) \qquad (4.105)$$

Table 4.1. Parameters for Bessel's equation applied to a hollow cylinder.

i	$g(x)$	x	n
1	$\psi_1(r)$	qr	$m + 1$
2	$\psi_2(r)$	qr	$m - 1$
3	$\psi_3(r)$	qr	m
4	$f(r)$	pr	m

$$\varepsilon_{zz} = u_{z,z}; \; \varepsilon_{\theta z} = \frac{1}{2}\left(u_{\theta,z} + \frac{1}{r}u_{z,\theta}\right)$$

and then in terms of the potential functions ϕ and ψ,

$$u_r = \phi_{,r} + \left(\frac{1}{r}\psi_{z,\theta} - \psi_{\theta,z}\right)$$

$$u_\theta = \frac{1}{r}\phi_{,\theta} + \left(\psi_{r,z} - \psi_{z,r}\right) \tag{4.106}$$

$$u_z = \phi_{,z} + \frac{1}{r}\left(\psi_\theta + r\psi_{\theta,r} - \psi_{r,\theta}\right),$$

resulting in

$$\sigma_{rr} = \left[2\mu\left[\phi_{,rr} - \frac{1}{r^2}\psi_{z,\theta} + \frac{1}{r}\psi_{z,\theta r} - \psi_{\theta,zr}\right]\right.$$
$$\left. + \lambda\left[\phi_{,rr} + \frac{1}{r^2}\phi_{,\theta\theta} + \frac{1}{r}\phi_{,r} + \phi_{,zz}\right]\right]e^{im\theta}e^{i(kz-\omega t)} \tag{4.107a}$$

$$\sigma_{r\theta} = \mu\left[\frac{2}{r}\phi_{,r\theta} + \frac{1}{r^2}\psi_{z,\theta\theta} - \frac{1}{r}\psi_{\theta,z\theta} - \frac{1}{r^2}\phi_{,\theta} - \frac{1}{r}\psi_{r,z} + \frac{1}{r}\psi_{z,r}\right.$$
$$\left. - \frac{1}{r^2}\phi_{,\theta} + \psi_{r,zr} - \psi_{z,rr}\right]e^{im\theta}e^{i(kz-\omega t)} \tag{4.107b}$$

$$\sigma_{rz} = \mu\left[2\phi_{,rz} + \frac{1}{r}\psi_{z,\theta z} - \psi_{\theta,zz} - \frac{1}{r^2}\psi_\theta + \frac{1}{r}\psi_{\theta,r} + \psi_{\theta,rr} + \frac{1}{r^2}\psi_{r,\theta}\right.$$
$$\left. - \frac{1}{r}\psi_{r,\theta r}\right]e^{im\theta}e^{i(kz-\omega t)} \tag{4.107c}$$

$$\psi = [-i(\psi_1 + \psi_2)\mathbf{e}_r - (\psi_1 - \psi_2)\mathbf{e}_\theta + \psi_3\mathbf{e}_z]e^{im\theta}e^{i(kz-\omega t)} \tag{4.108a}$$

$$\psi_1 = A_1 Z_{m+1}(qr) + B_1 W_{m+1}(qr)$$

$$\psi_2 = A_2 Z_{m-1}(qr) + B_2 W_{m-1}(qr)$$

$$\psi_3 = A_3 Z_m(qr) + B_3 W_m(qr)$$

$$\phi = [A_4 Z_m(pr) + B_4 W_m(pr)]e^{im\theta}e^{i(kz-\omega t)}. \tag{4.108b}$$

We have six stress BCs, but there are eight constants A_i and B_i. Thus, two more equations are necessary. Recalling that ψ in the Helmholtz decomposition is associated only with distortion, not volume change, it must be divergence free,

$$\nabla \cdot \psi = 0, \tag{4.109}$$

which is the gage invariance again, leading to

$$(1 + m)\psi_1 + r\psi_1' + (1 - m)\psi_2 + r\psi_2' - kr\psi_3 = 0. \tag{4.110}$$

We apply gage invariance at the two boundaries to obtain the additional equations. In total, there is a system of eight equations that can be written in matrix form as

$$\begin{bmatrix} \mathbb{G}_{11} & \cdots & \mathbb{G}_{18} \\ \vdots & \ddots & \vdots \\ \mathbb{G}_{81} & \cdots & \mathbb{G}_{88} \end{bmatrix} \begin{Bmatrix} A_1 \\ \vdots \\ B_4 \end{Bmatrix} = \begin{Bmatrix} 0 \\ \vdots \\ 0 \end{Bmatrix}. \tag{4.111}$$

Comparing to the general plate wave problem, $[\mathbb{G}] = [\mathbb{G}_{8\times8}]$ and $\{\mathbb{x}\} = (A_1 \cdots B_4)^T$. Readers interested in writing code to determine dispersion curves should note that in addition to the notation being different from Rose [1], the order of coefficients is different, and there are some other differences (mostly signs).

From $\sigma_{rr}(r = r_1) = 0$ we obtain

$$\mathbb{G}_{11} = 2\mu i k Z_{m+1}'(q r_1)$$

$$\mathbb{G}_{12} = 2\mu i k W_{m+1}'(q r_1)$$

$$\mathbb{G}_{13} = -2\mu i k Z_{m-1}'(q r_1)$$

$$\mathbb{G}_{14} = -2\mu i k W_{m-1}'(q r_1)$$

$$\mathbb{G}_{15} = \frac{2\mu i m}{r_1}\left[Z_m'(q r_1) - \frac{1}{r_1}Z_m(q r_1) \right]$$

$$\mathbb{G}_{16} = \frac{2\mu i m}{r_1}\left[W_m'(q r_1) - \frac{1}{r_1}W_m(q r_1) \right]$$

$$\mathbb{G}_{17} = -\lambda k_L^2 Z_m(p r_1) + 2\mu Z_m''(p r_1)$$

$$\mathbb{G}_{18} = -\lambda k_L^2 W_m(p r_1) + 2\mu W_m''(p r_1).$$

From $\sigma_{rr}(r = r_2) = 0$ we obtain the coefficients in the fourth row, $\mathbb{G}_{4J}$, by simply replacing r_1 with r_2. From $\sigma_{r\theta}(r = r_1) = 0$,

$$\mathbb{G}_{21} = -k\left[\frac{m + 1}{r_1}Z_{m+1}(q r_1) - Z_{m+1}'(q r_1) \right]$$

$$\mathbb{G}_{22} = -k\left[\frac{m + 1}{r_1}W_{m+1}(q r_1) - W_{m+1}'(q r_1) \right]$$

$$\mathbb{G}_{23} = k\left[\frac{m - 1}{r_1}Z_{m-1}(q r_1) + Z_{m-1}'(q r_1) \right]$$

$$\mathbb{G}_{24} = k\left[\frac{m - 1}{r_1}W_{m-1}(q r_1) + W_{m-1}'(q r_1) \right]$$

$$\mathbb{G}_{25} = -q^2 Z_m(qr_1) - 2Z_m''(qr_1)$$

$$\mathbb{G}_{26} = -q^2 W_m(qr_1) - 2W_m''(qr_1)$$

$$\mathbb{G}_{27} = \frac{2im}{r_1}\left[Z_m'(pr_1) - \frac{1}{r_1}Z_m(pr_1)\right]$$

$$\mathbb{G}_{28} = \frac{2im}{r_1}\left[W_m'(pr_1) - \frac{1}{r_1}W_m(pr_1)\right],$$

while the coefficients in the fifth row, $\mathbb{G}_{5J}$, come from simply replacing r_1 with r_2 in row two. From $\sigma_{rz}(r = r_1) = 0$,

$$\mathbb{G}_{31} = \left[k_T^2 - 2k^2 - \frac{m(m+1)}{r_1^2}\right]Z_{m+1}(qr_1) - \frac{m}{r_1}qZ_{m+1}'(qr_1)$$

$$\mathbb{G}_{32} = \left[k_T^2 - 2k^2 - \frac{m(m+1)}{r_1^2}\right]W_{m+1}(qr_1) - \frac{m}{r_1}qW_{m+1}'(qr_1)$$

$$\mathbb{G}_{33} = \left[-k_T^2 + 2k^2 + \frac{m(m-1)}{r_1^2}\right]Z_{m-1}(qr_1) - \frac{m}{r_1}qZ_{m-1}'(qr_1)$$

$$\mathbb{G}_{34} = \left[-k_T^2 + 2k^2 + \frac{m(m-1)}{r_1^2}\right]W_{m-1}(qr_1) - \frac{m}{r_1}qW_{m-1}'(qr_1)$$

$$\mathbb{G}_{35} = -\frac{km}{r_1}Z_m(qr_1)$$

$$\mathbb{G}_{36} = -\frac{km}{r_1}W_m(qr_1)$$

$$\mathbb{G}_{37} = 2ikpZ_m'(pr_1)$$

$$\mathbb{G}_{38} = 2ikpW_m'(pr_1)$$

and as before, the coefficients in the sixth row, $\mathbb{G}_{6J}$, come from simply replacing r_1 with r_2 in row three. Gage invariance at $r = r_1$ gives the coefficients

$$\mathbb{G}_{71} = (1 + m)Z_{m+1}(qr_1) + r_1 qZ_{m+1}'(qr_1)$$

$$\mathbb{G}_{72} = (1 + m)W_{m+1}(qr_1) + r_1 qW_{m+1}'(qr_1)$$

$$\mathbb{G}_{73} = (1 - m)Z_{m-1}(qr_1) + r_1 qZ_{m-1}'(qr_1)$$

$$\mathbb{G}_{74} = (1 - m)W_{m-1}(qr_1) + r_1 qW_{m-1}'(qr_1)$$

$$\mathbb{G}_{75} = -kr_1 Z_m(qr_1)$$

$$\mathbb{G}_{76} = -kr_1 W_m(qr_1)$$

$$\mathbb{G}_{77} = \mathbb{G}_{78} = 0.$$

Simply replace r_1 with r_2 in row seven to obtain row eight, $\mathbb{G}_{8J}$. The characteristic equation, det $[\mathbb{G}_{8\times8}] = 0$ yields the eigenvalues, giving the wavenumbers and phase speeds. The eigenvectors are the wavestructures, which can be plotted radially to visualize the wavestructure analogous to a plate or as an angular profile to visualize the effect of circumferential order m.

4.1.4 Arbitrary cross-sections

Up to this point the geometric domain of the waveguide (plate or hollow circular cylinder) was simple enough that equations could be formulated to generate a global matrix by satisfying boundary conditions and continuity conditions. Now we consider waveguides having an arbitrary cross-sectional geometry as shown in figure 4.4 that needs to be discretized for numerical solution. The waveguides are required to be uniform along the length in which the waves propagate, although periodicity can be handled by imposing Floquet periodicity conditions, as described by Hakoda *et al* [9]. To be uniform along the length, we mean that the geometry (i.e. the waveguide is prismatic), the material is homogeneous (along the length, although not necessarily in the cross-section), and the boundary conditions do not change. The uniformity along the length enables a semi-analytical finite element (SAFE) analysis of free wave propagation where the wavefield in the wave propagation direction (x_1 in our case) is harmonic.

Many authors have formulated their own elements, assembly, and solution as described in Rose's chapter 9 [1]. The SAFE formalism starts with equation (2.52), although we ignore the body force term, leaving

$$C_{ijkl} u_{l,\,kj} = \rho \ddot{u}_i$$

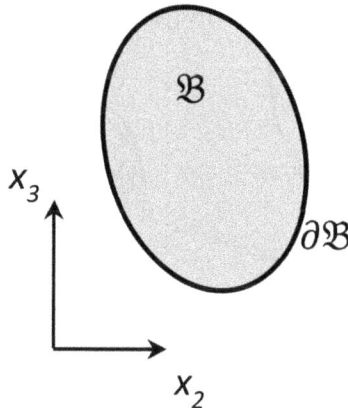

Figure 4.4. Domain and boundary within the cross-section of a prismatic waveguide.

for which we assume the displacement field can be written as

$$u_i(x_1, x_2, x_3, t) = U_i(x_2, x_3)e^{i(kx_1 - \omega t)}.$$

We wish to determine the unknown wavenumbers k and wavestructures $U_i(x_2, x_3)$ for a specified frequency ω. Both k and $U_i(x_2, x_3)$ are in general complex-valued for real ω. We now substitute this displacement field into the governing differential equation in the domain $\mathfrak{B}$ as well as into the traction-free boundary conditions on $\partial\mathfrak{B}$ to obtain

$$C_{ijkl}U_{k,lj} + ikC_{i1kl}U_{k,l} + ikC_{ijk1}U_{k,j} - k^2 C_{i1k1}U_k + \rho\omega^2 \delta_{ik}U_k = 0 \text{ in } \mathfrak{B}$$

$$C_{jikl}U_{k,l}n_j + ikC_{jik1}U_k n_j = 0 \text{ on } \partial\mathfrak{B}.$$

This eigenvalue problem is efficiently solved using the weak form in the commercially available Comsol Multiphysics software (Comsol, Inc.) as described by Hakoda *et al* [10], who present example problems for a rail cross-section, plate, and pipe. Lamb and SH waves can be found with a 1D finite element model of a plate, axisymmetric axial and torsional waves can be found with a 1D finite element model of a pipe, although flexural modes require a 2D model.

4.2 Nonlinear BVPs

Now we progress from linear problems to formulating the nonlinear boundary value problem for guided waves. We do so in two steps: first implementing a perturbation solution for second harmonic generation and then generalizing that approach in such a way to cleanly handle wave mixing problems using wave interactions. We consider an infinitely long prismatic waveguide (i.e. the cross-section does not change along the length) that will be denoted as domain $\mathfrak{B}$, which has a lateral boundary $\partial\mathfrak{B}_o$ in the reference configuration that is traction-free, as illustrated in figure 4.5.

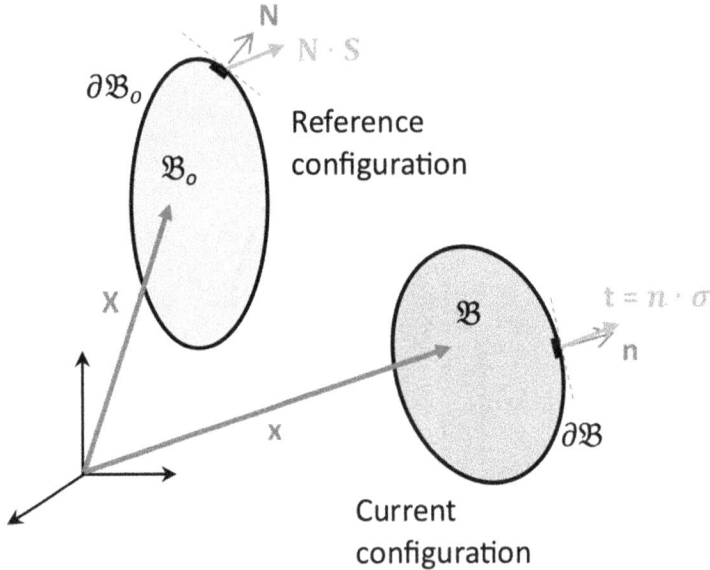

Figure 4.5. Domain and boundary of a prismatic waveguide.

4.2.1 Regular perturbation method

Starting with the equation of motion in chapter 2,

$$\nabla_X \cdot \mathbf{T}_{PK1} + \rho_o \mathbf{b}_o = \rho_o \dot{\mathbf{v}}, \qquad (2.26)$$

where the body force term has been carried along to show that part of the nonlinearity acts like a body force. But now we will neglect the actual body force, write $\mathbf{x}$ in terms of $\mathbf{u}$, and use the shorter notation $\mathbf{S}$ for the first Piola–Kirchhoff stress tensor to write the governing differential equation (GDE):

$$\nabla_X \cdot \mathbf{S} - \rho_o \dot{\mathbf{v}} = 0 \text{ in } \mathfrak{B}. \qquad (4.112)$$

The lateral boundary of the waveguide is traction-free:

$$\mathbf{N} \cdot \mathbf{S} = 0 \text{ on } \partial \mathfrak{B}_o. \qquad (4.113)$$

Decompose the stress tensor into linear and nonlinear parts $\mathbf{S} = \mathbf{S}_L + \mathbf{S}_{NL}$ as in example 2.4. In addition, decompose the displacement field into linear and nonlinear parts $\mathbf{u} = \mathbf{u}_L + \mathbf{u}_{NL}$, with the assumption that $|\mathbf{u}_{NL}| \ll |\mathbf{u}_L|$. It may be obvious that $\mathbf{u}_L \to \mathbf{S}_L$ and $\mathbf{u}_{NL} \to \mathbf{S}_{NL}$. The boundary value problem may now be written as

$$\nabla_X \cdot \mathbf{S}_L - \rho_o \dot{\mathbf{v}} = -\nabla_X \cdot \mathbf{S}_{NL} \text{ in } \mathfrak{B} \qquad (4.114)$$

$$\mathbf{N} \cdot \mathbf{S}_L = -\mathbf{N} \cdot \mathbf{S}_{NL} \text{ on } \partial \mathfrak{B}_o. \qquad (4.115)$$

Apply a two-step procedure to solve equations (4.114) and (4.115). Step one is to solve the linear problem, where both GDE and BCs are homogeneous and solutions have already been discussed, with the dispersion curves shown in chapter 5. Step two leverages $|\mathbf{u}_{NL}| \ll |\mathbf{u}_L|$ by assuming that $\mathbf{S}_{NL}$ can be approximated by the linear solution $\mathbf{u}_L$ in order to compute $\mathbf{u}_{NL}$ (in other words, we use $\mathbf{u}_L$ to compute $\mathbf{S}_{NL}$, and then by solving the BVP we obtain $\mathbf{u}_{NL}$). In so doing we are effectively ignoring the interactions between $\mathbf{u}_{NL}$ and $\mathbf{u}_L$ as well as those between $\mathbf{u}_{NL}$ and itself. Had we implemented a formal power series expansion in the perturbation solution (as was done in example 3.2) the result would have been the same. The actual solution is described in chapter 6 for a plate. In the forthcoming section we implement a framework applicable for higher harmonic generation and wave mixing.

4.2.2 Wave interactions

Chillara and Lissenden [11] organized the nonlinear BVP into a framework that simplifies the bookkeeping for wave mixing problems by specifically accounting for the wave interactions. Mixing of two or more distinct waves can formally be treated as a mutual interaction problem, i.e. wave mixing is a nonlinear phenomenon described by the mutual interaction of the primary waves. We may view self-interactions of primary waves that lead to higher harmonic generation at integer multiples of the primary (or driving) frequency as a special case of mutual interaction.

Let's consider two interacting primary wavefields having frequencies ω_a and ω_b and associated displacement fields $\mathbf{u}_a$ and $\mathbf{u}_b$. Since it is convenient to write stress components in terms of the displacement gradient as opposed to strain or displacement, we use $\mathbf{H}_a$ and $\mathbf{H}_b$. The resulting displacement and displacement gradients for interacting waves a and b are

$$\mathbf{u} = \mathbf{u}_a + \mathbf{u}_b + \mathbf{u}_{aa} + \mathbf{u}_{bb} + \mathbf{u}_{ab} \tag{4.116}$$

$$\mathbf{H} = \mathbf{H}_a + \mathbf{H}_b + \mathbf{H}_{aa} + \mathbf{H}_{bb} + \mathbf{H}_{ab}, \tag{4.117}$$

where a single subscript identifies the primary wavefield, repeated subscripts denote self-interactions, and dissimilar double subscripts denote the secondary wavefield from mutual interaction between the distinct waves. For clarity of presentation, we limit the interactions to second order here. This notation can be extended to third order and higher interactions and more than two interacting waves. The resulting first Piola–Kirchhoff stress can be decomposed into linear and nonlinear components that are written as an explicit function of the displacement gradient component:

$$\mathbf{S}(\mathbf{H}) = \mathbf{S}^L(\mathbf{H}_a) + \mathbf{S}^L(\mathbf{H}_b) + \mathbf{S}^L(\mathbf{H}_{aa}) + \mathbf{S}^L(\mathbf{H}_{bb}) + \mathbf{S}^L(\mathbf{H}_{ab}) + \mathbf{S}^{NL}(\mathbf{H}_a + \mathbf{H}_b). \tag{4.118}$$

The nonlinear part can be further subdivided based on the three types of wave interaction:

$$\mathbf{S}^{NL}(\mathbf{H}_a + \mathbf{H}_b) = \mathbf{S}^{NL}(\mathbf{H}_a, \mathbf{H}_a, 2) + \mathbf{S}^{NL}(\mathbf{H}_b, \mathbf{H}_b, 2) + \mathbf{S}^{NL}(\mathbf{H}_a, \mathbf{H}_b, 2). \tag{4.119}$$

The term $\mathbf{S}^{NL}(\mathbf{H}_a, \mathbf{H}_b, 2)$ designates the nonlinear terms of order 2 in the displacement gradient due to mutual interaction between waves a and b. Likewise, the term $\mathbf{S}^{NL}(\mathbf{H}_a, \mathbf{H}_a, 2)$ designates the nonlinear terms of order 2 in the displacement gradient due to self-interaction of waves a. Finally, the term $\mathbf{S}^{NL}(\mathbf{H}_b, \mathbf{H}_b, 2)$ designates self-interaction of waves b.

From example 2.4 we recall that the linear terms are

$$\mathbf{S}^L(\mathbf{H}) = \frac{\lambda}{2} tr(\mathbf{H} + \mathbf{H}^T)\, \mathbf{I} + \mu(\mathbf{H} + \mathbf{H}^T) \tag{4.120}$$

from which each component of $\mathbf{S}^L(\mathbf{H})$ can be computed. The self-interaction terms are also given in example 2.4 (as quadratic in terms of strain)

$$\begin{aligned}
\mathbf{S}^{NL}(\mathbf{H}_a, \mathbf{H}_a, 2) = {}& \frac{\lambda}{2} tr\left(\mathbf{H}_a + \mathbf{H}_a^T\right)\mathbf{H}_a + \frac{\lambda}{2} tr\left(\mathbf{H}_a^T \mathbf{H}_a\right)\mathbf{I} + \mu\left(\mathbf{H}_a \mathbf{H}_a + \mathbf{H}_a \mathbf{H}_a^T + \mathbf{H}_a^T \mathbf{H}_a\right) \\
& + \frac{\mathscr{A}}{4}\left(\mathbf{H}_a \mathbf{H}_a + \mathbf{H}_a \mathbf{H}_a^T + \mathbf{H}_a^T \mathbf{H}_a + \mathbf{H}_a^T \mathbf{H}_a^T\right) \\
& + \frac{\mathscr{B}}{4} tr\left(\mathbf{H}_a \mathbf{H}_a + \mathbf{H}_a \mathbf{H}_a^T + \mathbf{H}_a^T \mathbf{H}_a + \mathbf{H}_a^T \mathbf{H}_a^T\right)\mathbf{I} + \mathscr{B} tr(\mathbf{H}_a)\left(\mathbf{H}_a + \mathbf{H}_a^T\right) \\
& + \mathscr{C}\left(tr(\mathbf{H}_a)\right)^2 \mathbf{I}
\end{aligned} \tag{4.121}$$

with $\mathbf{S}^{NL}(\mathbf{H}_b, \mathbf{H}_b, 2)$ being analogous. The mutual interaction term is obtained from equation (4.121) by replacing the subscripts (a and a) in each term, first with (a and

b) and then with (b and a), $\mathbf{S}^{NL}(\mathbf{H}_a, \mathbf{H}_b, 2) = (ab + ba)/2$. The sum must be divided by 2 so that it can also be used to compute $\mathbf{S}^{NL}(\mathbf{H}_a, \mathbf{H}_a, 2)$,

$$
\begin{aligned}
\mathbf{S}^{NL}(\mathbf{H}_a, \mathbf{H}_b, 2) =& \frac{\lambda}{4}\mathrm{tr}\left(\mathbf{H}_a + \mathbf{H}_a^T\right)\mathbf{H}_b + \frac{\lambda}{4}\mathrm{tr}\left(\mathbf{H}_b + \mathbf{H}_b^T\right)\mathbf{H}_a + \frac{\lambda}{4}\mathrm{tr}\left(\mathbf{H}_a^T\mathbf{H}_b + \mathbf{H}_b^T\mathbf{H}_a\right)\mathbf{I} \\
&+ \frac{\mu}{2}\left[\mathbf{H}_a\mathbf{H}_b + \mathbf{H}_b\mathbf{H}_a + \mathbf{H}_a\mathbf{H}_b^T + \mathbf{H}_b\mathbf{H}_a^T + \mathbf{H}_a^T\mathbf{H}_b + \mathbf{H}_b^T\mathbf{H}_a\right] \\
&+ \frac{\mathscr{A}}{8}\left[\mathbf{H}_a\mathbf{H}_b + \mathbf{H}_b\mathbf{H}_a + \mathbf{H}_a\mathbf{H}_b^T + \mathbf{H}_b\mathbf{H}_a^T + \mathbf{H}_a^T\mathbf{H}_b + \mathbf{H}_b^T\mathbf{H}_a + \mathbf{H}_a^T\mathbf{H}_b^T + \mathbf{H}_b^T\mathbf{H}_a^T\right] \quad (4.122)\\
&+ \frac{\mathscr{B}}{8}\mathrm{tr}\left(\mathbf{H}_a\mathbf{H}_b + \mathbf{H}_b\mathbf{H}_a + \mathbf{H}_a\mathbf{H}_b^T + \mathbf{H}_b\mathbf{H}_a^T + \mathbf{H}_a^T\mathbf{H}_b + \mathbf{H}_b^T\mathbf{H}_a + \mathbf{H}_a^T\mathbf{H}_b^T + \mathbf{H}_b^T\mathbf{H}_a^T\right)\mathbf{I} \\
&+ \frac{\mathscr{B}}{2}\mathrm{tr}(\mathbf{H}_a)\left(\mathbf{H}_b + \mathbf{H}_b^T\right) + \frac{\mathscr{B}}{2}\mathrm{tr}(\mathbf{H}_b)\left(\mathbf{H}_a + \mathbf{H}_a^T\right) + \mathscr{C}\mathrm{tr}(\mathbf{H}_a)\mathrm{tr}(\mathbf{H}_b)\mathbf{I}.
\end{aligned}
$$

The BVP given by (4.114) and (4.115) can now be divided into five separate BVPs.
Primary waves a:

$$\nabla_X \cdot \mathbf{S}^L(\mathbf{H}_a) - \rho_o\ddot{\mathbf{u}}_a = 0 \text{ in } \mathscr{B} \qquad (4.123)$$

$$\mathbf{N} \cdot \mathbf{S}_L(\mathbf{H}_a) = 0 \text{ on } \partial\mathscr{B}_o. \qquad (4.124)$$

Primary waves b:

$$\nabla_X \cdot \mathbf{S}^L(\mathbf{H}_b) - \rho_o\ddot{\mathbf{u}}_b = 0 \text{ in } \mathscr{B} \qquad (4.125)$$

$$\mathbf{N} \cdot \mathbf{S}^L(\mathbf{H}_b) = 0 \text{ on } \partial\mathscr{B}_o \qquad (4.126)$$

Self-interaction of waves a:

$$\nabla_X \cdot \mathbf{S}^L(\mathbf{H}_{aa}) - \rho_o\ddot{\mathbf{u}}_{aa} = -\nabla_X \cdot \mathbf{S}^{NL}(\mathbf{H}_a, \mathbf{H}_a, 2) \text{ in } \mathscr{B} \qquad (4.127)$$

$$\mathbf{N} \cdot \mathbf{S}^L(\mathbf{H}_{aa}) = -\mathbf{N} \cdot \mathbf{S}^{NL}(\mathbf{H}_a, \mathbf{H}_a, 2) \text{ on } \partial\mathscr{B}_o. \qquad (4.128)$$

Self-interaction of waves b:

$$\nabla_X \cdot \mathbf{S}^L(\mathbf{H}_{bb}) - \rho_o\ddot{\mathbf{u}}_{bb} = -\nabla_X \cdot \mathbf{S}^{NL}(\mathbf{H}_b, \mathbf{H}_b, 2) \text{ in } \mathscr{B} \qquad (4.129)$$

$$\mathbf{N} \cdot \mathbf{S}^L(\mathbf{H}_{bb}) = -\mathbf{N} \cdot \mathbf{S}^{NL}(\mathbf{H}_b, \mathbf{H}_b, 2) \text{ on } \partial\mathscr{B}_o \qquad (4.130)$$

Mutual interaction of waves a and b:

$$\nabla_X \cdot \mathbf{S}^L(\mathbf{H}_{ab}) - \rho_o\ddot{\mathbf{u}}_{ab} = -\nabla_X \cdot \mathbf{S}^{NL}(\mathbf{H}_a, \mathbf{H}_b, 2) \text{ in } \mathscr{B} \qquad (4.131)$$

$$\mathbf{N} \cdot \mathbf{S}^L(\mathbf{H}_{ab}) = -\mathbf{N} \cdot \mathbf{S}^{NL}(\mathbf{H}_a, \mathbf{H}_b, 2) \text{ on } \partial\mathscr{B}_o. \qquad (4.132)$$

Example 4.1. Third order interactions.

This example demonstrates how to generalize the wave interactions approach by formulating the BVP for the interaction of two distinct primary waves a and b to third order. We follow Chillara and Lissenden [11] and Liu *et al* [12]. Suppose that we are interested in determining the nonlinearity from the interaction of two collimated planar waves a and b and we recognize that there will be second harmonics generated and that they will interact with the primary waves in addition to the third order self-interactions. Therefore, we decompose the full wavefield as

$$\mathbf{u} = \mathbf{u}_a + \mathbf{u}_b + \mathbf{u}_{aa} + \mathbf{u}_{bb} + \mathbf{u}_{ab} + \mathbf{u}_{aaa} + \mathbf{u}_{bbb} + \mathbf{u}_{aab} + \mathbf{u}_{abb}$$

and presume that all second and third order terms are significantly smaller than the primary waves ($\mathbf{u}_a$ and $\mathbf{u}_b$) that generate them at the driving frequencies of ω_a and ω_b, respectively. The nonlinear terms are described as:
- Self-interactions at quadratic nonlinearity ($\mathbf{u}_{aa}$ and $\mathbf{u}_{bb}$) at frequencies $2\omega_a$ and $2\omega_b$, respectively.
- Mutual interactions at quadratic nonlinearity ($\mathbf{u}_{ab}$) at frequencies $|\omega_a \pm \omega_b|$.
- Self-interactions at cubic nonlinearity ($\mathbf{u}_{aaa}$ and $\mathbf{u}_{bbb}$), at frequencies of $3\omega_a$ and $3\omega_b$, respectively.
- Mutual interactions at cubic nonlinearity ($\mathbf{u}_{aab}$ and $\mathbf{u}_{abb}$) at frequencies $|2\omega_a \pm \omega_b|$ and $|\omega_a \pm 2\omega_b|$, respectively.

Extrapolating (4.119) to third order enables decomposition of the first Piola–Kirchhoff stress tensor,

$$\mathbf{S}(\mathbf{H}) = \mathbf{S}^L(\mathbf{H}) + \mathbf{S}^{NL}(\mathbf{H})$$

$$\mathbf{S}^L(\mathbf{H}) = \mathbf{S}^L(\mathbf{H}_a) + \mathbf{S}^L(\mathbf{H}_b) + \mathbf{S}^L(\mathbf{H}_{aa}) + \mathbf{S}^L(\mathbf{H}_{bb}) + \mathbf{S}^L(\mathbf{H}_{ab}) + \mathbf{S}^L(\mathbf{H}_{aaa})$$
$$+ \mathbf{S}^L(\mathbf{H}_{bbb}) + \mathbf{S}^L(\mathbf{H}_{aab}) + \mathbf{S}^L(\mathbf{H}_{abb})$$

$$\mathbf{S}^{NL}(\mathbf{H}) = \mathbf{S}^{NL}(\mathbf{H}_a + \mathbf{H}_b) = \mathbf{S}^Q(\mathbf{H}_a, \mathbf{H}_b, 2, 0) + \mathbf{S}^Q(\mathbf{H}_a, \mathbf{H}_b, 0, 2)$$
$$+ \mathbf{S}^Q(\mathbf{H}_a, \mathbf{H}_b, 1, 1) + \mathbf{S}^C(\mathbf{H}_a, \mathbf{H}_b, 3, 0) + \mathbf{S}^C(\mathbf{H}_a, \mathbf{H}_b, 0, 3)$$
$$+ \mathbf{S}^C(\mathbf{H}_a, \mathbf{H}_b, 2, 1) + \mathbf{S}^C(\mathbf{H}_a, \mathbf{H}_b, 1, 2)$$

where the decomposition of the displacement gradient $\mathbf{H}$ follows that of the displacement $\mathbf{u}$. The notation for the seven nonlinear stress interaction terms is systematic and precise; a new notation is introduced due to the large quantity of terms when interactions above second order are permitted. The superscripts Q and C denote quadratic and cubic nonlinearity with respect to displacement gradient and these terms are given in example 2.4 for isotropic materials. The argument for the nonlinear stress explicitly denotes the interaction between wavefields a and b through $\mathbf{H}_a$ and $\mathbf{H}_b$. The indices in the argument dictate the order of the interaction, with the first index referring to $\mathbf{H}_a$ and the second index referring to $\mathbf{H}_b$. If one of the indices is zero then it is a self-interaction, otherwise it is a mutual interaction. For example, the term $\mathbf{S}^C(\mathbf{H}_a, \mathbf{H}_b, 0, 3)$ represents self-interaction of wavefield b.

The BVP can now be divided into eight separate BVPs.

Primary waves (which could themselves be decomposed into nine linear problems):

$$\nabla_X \cdot \mathbf{S}^L(\mathbf{H}) - \rho_o \ddot{\mathbf{u}} = 0 \text{ in } \mathcal{B}$$

$$\mathbf{N} \cdot \mathbf{S}^L(\mathbf{H}) = 0 \text{ on } \partial \mathcal{B}_o.$$

Self-interaction of waves a at quadratic nonlinearity:

$$\nabla_X \cdot \mathbf{S}^L(\mathbf{H}_{aa}) - \rho_o \ddot{\mathbf{u}}_{aa} = -\nabla_X \cdot \mathbf{S}^Q(\mathbf{H}_a, \mathbf{H}_b, 2, 0) \text{ in } \mathfrak{B}$$

$$\mathbf{N} \cdot \mathbf{S}^L(\mathbf{H}_{aa}) = -\mathbf{N} \cdot \mathbf{S}^Q(\mathbf{H}_a, \mathbf{H}_b, 2, 0) \text{ on } \partial\mathfrak{B}_o.$$

Self-interaction of waves b at quadratic nonlinearity:

$$\nabla_X \cdot \mathbf{S}^L(\mathbf{H}_{bb}) - \rho_o \ddot{\mathbf{u}}_{bb} = -\nabla_X \cdot \mathbf{S}^Q(\mathbf{H}_a, \mathbf{H}_b, 0, 2) \text{ in } \mathfrak{B}$$

$$\mathbf{N} \cdot \mathbf{S}^L(\mathbf{H}_{bb}) = -\mathbf{N} \cdot \mathbf{S}^Q(\mathbf{H}_a, \mathbf{H}_b, 0, 2) \text{ on } \partial\mathfrak{B}_o.$$

Mutual interaction of waves a and b at quadratic nonlinearity:

$$\nabla_X \cdot \mathbf{S}^L(\mathbf{H}_{ab}) - \rho_o \ddot{\mathbf{u}}_{ab} = -\nabla_X \cdot \mathbf{S}^Q(\mathbf{H}_a, \mathbf{H}_b, 1, 1) \text{ in } \mathfrak{B}$$

$$\mathbf{N} \cdot \mathbf{S}^L(\mathbf{H}_{ab}) = -\mathbf{N} \cdot \mathbf{S}^Q(\mathbf{H}_a, \mathbf{H}_b, 1, 1) \text{ on } \partial\mathfrak{B}_o.$$

Self-interaction of waves a at cubic nonlinearity:

$$\nabla_X \cdot \mathbf{S}^L(\mathbf{H}_{aaa}) - \rho_o \ddot{\mathbf{u}}_{aaa} = -\nabla_X \cdot \mathbf{S}^C(\mathbf{H}_a, \mathbf{H}_b, 3, 0) \text{ in } \mathfrak{B}$$

$$\mathbf{N} \cdot \mathbf{S}^L(\mathbf{H}_{aaa}) = -\mathbf{N} \cdot \mathbf{S}^C(\mathbf{H}_a, \mathbf{H}_b, 3, 0) \text{ on } \partial\mathfrak{B}_o.$$

Self-interaction of waves b at cubic nonlinearity:

$$\nabla_X \cdot \mathbf{S}^L(\mathbf{H}_{bbb}) - \rho_o \ddot{\mathbf{u}}_{bbb} = -\nabla_X \cdot \mathbf{S}^C(\mathbf{H}_a, \mathbf{H}_b, 0, 3) \text{ in } \mathfrak{B}$$

$$\mathbf{N} \cdot \mathbf{S}^L(\mathbf{H}_{bbb}) = -\mathbf{N} \cdot \mathbf{S}^C(\mathbf{H}_a, \mathbf{H}_b, 0, 3) \text{ on } \partial\mathfrak{B}_o.$$

Mutual interaction of waves a and b at cubic nonlinearity (a interacting twice):

$$\nabla_X \cdot \mathbf{S}^L(\mathbf{H}_{aab}) - \rho_o \ddot{\mathbf{u}}_{aab} = -\nabla_X \cdot \mathbf{S}^C(\mathbf{H}_a, \mathbf{H}_b, 2, 1) \text{ in } \mathfrak{B}$$

$$\mathbf{N} \cdot \mathbf{S}^L(\mathbf{H}_{aab}) = -\mathbf{N} \cdot \mathbf{S}^C(\mathbf{H}_a, \mathbf{H}_b, 2, 1) \text{ on } \partial\mathfrak{B}_o.$$

Mutual interaction of waves a and b at cubic nonlinearity (b interacting twice):

$$\nabla_X \cdot \mathbf{S}^L(\mathbf{H}_{abb}) - \rho_o \ddot{\mathbf{u}}_{abb} = -\nabla_X \cdot \mathbf{S}^C(\mathbf{H}_a, \mathbf{H}_b, 1, 2) \text{ in } \mathfrak{B}$$

$$\mathbf{N} \cdot \mathbf{S}^L(\mathbf{H}_{abb}) = -\mathbf{N} \cdot \mathbf{S}^C(\mathbf{H}_a, \mathbf{H}_b, 1, 2) \text{ on } \partial\mathfrak{B}_o.$$

Here we see that good bookkeeping greatly facilitates solving wave interaction problems to third order (and beyond if there is ever a reason to do so) without 'forgetting' to include some of the terms. Naturally, the solution depends upon knowing all of the nonlinear stress terms. Equation (4.121) gives $\mathbf{S}^Q(\mathbf{H}_a, \mathbf{H}_b, 2, 0)$ and $\mathbf{S}^Q(\mathbf{H}_a, \mathbf{H}_b, 0, 2)$, while equation (4.122) gives $\mathbf{S}^Q(\mathbf{H}_a, \mathbf{H}_b, 1, 1)$. The cubic self-interaction terms $\mathbf{S}^C(\mathbf{H}_a, \mathbf{H}_b, 3, 0)$ and $\mathbf{S}^C(\mathbf{H}_a, \mathbf{H}_b, 0, 3)$ are worked out in example 2.4, but it is left to the reader to work out the cubic mutual interaction terms $\mathbf{S}^C(\mathbf{H}_a, \mathbf{H}_b, 2, 1)$ and $\mathbf{S}^C(\mathbf{H}_a, \mathbf{H}_b, 1, 2)$ if sufficiently motivated.

While it may be obvious to the experienced reader, this example problem solves the important practical problem of third harmonic generation due to self-interaction of SH waves in a plate. This is a much-simplified problem without mutual interactions because, as was demonstrated by the formulation of equation (3.26), pure shear waves cannot generate second harmonics in isotropic materials. Thus, for self-interaction of SH waves in a plate we simply have the following:

1) Primary waves:

$$\nabla_X \cdot \mathbf{S}^L(\mathbf{H}) - \rho_o \ddot{\mathbf{u}} = 0 \text{ in } \mathcal{B}$$

$$\mathbf{N} \cdot \mathbf{S}^L(\mathbf{H}) = 0 \text{ on } \partial\mathcal{B}_o.$$

2) Self-interaction of waves a:

$$\nabla_X \cdot \mathbf{S}^L(\mathbf{H}_{aaa}) - \rho_o \ddot{\mathbf{u}}_{aaa} = -\nabla_X \cdot \mathbf{S}^C(\mathbf{H}_a, \mathbf{H}_b, 3, 0) \text{ in } \mathcal{B}$$

$$\mathbf{N} \cdot \mathbf{S}^L(\mathbf{H}_{aaa}) = -\mathbf{N} \cdot \mathbf{S}^C(\mathbf{H}_a, \mathbf{H}_b, 3, 0) \text{ on } \partial\mathcal{B}_o.$$

Likewise, we may recall that $\mathscr{A}$ is the only TOEC and $\mathscr{G}$ is the only FOEC to play a role in pure shear wave problems (see example 3.5). Thus, modeling third harmonic generation from self-interaction of SH waves only requires that we know linear elastic properties and $\mathscr{A}$ and $\mathscr{G}$.

4.3 Closure

The boundary value problems for linear and nonlinear guided wave problems were formulated. Based on these formulations the linear solutions, i.e. the dispersion curves and wavestructures, will be given in chapter 5. The nonlinear solution for plate problems will be given in chapter 6.

References

[1] Rose J L 2014 *Ultrasonic Guided Waves in Solid Media* (Cambridge: Cambridge University Press)
[2] Rayleigh J W S 1945 *Theory of Sound* (New York: Dover)
[3] Viktorov I A 1967 *Rayleigh and Lamb Waves—Physical Theory and Applications* (New York: Plenum)
[4] Lamb H 1917 On waves in an elastic plate *Proc. R. Soc. Lond.* A **93** 114–28
[5] Solie L P and Auld B A 1973 Elastic waves in free anisotropic plates *J. Acoust. Soc. Am.* **54** 50–65
[6] Lowe M J S 1995 Matrix techniques for modeling ultrasonic waves in multilayered media *IEEE Trans. Ultrason. Ferroelectr. Freq. Control* **42** 525–42
[7] Gazis D C 1959 Three-dimensional investigation of the propagation of waves in hollow circular cylinders. I. Analytical foundation *J. Acoust. Soc. Am.* **31** 568–73
[8] Gazis D C 1959 Three-dimensional investigation of the propagation of waves in hollow circular cylinders. II. Numerical results *J. Acoust. Soc. Am.* **31** 573–8
[9] Hakoda C, Rose J, Shokouhi P and Lissenden C 2018 Using Floquet periodicity to easily calculate dispersion curves and wave structures of homogeneous waveguides *AIP Conf. Proc.* **37** 020016

[10] Hakoda C, Lissenden C and Rose J L 2018 Weak form implementation of the semi-analytical finite element (SAFE) method for a variety of elastodynamic waveguides *AIP Conf. Proc.* **37** 230001

[11] Chillara V K and Lissenden C J 2012 Interaction of guided wave modes in isotropic weakly nonlinear elastic plates: higher harmonic generation *J. Appl. Phys.* **111** 124909

[12] Liu Y, Chillara V K, Lissenden C J and Rose J L 2013 Third harmonic shear horizontal and Rayleigh Lamb waves in weakly nonlinear plates *J. Appl. Phys.* **114** 114908

Chapter 5

Ultrasonic guided waves—linear features

Despite our inability to see elastic waves like we see water waves, it is quite useful to visualize their characteristics, which we do through dispersion curves and wave-structures. Having formulated the boundary value problems in chapter 4, it is now time to plot frequency–wavenumber, phase velocity, and group velocity dispersion curves for waves in different waveguides—and this just scratches the surface. Since the previous chapters have focused more on the mathematics than the physics of wave propagation, we start with a summary of the physical characteristics of waves.

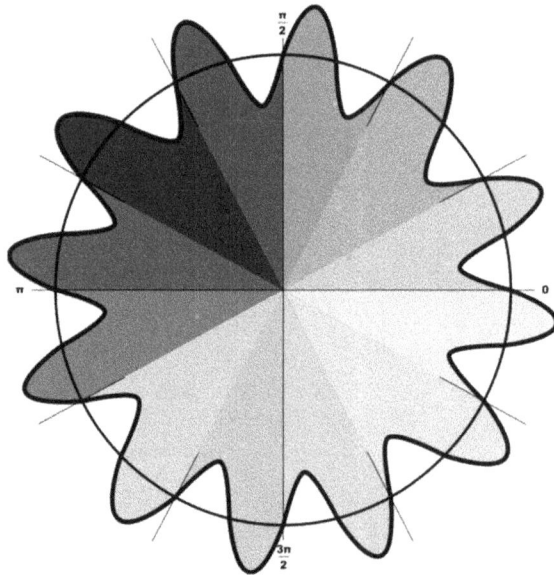

Chapter 4 formulated the boundary value problems that describe elastic wave propagation in waveguides for both linear and nonlinear domains. The linear problems were reduced to eigen-problems, whose solution provides the deterministic features of the guided waves. The eigenvalues give the wavenumbers and wave speeds, and the eigenvectors represent the displacement profiles (i.e. the wavestructure or mode shapes) in the cross-section of the waveguide, which is required to not vary along its length. In this chapter we investigate the solutions in the linear domain to prepare us for the solutions to the nonlinear problems that are given in chapters 6–8.

5.1 Physical characteristics of waves

We start by reviewing the fundamental physical characteristics of wave propagation, which we have managed to parlay until now through our prior focus on the mathematics. Elastic waves are mechanical disturbances that propagate in solid media. Elastic waves having infinitesimal amplitudes on the order of nanometers tend to propagate as harmonic waves in the crystal lattice, with an exception being the finite amplitude broadband pulse-generated waves, as described in chapter 9, for example. Harmonic waves are described mathematically by their transient displacement field,

$$u_i(x_j,\, t) = A_i[\cos{(kx_1 - \omega t)} + \mathrm{i} \sin{(kx_1 - \omega t)}], \qquad (5.1)$$

or alternatively, we use Euler's formula to write it more compactly in terms of the exponential function as

$$u_i(x_j,\, t) = A_i \mathrm{e}^{\mathrm{i}(kx_1 - \omega t)} \qquad (5.2)$$

for waves propagating in the x_1 direction. For clarity we distinguish between the imaginary unit i (not italic) and an index for a tensor or vector i (italic). Clearly the displacement field is complex-valued, with the physical part being the real part. The imaginary part is 90° out of phase with the real part. The leading coefficient is the wave amplitude A_i. The argument $(kx_1 - \omega t)$ is known as the phase. The argument in equations (5.1) and (5.2) represents waves propagating in the $+x_1$ direction with increasing time t since x_1 must increase with increasing t to keep the phase constant. The time period of one wave cycle is T, which corresponds to the wavelength λ in the spatial domain. The frequency f, in hertz, is the inverse of T, and the circular frequency is $\omega = 2\pi f$, in rad s^{-1}. Finally, the wavenumber k is the number of wavelengths in the unit circle, i.e. $k = 2\pi/\lambda$, in rad m^{-1} (see the illustration at the beginning of this chapter). Although wavenumber is a scalar in the direction of wave propagation, we may describe wavenumber relative to the relevant coordinate system by the wavevector.

5.1.1 Phase velocity

If we think of a mechanical disturbance as a wave packet having a finite number of cycles, as shown in figure 5.1, then we can define the speed at which one phase of that packet travels to be its wavelength divided by the time period:

$$c_p = \frac{\lambda}{T} = \lambda f. \qquad (5.3)$$

The name phase velocity implies that this speed can be written in terms of the relevant coordinate system although this is rarely done. Since the wavefield is

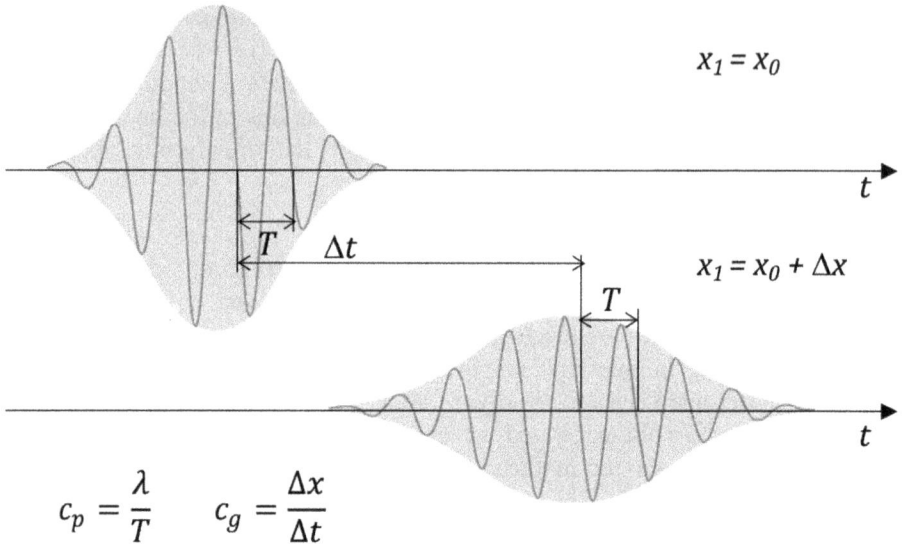

$$c_p = \frac{\lambda}{T} \qquad c_g = \frac{\Delta x}{\Delta t}$$

Figure 5.1. Wave speeds in terms of phase velocity and group velocity.

directly represented by the displacement field, which is written in terms of the wavenumber, it is useful to relate the phase velocity to the wavenumber,

$$c_p = \lambda f = \frac{2\pi}{k} \frac{\omega}{2\pi} = \frac{\omega}{k}, \tag{5.4}$$

which is also apparent from the phase $(kx_1 - \omega t)$.

5.1.2 Wavestructure

We use the term wavestructure to denote the modal displacement field in the cross-section of the waveguide. This is a representation of the eigenvector. The wavestructure is analogous to a mode shape in mechanical vibrations in that it does not quantify the magnitude of the displacement field, only the shape. For a half-space or plate, the displacement field for planar waves is only a function of the x_3 direction, thus we often denote the wavestructure $U_i(x_3)$, giving the displacement field

$$u_i(x_j, t) = U_i(x_3)e^{i(kx_1 - \omega t)}. \tag{5.5}$$

It is customary to normalize the wavestructures either by setting the maximum value equal to unity or to make the modes orthonormal, as will be discussed in chapter 6. If the $U_i(x_3)$ do not represent the actual amplitude of the waves then we should add scaling factor A to equation (5.5). There are many instances when particle velocities and stresses need to be computed from the displacement wavestructures. Consider

$$v_i(x_j, t) = V_i(x_3)e^{i(kx_1 - \omega t)} \tag{5.6}$$

$$\sigma_{ij}(x_j, t) = \Sigma_{ij}(x_3)e^{i(kx_1 - \omega t)}, \tag{5.7}$$

where $V_i(x_3)$ and $\Sigma_{ij}(x_3)$ are computed from $U_i(x_3)$ as follows:

$$V_i(x_3) = -i\omega U_i(x_3) \tag{5.8}$$

$$\begin{aligned}
\Sigma_{11}(x_3) &= (\lambda + 2\mu)ik U_1 + \lambda U_{3,3} \\
\Sigma_{12}(x_3) &= \mu ik U_2 \\
\Sigma_{13}(x_3) &= \mu U_{1,3} + \mu ik U_3
\end{aligned} \tag{5.9}$$

$$\begin{aligned}
\Sigma_{33}(x_3) &= \lambda ik U_1 + (\lambda + 2\mu) U_{3,3} \\
\Sigma_{23}(x_3) &= \mu U_{2,3} \\
\Sigma_{13}(x_3) &= \mu U_{1,3} + \mu ik U_3.
\end{aligned} \tag{5.10}$$

Equation (5.9) gives the stress components acting in the wave propagation direction and equation (5.10) gives the tractions on a surface with unit normal in the x_3 direction. Both equations (5.9) and (5.10) are restricted to isotropic materials, but similar relations can be obtained for anisotropic materials.

5.1.3 Group velocity

If each phase of the wave packet in figure 5.1 has the same frequency, as is the case for a continuous wave (not a wave packet), then the wave packet itself travels at the phase velocity. However, for a finite number of cycles the packet has a frequency bandwidth and we can analyse the velocity of the wave packet, or group of phases. For simplicity, let's analyse two phases having frequencies that are only slightly different, $\omega_1 > \omega_2$. The real part of the displacement field is the superposition

$$u_1 = A_1 \cos(k_1 x_1 - \omega_1 t) + A_1 \cos(k_2 x_1 - \omega_2 t), \tag{5.11}$$

which can be manipulated using the obscure product-sum trigonometric identity,

$$\cos\alpha + \cos\beta = 2\cos\left(\frac{\alpha + \beta}{2}\right)\cos\left(\frac{\alpha - \beta}{2}\right),$$

to obtain

$$u_1 = 2A_1 \cos\left(\frac{(k_1 x_1 - \omega_1 t) + (k_2 x_1 - \omega_2 t)}{2}\right)\cos\left(\frac{(k_1 x_1 - \omega_1 t) - (k_2 x_1 - \omega_2 t)}{2}\right)$$

$$u_1 = 2A_1 \cos\left(\frac{(k_1 + k_2)x_1 - (\omega_1 + \omega_2)t}{2}\right)\cos\left(\frac{(k_1 - k_2)x_1 - (\omega_1 - \omega_2)t}{2}\right)$$

$$u_1 = 2A_1 \cos\left(k_{\text{avg}} x_1 - \omega_{\text{avg}} t\right)\cos\left(\frac{\Delta k x_1 - \Delta\omega t}{2}\right). \tag{5.12}$$

We understand equation (5.12) to be modulation of the high frequency component ω_{avg} with the low frequency component $\Delta\omega$. The phase velocity of the high frequency component is $c_{\text{avg}} = \frac{\omega_{\text{avg}}}{k_{\text{avg}}}$ and the phase velocity of the low frequency component is $c_\Delta = \frac{\Delta\omega}{\Delta k}$. Since ω_1 and ω_2 are almost equal $c_{avg} \rightarrow c_p$ and the high frequency component represents the phase. The low frequency component represents the propagation of the envelope or group,

$$c_g = c_\Delta = \frac{\Delta\omega}{\Delta k}, \tag{5.13}$$

which in the limit becomes

$$c_g = \frac{d\omega}{dk}. \tag{5.14}$$

Example 5.1. Group velocity calculation.

Consider two finite-duration waves having similar frequencies and wavenumbers. Estimate the group velocity and show the modulation.

Take $\omega_1 = 1 \times 10^6$ rad s^{-1}, $k_1 = 200$ rad m^{-1}.
Take $\omega_2 = 0.96 \times 10^6$ rad s^{-1}, $k_2 = 190$ rad m^{-1}.
The phase velocities are $c_1 = \frac{1 \times 10^6}{200} = 5000$ m s^{-1}
and $c_2 = \frac{0.96 \times 10^6}{190} = 5053$ m s^{-1},

$c_{avg} = \frac{\omega_{avg}}{k_{avg}} = \frac{0.98 \times 10^6}{195} = 5026$ m s^{-1},

$c_g = c_\Delta = \frac{\Delta\omega}{\Delta k} = \frac{0.04 \times 10^6}{10} = 4000$ m s^{-1}.

The modulation of the two waves, showing the envelopes, is depicted in figure 5.2.

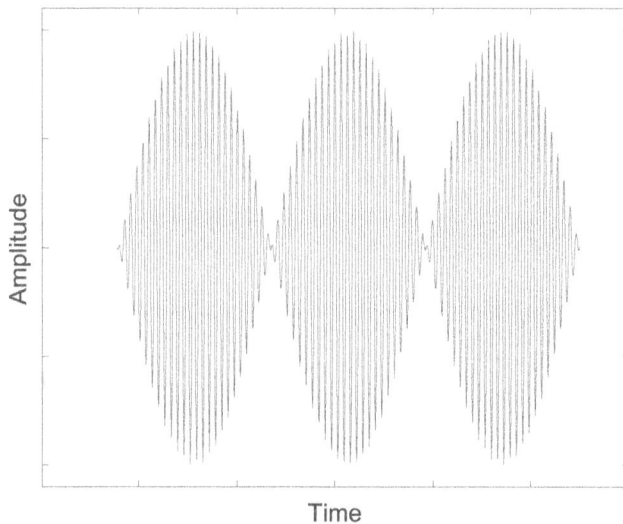

Figure 5.2. Modulation of two waves.

The previously described fundamental wave characteristics can be employed to write equation (5.14) in a variety of different ways [1], each requiring computation of a derivative. Computing derivatives is usually straightforward, but here there are challenges that stem from the multi-modal nature of guided waves. We will soon see that there exist multiple modes at a prescribed frequency, each having different wavenumbers, phase velocities, and wavestructures. The sorting of these modes becomes problematic when modes cross or veer away from one another because there is not a definitive characteristic defining a specific mode. Another approach to the group velocity is to relate it to the speed of the energy flux. In fact, the speed of the energy flux has been shown to be equal to the group speed (see Biot [2] and Achenbach [3]). Thus, we can conceptualize that the envelope in the above analysis provides the energy transport. We seek an expression for the group velocity that does not involve a problematic derivative.

We start by balancing the kinetic energy and the potential energy. Instead of writing the potential energy in terms of the body forces and surface tractions we use the elastic strain energy associated with them:

$$T = \int_V \frac{1}{2}\rho \dot{u}_i \dot{u}_i dV \tag{5.15}$$

$$V = U = \int_V \frac{1}{2}\sigma_{ij}\varepsilon_{ij} dV. \tag{5.16}$$

We introduce the complex conjugate (denoted with an asterisk*) to force the energies to be real:

$$\int_V \frac{1}{2}\rho \dot{u}_i \dot{u}_i^* dV = \int_V \frac{1}{2}\sigma_{ij}\varepsilon_{ij}^* dV. \tag{5.17}$$

Since we have a time harmonic displacement field ($\sim e^{-i\omega t}$), $\dot{u}_i = -i\omega u_i$, and we can eliminate the time derivative

$$\int_V \rho(-i\omega u_i)(i\omega u_i^*)dV = \int_V \sigma_{ij}\varepsilon_{ij}^* dV \tag{5.18}$$

giving

$$\omega^2 = \frac{\int_V \sigma_{ij}\varepsilon_{ij}^* dV}{\int_V \rho u_l u_l^* dV}. \tag{5.19}$$

We want an expression for group velocity, so take the derivative with respect to wavenumber,

$$2\omega \frac{d\omega}{dk} = \frac{\partial}{\partial k}\left[\frac{\int_V \sigma_{ij}\varepsilon_{ij}^* dV}{\int_V \rho u_l u_l^* dV}\right] \tag{5.20}$$

and now the group velocity can be expressed as

$$c_g = \frac{d\omega}{dk} = \frac{1}{2\omega}\frac{\partial}{\partial k}\left[\frac{\int_V \sigma_{ij}\varepsilon_{ij}^* dV}{\int_V \rho u_l u_l^* dV}\right], \tag{5.21}$$

which applies to any linear elastic material, and becomes powerful when we apply the displacement, strain, and stress fields determined by the wavestructure. The wavestructure is given by $U_i(x_3)$ in equation (5.5) for a plate. Use the infinitesimal strain-displacement relation equation (2.8) and Hooke's law equation (2.42) to write stress and strain in terms of $U_i(x_3)$ and its derivatives with respect to x_3. Thus, we can take the derivatives in general one time, resulting in an expression for c_g that is easily computed. We will implement equation (5.21) for Lamb waves in section 5.3.

5.1.4 Attenuation

The concept of attenuation due to internal damping, scattering, and diffraction was introduced in section 2.3.4. We now address attenuation in guided waves attributable to internal damping, whereby the material is viscoelastic (i.e. time-dependent elastic). We commonly use mechanical analogs to represent viscoelastic material behavior, where a spring represents elasticity, and a dashpot represents viscous damping. The simplest configurations for a single spring and a single dashpot are series and parallel. The series configuration is known as the Maxwell model, having the 1D stress–strain relation $\sigma + \frac{\eta}{E}\dot\sigma = \eta\dot\varepsilon$, and the parallel config-uration is known as the Kelvin–Voigt model, having the 1D stress–strain relation $\sigma = E\varepsilon + \eta\dot\varepsilon$, where E and η are the Young's modulus and viscosity, respectively.

When analysing time harmonic waves, we have

$$\varepsilon = \varepsilon_o e^{i\omega t}, \ \sigma = \sigma_o e^{i\omega t} \tag{5.22}$$

and we can write the stress–strain relation as

$$\sigma_o = \frac{(\omega\eta)^2 E + i\omega\eta E^2}{E^2 + (\omega\eta)^2}\varepsilon_o \tag{5.23}$$

$$\sigma_o = (E + i\omega\eta)\varepsilon_o \tag{5.24}$$

for the Maxwell and Kelvin–Voigt models, respectively. Both models can be represented by a complex elasticity (or stiffness), $C^*=C' + iC''$, and generalized in 3D to a complex elasticity tensor, $\mathbf{C}^*=\mathbf{C}' + i\mathbf{C}''$, where $\mathbf{C}^*$ simply means that it is complex (not the complex conjugate, which is the meaning of * everywhere else in this book) and $\mathbf{C}'$ and $\mathbf{C}''$ represent the storage and loss moduli, respectively. The viscoelastic correspondence principle enables the complex solution to a viscoelastic problem to be obtained from the real solution of an elastic problem. In the Kelvin–Voigt model the storage modulus is the elasticity tensor, $\mathbf{C}' = \mathbf{C}$, and the loss modulus is the frequency times the viscosity tensor, $\mathbf{C}'' = \omega\eta$. The viscosity tensor generally has the same form as the elasticity tensor.

In the so-called hysteretic model the loss modulus is taken to be independent of the frequency, and is just the viscosity tensor [4]. Thus, the viscosity tensor only needs to be measured at one frequency. Waves propagating in waveguides comprised of materials having internal damping have a complex wavenumber k, resulting in the displacement field:

$$u(x_1, x_3, t) = U(x_3)e^{i(kx_1-\omega t)} = U(x_3)e^{-\alpha x_1}e^{i(k_r x_1-\omega t)}, \tag{5.25}$$

where k_r and $k_i = \alpha$ are the real and imaginary parts of k, respectively. Thus, we need to plot both phase velocity and attenuation dispersion curves for lossy plates, where the attenuation α is assigned units of nepers/m (or decibels/m). Destrade *et al* [5] studied the Kelvin–Voigt model for nonlinear waves and show that the simplicity of the Kelvin–Voigt model does not translate into a simple analysis of nonlinear waves.

5.2 Rayleigh waves

The Rayleigh wave problem was actually solved in chapter 4. The eigen-problem is given in equation (4.21). Setting the determinant of the matrix equal to zero gave the characteristic equation (4.22), which was solved for the Rayleigh wave speed c_R in equation (4.25). The wavenumber $k = \frac{\omega}{c_R}$. The coefficients A_1 and B_1 give the eigenvectors and are related through (4.27). The displacement field is specified by equations (4.16) and (4.17), which give the wavestructure to be

$$U_1(x_3) = ikA_1 e^{ipx_3} - iqB_1 e^{iqx_3} \tag{5.26}$$

$$U_3(x_3) = ipA_1 e^{ipx_3} + ikB_1 e^{iqx_3}. \tag{5.27}$$

From the definitions of p and q in equation (4.9) we recognize that since $c_R < c_T < c_L$, both p and q are positive imaginary numbers, and thus the arguments of the exponential functions are negative real numbers and decay exponentially with distance from the surface. Readers are encouraged to prove to themselves that if we take $B_1 = 1$, then U_1 is real and U_3 is imaginary. Moreover, p and q are frequency-dependent, making the wavestructure also frequency-dependent.

The dispersion curves and wavestructure for Rayleigh waves are shown in figure 5.3. In ω–k space the curve is a straight line having a slope of c_R. The phase velocity dispersion curve has a constant value $c_p = c_R$, indicating that Rayleigh waves are nondispersive, that is to say that their speed is not a function of frequency. The dispersion curves are plotted for three values of Poisson's ratio to show that there is limited variability for the typical range $0 < \nu < 1/2$. The wavestructure plot shows that most of the energy is within one wavelength of the surface. Not all surface waves are Rayleigh waves, and some surface waves are dispersive.

5.3 Waves in plates

5.3.1 Shear-horizontal (SH) waves

We saw from cases 3 and 4 of the general formulation of the plate problem in chapter 4 that SH waves are naturally decoupled from Lamb waves if the plate is isotropic. In the general formulation the global matrix was formed by satisfying the traction-free boundary conditions and the gauge invariance on the plate surface in equation (4.38), $[\mathbb{G}]\{x\} = \{0\}$. This is the eigen-problem, which can be solved by setting $\det[\mathbb{G}] = 0$. Root finding needs to be done carefully because $\mathbb{G}$ is complex and the roots may be real, imaginary or complex. See Lowe [6] for details on root finding. In the case of an isotropic plate we are able to solve sub-determinants as was described in chapter 4. For SH waves

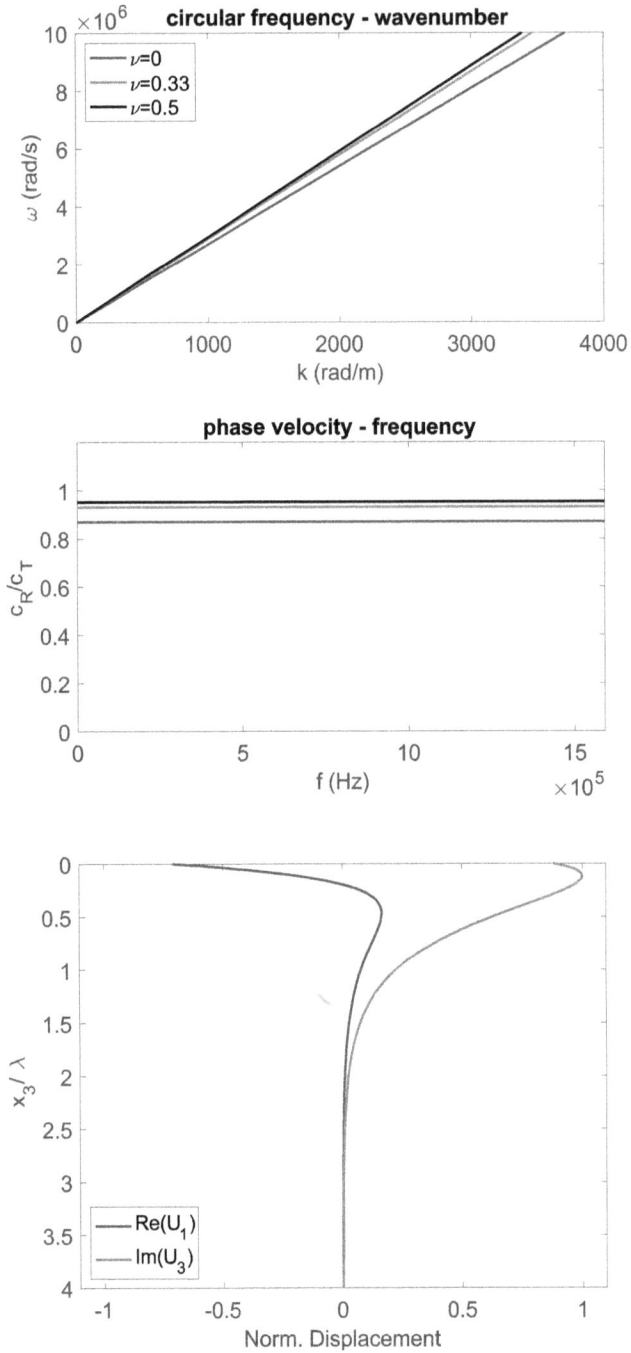

Figure 5.3. ω–k and c_p–f dispersion curves and wavestructure for Rayleigh waves for aluminum half-space.

the dispersion relations are simple enough to solve analytically, as was also done in chapter 4. Thus, we just need to present and describe the SH wave solutions.

Three dispersion curves for SH waves are plotted in figure 5.4 for an aluminum plate, where symmetric modes and antisymmetric modes are colored blue and red, respectively. The properties for aluminum are taken to be

$$\rho = 2700 \text{ kg m}^{-3},\ E = 69 \text{ GPa},\ \nu = 0.33,$$

where Lame's constant and the shear modulus are

$$\lambda = 50.35 \text{ GPa},\ \mu = 25.94 \text{ GPa},$$

respectively, giving the bulk wave speeds to be

$$c_L = 6153 \text{ m s}^{-1},\ c_T = 3100 \text{ m s}^{-1}.$$

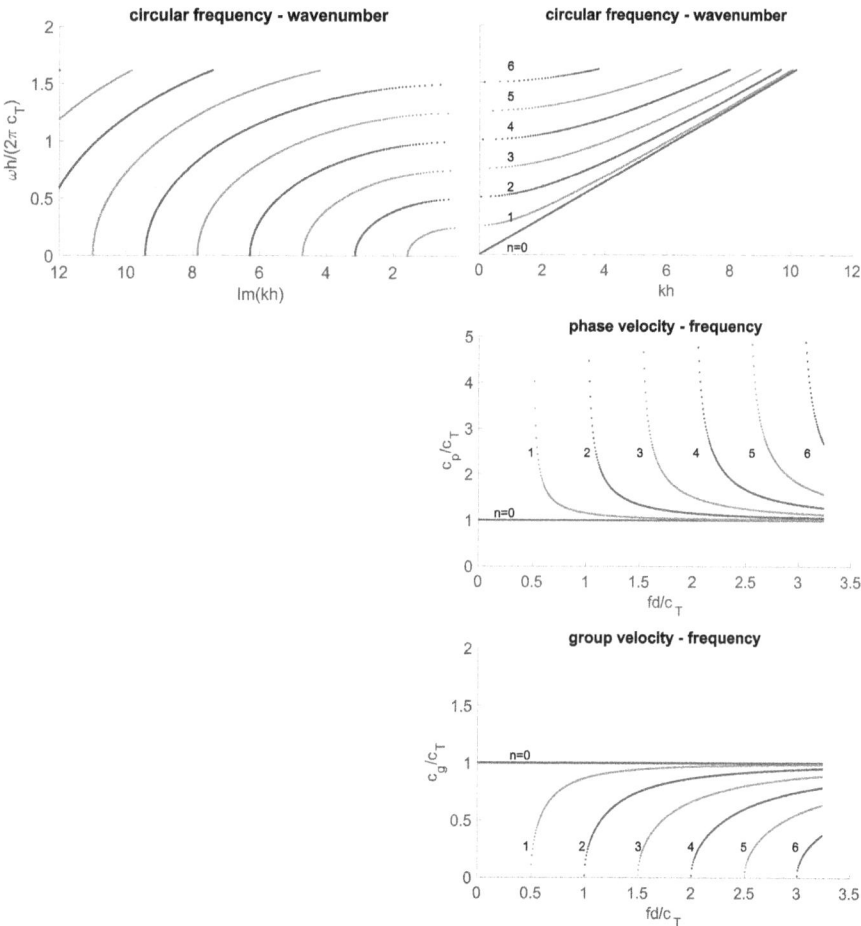

Figure 5.4. ω–k, c_p–fd, and c_g–fd dispersion curves for SH waves.

The nondimensionalized ω–k dispersion curves are shown on the right-hand side of figure 5.4 for real wavenumbers, which correspond to propagating modes. The left portion of figure 5.4 shows the imaginary wavenumbers, which correspond to evanescent wave modes that are unable to propagate any appreciable distance (i.e. the e^{ikx_1} term becomes a decaying exponential instead of harmonic). The ω–k dispersion curves clearly show the mode cutoff frequencies at $kh = 0$, below which that mode cannot propagate. Mode numbers are just names that are assigned to the modes for identification purposes. It is common to assign mode numbers (SHn, where $n = 0, 1, 2, 3, ...$) by simply progressing upward along the frequency axis, although for SH waves the mode numbers also correspond to the integer in the argument of the sine/cosine function in the wavestructure (equations (4.56) and (4.57)),

$$U_1 = U_3 = 0$$
$$\text{symmetric} \quad U_2(x_3) = A_2 \cos(qx_3)$$
$$\text{antisymmetric} \; U_2(x_3) = A_1 \sin(qx_3),$$

where $q = n\pi/2h$. The SH wavestructure depends on the mode, but not the frequency. Rose [1] shows that the group velocity is

$$c_G = c_T \sqrt{1 - \frac{(n/2)^2}{(fd/c_T)^2}} \, . \tag{5.28}$$

Equation (4.63) reveals that the ω–k dispersion curves are hyperbolas for real wavenumbers and circular for imaginary wavenumbers. The exception is the fundamental mode ($n = 0$), which is a straight line, has no cutoff, and is non-dispersive. The phase velocity dispersion curves and group velocity dispersion curves in figure 5.4 are nondimensionalized with respect to the shear wave speed c_T. The abscissa is the fd product divided by c_T, where $d = 2h$ is the plate thickness. SH waves depend only on c_T, and not c_L, and c_T depends only on the shear modulus and mass density. Thus, figure 5.4 applies to a plate of any thickness composed of any isotropic material. It is important to realize that both the phase speed and group speed depend on plate thickness. As evident from equation (5.4), at the cutoff frequencies, where $k = 0$, the phase velocity goes to infinity. The cutoff fd-products for SH waves are $(fd)_n = \frac{nc_T}{2}$ for $n = 0, 1, 2, ...$ As $fd \to \infty$ both c_p and $c_g \to c_T$ for all modes. Finally, we observe that $c_p \geqslant c_T$ and $c_g \leqslant c_T$ are always true for all modes.

5.3.2 Lamb waves

The Lamb wave eigen-problem described in chapter 4 results in the characteristic equation being transcendental, equations (4.42) and (4.45) for symmetric modes and antisymmetric modes, respectively. A popular approach is to sweep through real frequencies and search for the wavenumbers. The solution requires a root finding algorithm capable of handling complex variables. The roots are real, imaginary, or complex, but only the real roots represent propagating modes. Indeed, there are pairs of roots, two real wavenumbers having opposite signs, as one propagates in the positive x_1 direction and one in the negative x_1 direction. The imaginary wavenumber pairs also have opposite signs, but do not propagate. Typically, it is the real positive wavenumbers that are of interest. Finding roots of a complex characteristic equation is not as straightforward as looking for zero crossings. The approach described by Lowe [6] works well. Once the wavenumbers are known, the phase velocity and group velocity can be computed for each wavenumber–frequency combination. The Lamb wave dispersion curves for an aluminum plate plotted in figure 5.5 are colored blue for symmetric modes and red for antisymmetric modes. The material properties specified in the previous section for SH waves were used. As for the SH waves, the cutoff frequencies are evident on the ordinate (where $k = 0$) of the ω–k dispersion curves, but the frequency values are more complicated. Rose's [1] analysis indicates that the mode cutoffs occur at

$$\text{symmetric modes } fd = \begin{cases} nc_T \\ nc_L/2 \end{cases} \tag{5.29}$$

$$\text{antisymmetric modes } fd = \begin{cases} nc_T/2 \\ nc_L \end{cases} \tag{5.30}$$

for $n = 1, 2, 3, \ldots$. The cutoff could be due to c_T or c_L for any mode, and for the symmetric modes, if $c_L/c_T \approx 2$ it is difficult to distinguish which one is the cause of the cutoff. This can be important if we want to utilize a specific wave motion at the cutoff frequency, which is essentially that of a standing wave reverberating in the x_3 direction. Note that in figure 5.5 the ordinate is $\frac{\omega h}{2\pi c_T} = \frac{fd}{2c_T}$.

The dispersion curves in figure 5.5 are plotted as a function of the frequency–thickness product fd/c_T (or $\omega h/\pi c_T$) making them applicable to any plate thickness. It is evident that the mode cutoffs shift down when the plate thickness increases. As for SH waves, the phase velocity and group velocity dispersion curves are non-dimensionalized by c_T. However, these curves depend on c_L, and thus the Young's modulus and Poisson's ratio of the material.

Lamb wave modes are identified by the symmetry or antisymmetry of the $U_1(x_3)$ wavestructure component and the order of their cutoff frequencies as Sn and An, where $n = 0, 1, 2, 3, \ldots$. The fundamental modes do not have cutoffs. At low frequencies the S0 and A0 modes approach longitudinal plate waves and flexural waves, respectively, while at high frequencies they both approach the Rayleigh wave speed, c_R. In the high frequency limit, all the other modes approach the shear wave speed, c_T.

The group velocity is plotted in figure 5.5 based on equation (5.21), which we have specialized for Lamb waves. Based on equations (4.43), the wavestructure components for symmetric modes are

$$U_1(x_3) = ikA \cos(px_3) - qD_2 \cos(qx_3)$$

$$U_2 = 0$$

$$U_3(x_3) = -pA \sin(px_3) + ikD_2 \sin(qx_3), \tag{5.31}$$

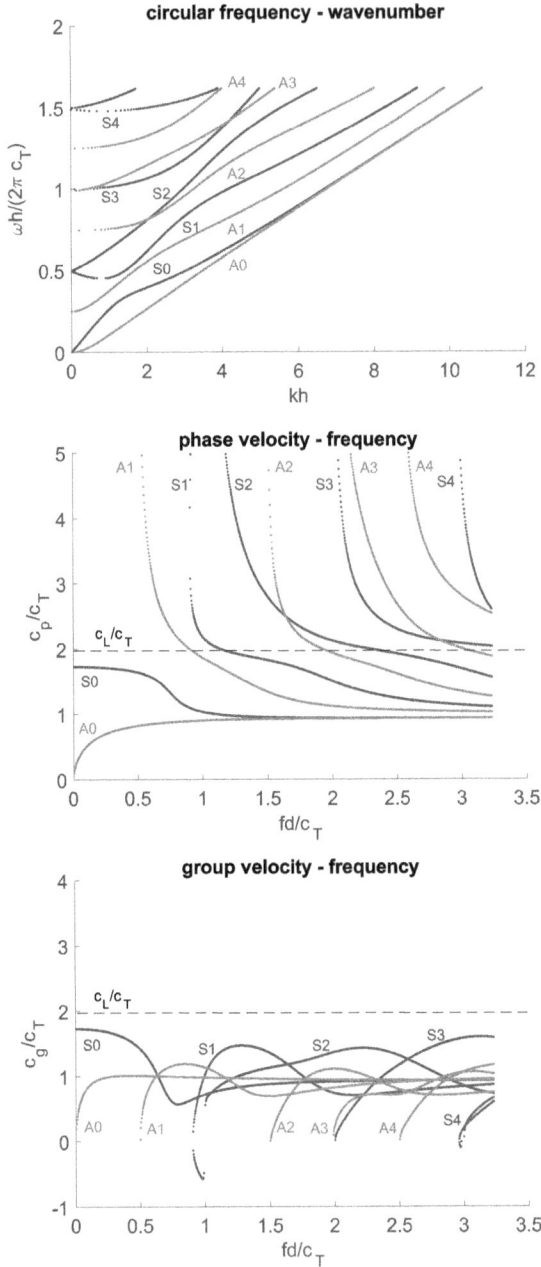

Figure 5.5. ω–k, c_p–fd, and c_g–fd dispersion curves for Lamb waves.

where the eigenvectors give the coefficients to be

$$A = D_2 \frac{2i\mu k q \cos(qh)}{[(\lambda + 2\mu)p^2 + \lambda k^2]\cos(ph)} \tag{5.32}$$

and for antisymmetric modes are

$$U_1(x_3) = ikB \sin(px_3) + qC_2 \sin(qx_3)$$

$$U_2 = 0$$

$$U_3(x_3) = pB \cos(px_3) + ikC_2 \cos(qx_3), \tag{5.33}$$

where

$$B = -C_2 \frac{2i\mu kq \sin(qh)}{[(\lambda + 2\mu)p^2 + \lambda k^2] \sin(ph)}. \tag{5.34}$$

For Lamb waves, where the plate is isotropic, Hakoda and Lissenden [7] indicate the simplified form of equation (5.21) for the group velocity to be

$$c_g = \frac{\int_{-h}^{h} [\xi_1 + \xi_2 + \xi_3] dx_3}{2\omega \int_{-h}^{h} \rho[U_i U_i^*] dx_3}, \tag{5.35}$$

$$\xi_1 = 2k_1[(\lambda + 2\mu)U_1 U_1^* + \mu(U_2 U_2^* + U_3 U_3^*)] \tag{5.36}$$

$$\xi_2 = -i\lambda \left(U_{3,3} U_1^* - U_1 U_{3,3}^* \right) \tag{5.37}$$

$$\xi_3 = -i\mu \left(U_{1,3} U_3^* - U_3 U_{1,3}^* \right). \tag{5.38}$$

The wavestructures for Lamb waves, $U_i(x_3)$, are given in equations (5.31) and (5.33). The necessary derivatives are

$$\text{symmetric modes} \quad \begin{aligned} U_{1,3} &= -ikpA \sin(px_3) + q^2 D_2 \sin(qx_3) \\ U_{3,3} &= -p^2 A \cos(px_3) + ikqD_2 \cos(qx_3) \end{aligned} \tag{5.39}$$

$$\text{antisymmetric modes} \quad \begin{aligned} U_{1,3} &= ikpB \cos(px_3) + q^2 C_2 \cos(qx_3) \\ U_{3,3} &= -p^2 B \sin(px_3) - ikqC_2 \sin(qx_3) \end{aligned}. \tag{5.40}$$

For the benefit of those who want to implement equation (5.21) in their own work, the signs on ξ_2 and ξ_3 in equations (5.37) and (5.38) are opposite those given by Hakoda and Lissenden [7] because they used the opposite sign on the exponential function, i.e. $u_i(x_i, t) = U_i(x_3)e^{-i(kx_1 - \omega t)}$.

Referring back to the group velocity dispersion curves in figure 5.5, two things are noteworthy. First, at a given frequency the mode having the fastest group velocity will give the first arriving signal, which is the easiest to analyse. Second, the S1 and S4 modes are unique in that they have a negative group velocity near the cutoff frequency.

Wavestructures are plotted in figure 5.6 at selected points on the dispersion curves in figure 5.5. To show that the U_1 and U_3 are out of phase, both the real part and the imaginary part of each are plotted. In all cases but one, U_1 is real and U_3 is imaginary. The A0 and S0 modes are plotted at frequencies of $fd/c_T = 0.1, 1, 2$ in the top two rows. The A1 and S1 modes are plotted at frequencies of $fd/c_T = 1, 2$ in the middle

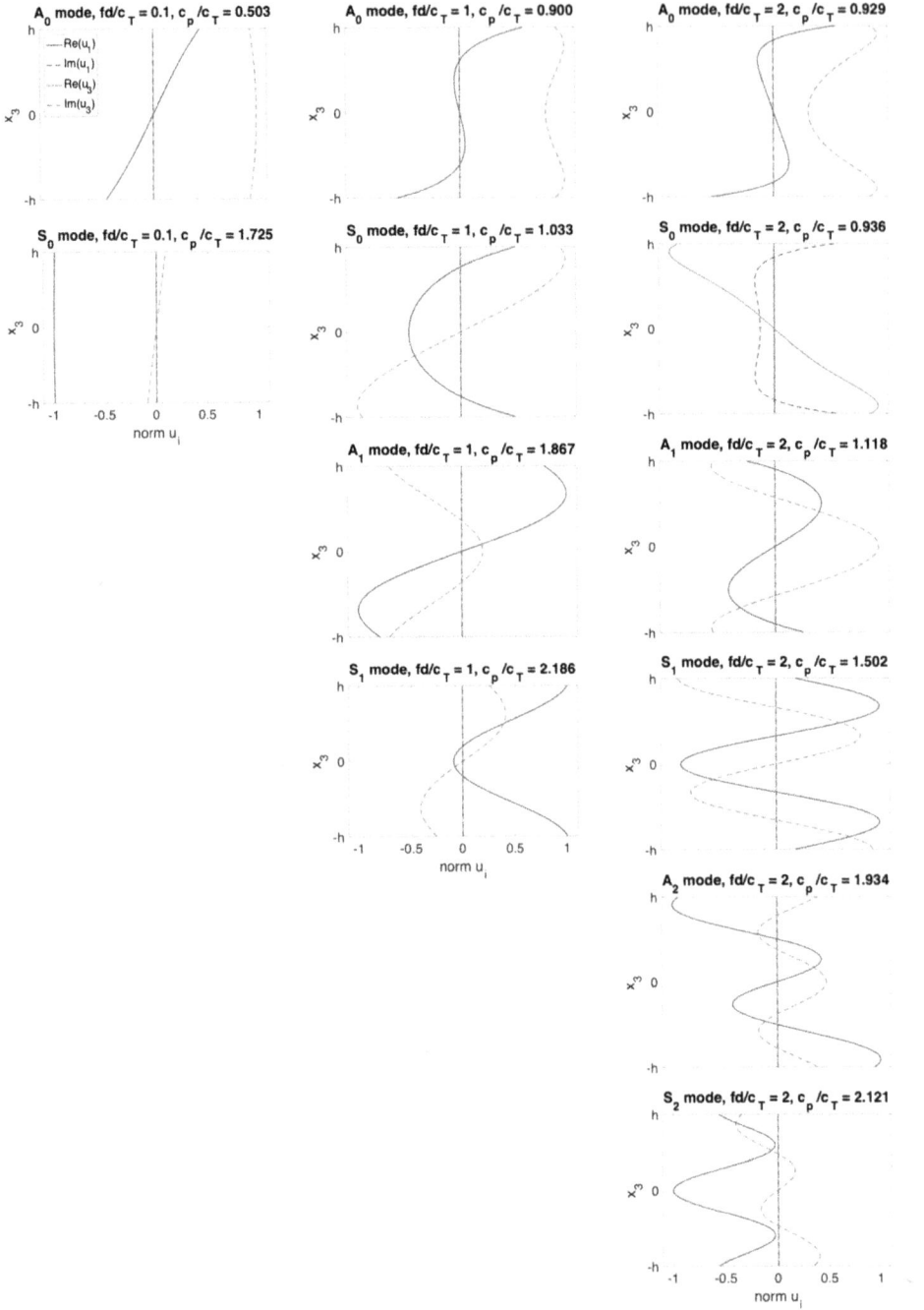

Figure 5.6. Wavestructures for selected Lamb wave modes and frequencies.

two rows, and the A2 and S2 modes are plotted at $fd/c_T = 2$ in the bottom subfigures. The U_1 and U_3 values are normalized such that the maximum value is unity. The symmetry/antisymmetry of the distributions about $x_3 = 0$ is evident.

The SH dispersion curves depend only on the shear modulus (and mass density) of the plate. The Lamb wave dispersion curves depend on both independent linear elastic properties, making it not possible to eliminate the effect of the material properties. How much effect do material properties have on the Lamb wave dispersion curves? Let's find out by taking advantage of the relatively small range of values that Poisson's ratio can have, i.e. $0 < \nu < 1/2$. Relevant material parameters are provided in table 5.1; by column they are mass density, Young's modulus, Poisson's ratio, Lame's constant, shear modulus, shear wave speed, longitudinal wave speed, and the wave speed ratio. The values in the last five columns are computed from the first three columns. As Poisson's ratio goes to $1/2$ the material becomes incompressible and Lame's constant and longitudinal wave speed become quite high. In addition to steel and aluminum, two fictitious materials having Poisson's ratios of 0.02 and 0.48 have been added. Dispersion curves are computed based on c_T and c_L and we notice the smallest difference between them occurs for the smallest Poisson's ratio and the largest difference comes from the largest Poisson's ratio. The resulting dispersion curves are shown in figure 5.7. Intuition might lead us to think that the large difference between c_T and c_L would have the largest effect on the dispersion curves, but actually quite the opposite is true. The small difference between c_T and c_L for small Poisson's ratio limits the low frequency plate wave speed and causes the S0 and S1 modes to veer apart at $fd/c_T \sim 0.7$. Prior to the veer the S0 mode is nondispersive and after the veer the S1 mode is nondispersive. The veer occurs when the phase velocity is equal to c_L. There is another veering of the S1 and S2 modes at $fd/c_T \sim 2.2$. The higher order symmetric modes can be described as highly dispersive after the cutoff, nondispersive at the plate wave speed between the veers, and then they asymptotically approach c_T. The group velocity dispersion curves for the symmetric modes are also unique as the mode that is nondispersive shifts from S0 to S1 to S2 and the group velocity is limited by c_L. We also can see that the A2, S1, and S3 modes each have a negative group velocity regime immediately after the cutoff frequency.

We can visualize the complete set of Lamb wave modes plotted on dispersion curves by including the evanescent modes from imaginary and complex wavenumber roots. Without sorting the modes this becomes difficult to digest. Thus, in figure 5.8 we have sorted real, imaginary, and complex roots for an aluminum plate, whereby *complex* here means that both the real and imaginary components are nonzero.

Table 5.1. Isotropic material parameters.

Material	ρ (kg m^{-3})	E (GPa)	ν (–)	λ (GPa)	μ (GPa)	c_T (m s^{-1})	c_L (m s^{-1})	c_L/c_T
Steel	7800	210	0.28	104	82.0	3243	5867	1.81
Aluminum	2700	69	0.33	50.4	25.9	3100	6153	1.98
Fictitious1	2700	69	0.02	1.41	33.8	3539	5057	1.43
Fictitious2	2700	69	0.48	559	23.3	2938	14 982	5.10

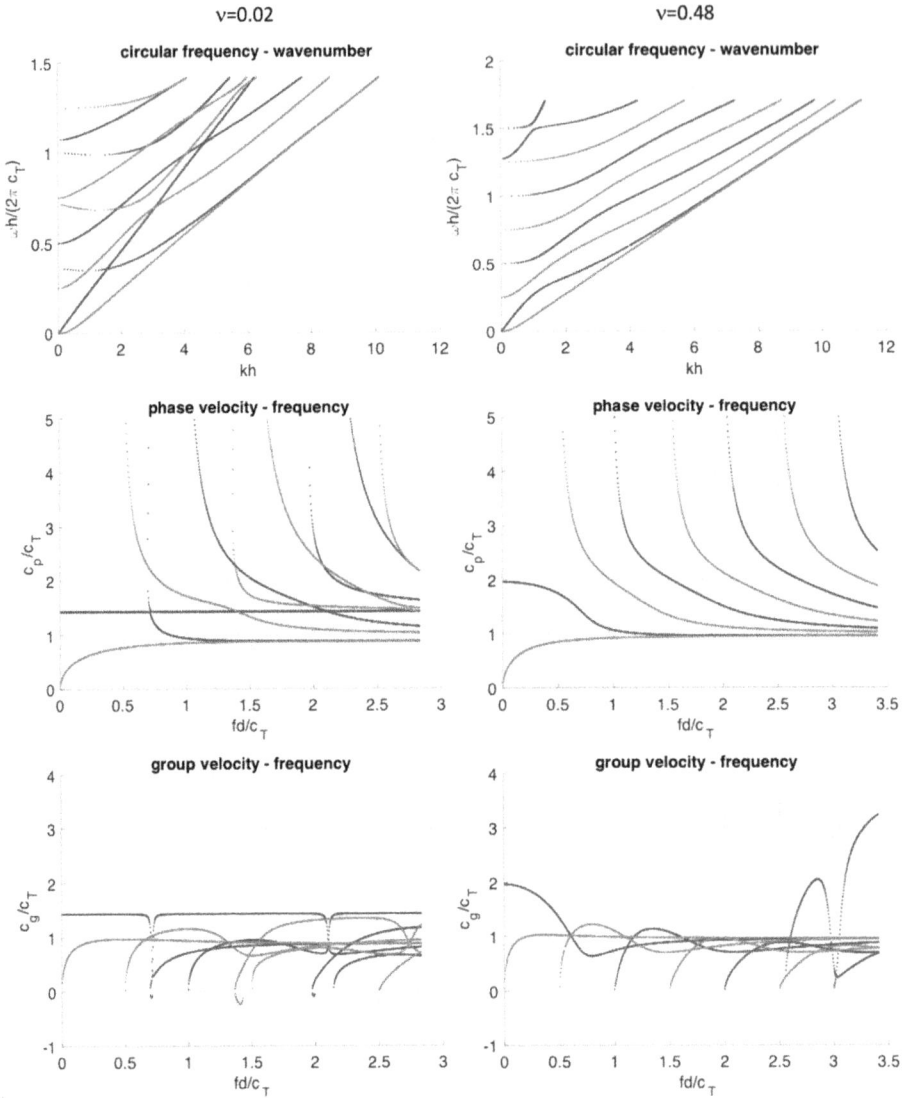

Figure 5.7. Dispersion curves for Lamb waves in materials having extreme properties: Fictitious1 with $v = 0.02$ (left) and Fictitious2 with $v = 0.48$ (right).

5.3.3 Anisotropic plates

Anisotropy can have a marked effect on wave propagation. We provide two examples of phase velocity dispersion curves in figure 5.9; one for a monocrystalline silicon wafer and the other for a unidirectional carbon fiber reinforced polymer (CFRP). Mass density and elastic stiffness coefficients are given in table 5.2. Both SH-like and Lamb-like wave modes are shown in figure 5.9. When the wave propagation direction is not in a material principal direction (e.g. at 45 degrees in the

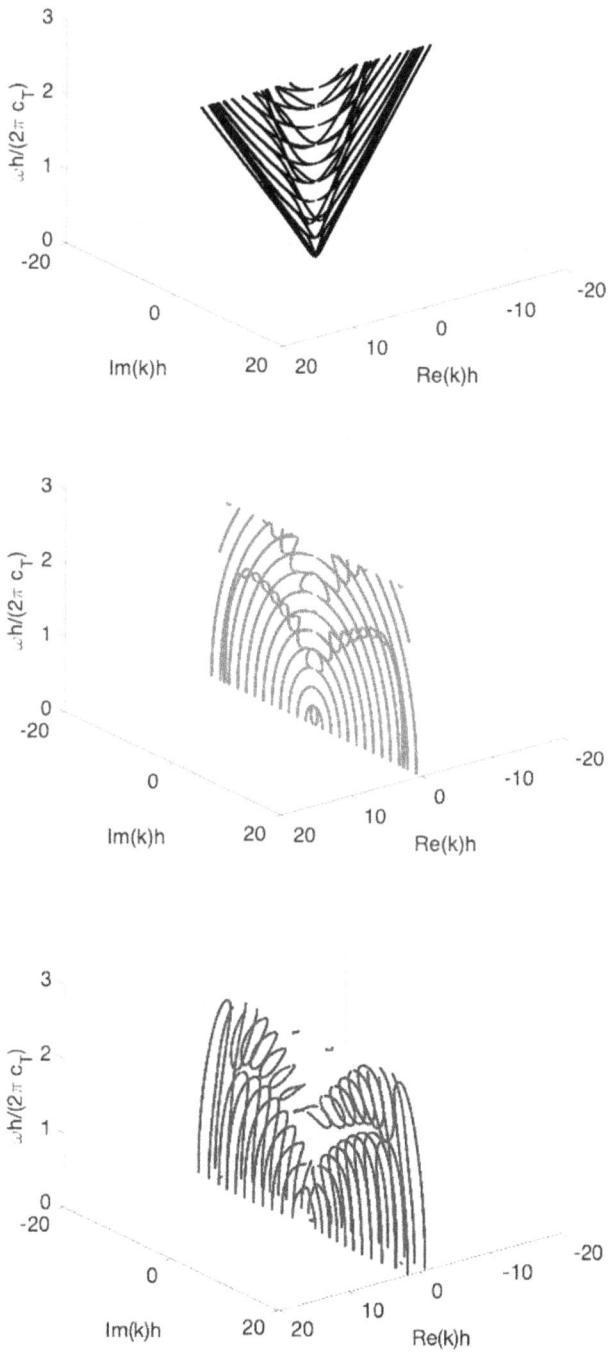

Figure 5.8. Nondimensionalized 3D dispersion curves for Lamb waves in an aluminum plate. Black = real, red = imaginary, blue = complex.

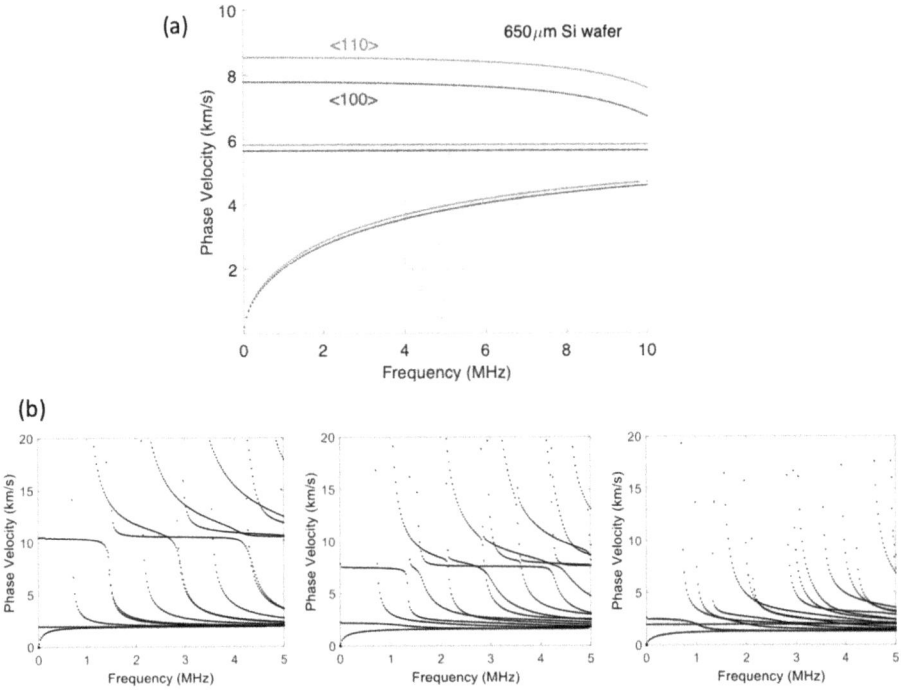

Figure 5.9. Dispersion curves for anisotropic plates: (a) 650 μm thick monocrystalline silicon wafer and (b) 1 mm thick unidirectional CFRP with wavevector at 0 (left), 45 (middle), and 90 (right) degrees to fiber direction.

Table 5.2. Anisotropic material parameters.

	ρ (kg m^{-3})	C_{11} (GPa)	$C_{22} = C_{33}$ (GPa)	$C_{12} = C_{13}$ (GPa)	C_{23} (GPa)	C_{44} (GPa)	$C_{55} = C_{66}$ (GPa)
Si	2330	165.7	165.7	63.9	63.9	79.6	79.6
CFRP	1608	178.2	14.44	8.347	8.119	3.161	6.100

CFRP) the SH-like and Lamb-like wave modes are coupled and there are no strictly nondispersive modes. As anticipated, the high anisotropy of the CFRP leads to large differences in the dispersion curves based on propagation direction.

5.3.4 Finite-width plates

Lamb waves propagate in an idealized plate that is infinitely wide and infinitely long. Thus, a plane strain model is used. But often real plates have a finite width. A 2D SAFE formulation is well suited for modeling finite width plates. Figure 5.10 shows the phase velocity dispersion curves for aluminum plates having width w to thickness d ratios of $w/d = 5, 10, 20$. Removing the requirement that none of the variables depend on x_2 changes the problem dramatically. While an infinitely wide plate has three fundamental

(a) $w/d = 20$

(d) wavestructures

B

(b) $w/d = 10$

C

(c) $w/d = 5$

D

E

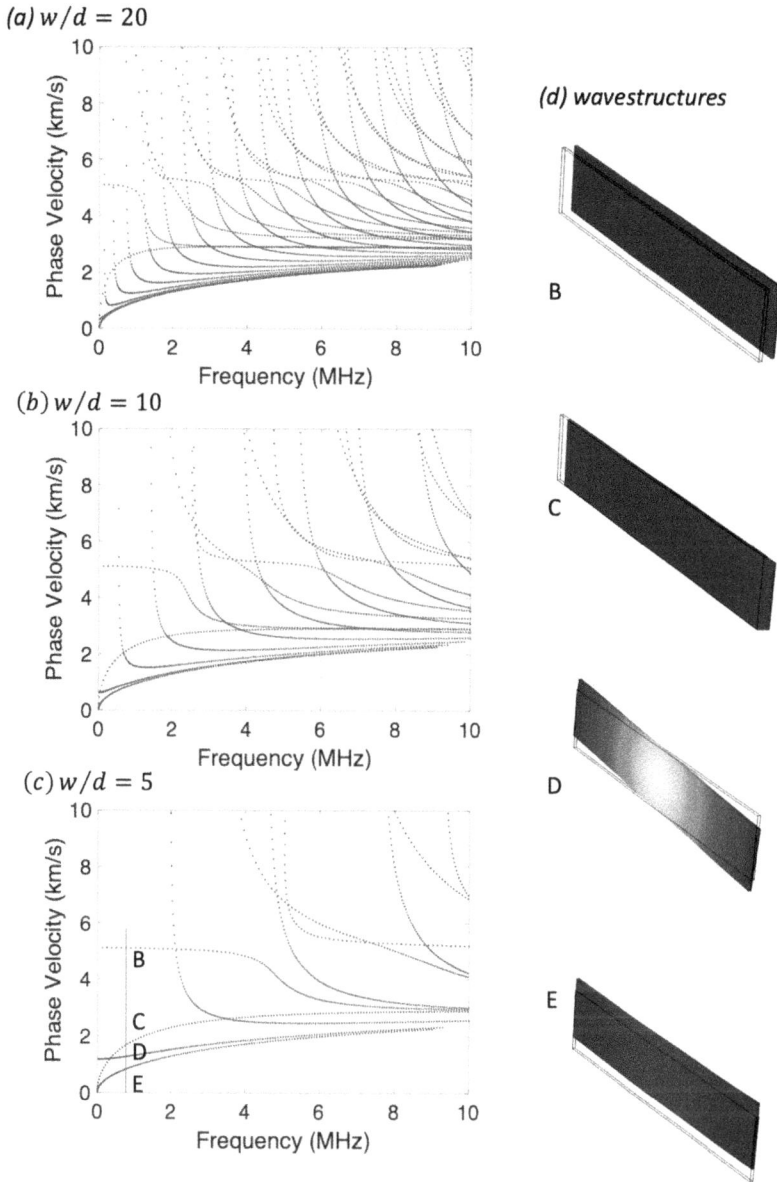

Figure 5.10. Phase velocity dispersion curves for finite width aluminum plates: (a) $w/d = 20$, (b) $w/d = 10$, (c) $w/d = 5$, (d) wavestructures for $w/d = 5$ at 1 MHz points B, C, D, E. The wire frame indicates reference position. $d = 0.1$ mm. Animation available at https://doi.org/10.1088/978-0-7503-4911-6.

wave modes S0, A0, and SH0 that are not cutoff, finite width plates have four fundamental wave modes whose wavestructures are shown in figure 5.10(d) at 1 MHz. The wavestructures are shown as the plate cross-section relative to the reference position, with the color representative of the displacement magnitude. The wavestructures with a solid blue color mean that the displacement magnitude is relatively uniform. The wave mode at point B is like the S0 lamb mode, an in-plane plate wave. The mode at point E is like the A0 lamb mode, a flexural mode at low frequencies. The mode at point D is a twisting mode that has some similarities to the SH0 mode. However, unlike the SH0 mode, the twisting mode is dispersive. Finally, the mode at point C is a new flexural mode like the A0 mode, but bending about the other axis of the plate.

5.4 Hollow cylinder waves

Guided waves propagating longitudinally in a hollow cylinder (i.e. in the z direction in figure 4.3) are first classified as axisymmetric or flexural. Referring back to the potential functions in equations (4.89) and (4.90), if $m = 0$ then the wavestructure is

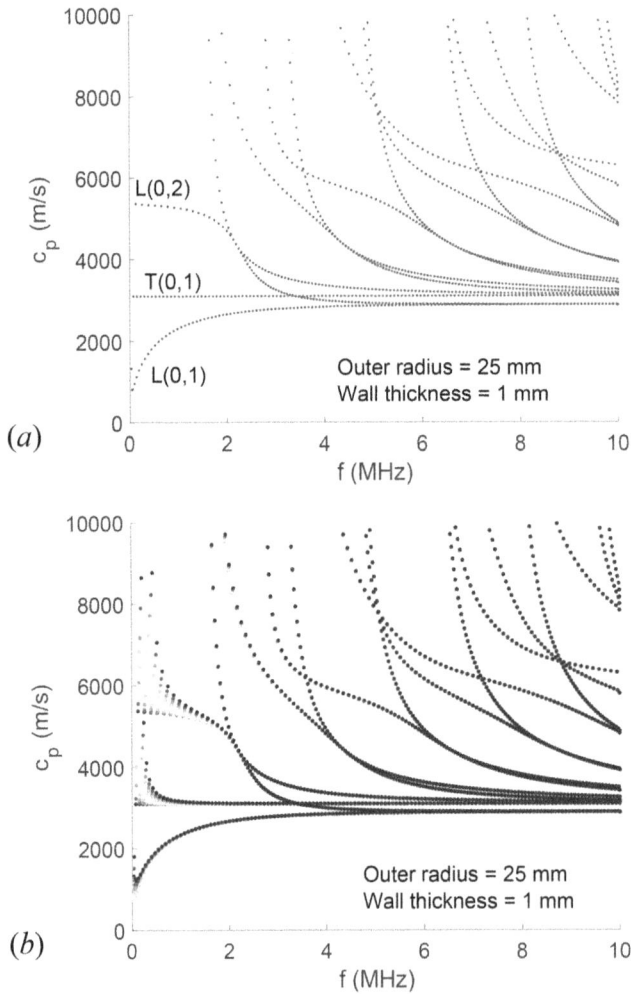

Figure 5.11. Phase velocity dispersion curves for hollow cylinders: (a) axisymmetric modes ($m = 0$) and (b) axisymmetric and flexural modes ($m = 0 - 10$).

axisymmetric, otherwise (i.e. $m \neq 0$) the wavestructure depends on the θ coordinate through the periodic $e^{im\theta}$ term and the waves are not axisymmetric, but rather are said to be flexural in nature. A further classification of waves in hollow cylinders is based on the polarity as either longitudinal, $L(m, n)$ or torsional, $T(m, n)$. Longitudinal waves are polarized in the r–z planes, torsional waves on the other hand are polarized in the circumferential direction. The first index, m, denotes the circumferential order, and the second index, n, denotes the family of the wave mode.

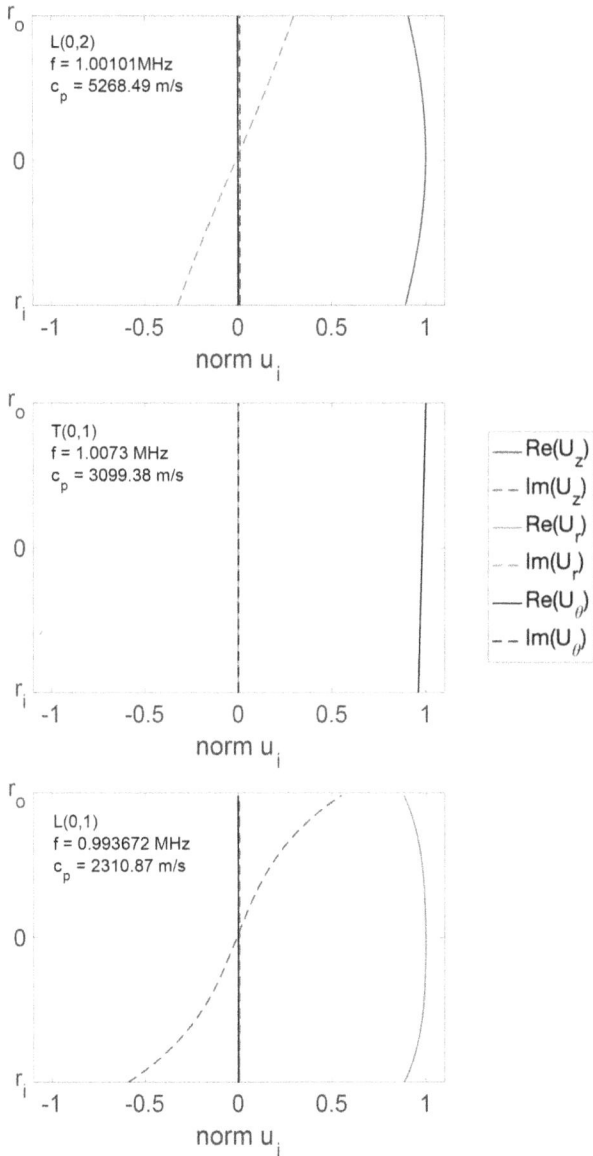

Figure 5.12. Wavestructures for fundamental modes in hollow cylinders at nominally 1 MHz.

The phase velocity dispersion curves for an aluminum pipe having a 25 mm outer radius and 1 mm wall thickness are shown in figure 5.11. In figure 5.11(a) only the axisymmetric waves are shown, while figure 5.11(b) shows circumferential orders 0–10. For this combination of radius and wall thickness the circumferential order m has very little effect above 1–2 MHz. This is not usually the case. Here, we chose the wall thickness and the material for comparison with the dispersion curves for a plate already presented. Except for low frequencies, the L(0, n) dispersion curves are comparable to the Lamb wave dispersion curves and the T(0, n) dispersion curves are comparable to the SH dispersion curves. To provide full disclosure, figure 5.11 (and figure 5.10) was generated using Floquet periodicity in COMSOL as described by Hakoda $et\ al$ [8]. It is worth pointing out that cylinders having smaller radius/thickness ratios than shown in figure 5.11 can have large deviations in the dispersion curves for fundamental axisymmetric modes at low frequencies (approaching zero), see for example Liu $et\ al$ [9].

The wavestructures of the three fundamental modes at ∼1 MHz are shown in figure 5.12. The similarity with the fundamental Lamb and SH modes in a plate is clear.

5.5 Other types of guided waves

There are many other types of guided waves. Love, Sezawa, Sholte, and Stoneley waves were mentioned in chapter 4 with respect to layered media. Laminated composites comprised of unidirectional CFRP plies are commonly used in aerostructures. Rods, beams, and rail are in common use. Features of structural components such as welds and edges also serve as waveguides. SAFE is a powerful tool for determining the dispersion curves and wavestructures for these types of guided waves.

5.6 Closure

Solutions to linear free wave propagation problems provide the foundation for the nonlinear ultrasonic guided wave solutions to be explained in chapters 6–8.

References

[1] Rose J L 2014 $Ultrasonic\ Guided\ Waves\ in\ Solid\ Media$ (Cambridge: Cambridge University Press)
[2] Biot M A 1957 General theorems on the equivalence of group velocity and energy transport $Phys.\ Rev.$ **105** 1129–37
[3] Achenbach J D 1975 $Wave\ Propagation\ in\ Elastic\ Solids$ (Amsterdam: North Holland)
[4] I, Bartoli A, Marzani F, Lanza , di Scalea E and Viola 2006 Modeling wave propagation in damped waveguides of arbitrary cross-section $J.\ Sound\ Vib.$ **295** 685–707
[5] Destrade M, Saccomandi G and Vianello M 2013 Proper formulation of viscous dissipation for nonlinear waves in solids $J.\ Acoust.\ Soc.\ Am.$ **133** 1255–9
[6] Lowe M J S 1995 Matrix techniques for modeling ultrasonic waves in multilayered media $IEEE\ Trans.\ Ultrason.\ Ferroelectr.\ Freq.\ Control$ **42** 525–42

[7] Hakoda C and Lissenden C J 2020 Application of a general expression for the group velocity vector of elastodynamic guided waves *J. Sound Vib.* **469** 115165

[8] Hakoda C, Rose J, Shokouhi P and Lissenden C 2018 Using Floquet periodicity to easily calculate dispersion curves and wave structures of homogeneous waveguides *AIP Conf. Proc.* **1949** 020016

[9] Liu Y, Khajeh E, Lissenden C J and Rose J L 2013 Interaction of torsional and longitudinal guided waves in weakly nonlinear circular cylinders *J. Acoust. Soc. Am.* **133** 2541–53

Part II

Modeling nonlinearity

IOP Publishing

Nonlinear Ultrasonic Guided Waves

Cliff J Lissenden

Chapter 6

Nonlinear analysis of plates

Helpless and hopeless would be appropriate feelings if there was no way to predict the propagation characteristics of nonlinear ultrasonic guided waves. If we were interested in second harmonics, we would know to look in the frequency spectrum at twice the driving frequency, but with multiple dispersive modes, which mode or modes would propagate and what (small!) amplitude should be expected? Fortunately, a model, in fact a 'just-right' model is able to provide the necessary understanding. It is a complex problem (pun intended) that has a complex, but understandable, solution. This chapter walks slowly through that solution starting with the complex reciprocity relation and proving that the wave modes are orthogonal to enable use of the normal mode expansion. Given weak nonlinearity, the perturbation method of successive approximations is an appropriate solution methodology that leads to the internal resonance conditions for cumulative growth of the secondary waves. The special case of second harmonic generation is used to demonstrate the solution procedure for clarity, but the method applies to wave mixing and higher harmonic generation as well. Interest in wave mixing and higher harmonics springs from two sources: (i) the options for generating internally resonant second harmonics in plates are extremely limited, as well as difficult to actuate, leading to a desire to expand the design space for nonlinear guided waves and (ii) measurement systems often contain nonlinearities at integer multiples of the excitation frequency, providing impetus to move the frequencies at which signals are received and analysed away from the second and third harmonics.

The normal mode expansion is an important part of modeling many aspects of guided wave propagation, such as actuator-induced wavefields (e.g. [1, 2]), wave scattering, and mode conversions at waveguide transitions (e.g. [3]). Rose [4] employs the normal mode expansion for the source influence (their chapter 13), phased arrays for pipes (their chapter 16), guided wave transducers (their chapter 19), and nonlinear guided wave propagation (their chapter 20). This chapter delves into the normal mode expansion to make clear its application for solving the nonlinear guided wave problem.

Let's summarize the situation. The equation of motion applied to a harmonic disturbance ($e^{-i\omega t}$) on a finite domain (the waveguide) leads to an eigen-problem. The eigenvalues are directly related to the wavenumber (and phase velocity) and the eigenvectors give the wavestructure (mode shapes). A mode is a set of eigen-solutions that are a function of frequency and have similar characteristics, such as the $U_i^{[m]}(x_3)$ profile of mode m being symmetric. Normal modes propagate freely at a fixed phase velocity, having a unique displaced shape that oscillates at a fixed frequency. Each mode is orthogonal to all other modes at that frequency. If the complete set of orthogonal modes is normalized, they are orthonormal and can be used in the normal mode expansion (also known as the eigenfunction expansion method) to solve forced wave problems, mode conversion problems, and nonlinear problems as will be described in this chapter.

The normal mode expansion relies upon the normality of each mode in the expansion, thus the orthogonality relation is critical, and must be derived. Likewise, the completeness of the eigenfunctions is necessary for the expansion to be exact, but often, completeness is simply assumed. The normal mode expansion enables us to solve the nonlinear guided wave problem by determining the expansion coefficients from the orthogonality relation. The solution depends upon whether or not the primary waves and secondary waves are internally resonant. We start by using elastic reciprocity to derive a relation that will enable us to determine the orthogonality of wave modes in a plate.

6.1 Reciprocity

We will follow Auld's approach (vol 2, chapter 10, section J) [5] to use elastic reciprocity to derive the orthogonality relations for a plate-like waveguide. Auld includes piezoelectricity in the formulation, but for clarity and brevity we will not. Interested readers may wish to read Achenbach's book [6] on reciprocity in elastodynamics. In general terms we take elastic reciprocity to be a derived relationship between two distinct fields, say for example between displacement field 1 and displacement field 2. This derived relation is then used to obtain information about field 2 from field 1.

We develop a reciprocity relation by considering two distinct wavefields 1 and 2 and manipulating the field equations in order to uncover a reasonably simple relationship between them. As introduced in chapter 2, the elastodynamics problem has three field equations (conservation of linear momentum, elastic constitutive law, and the strain-displacement relation) and three variables (stress, strain, and displacement):

$$\sigma_{ij,j} + \rho b_i = \rho \dot{v}_i \tag{6.1}$$

$$\sigma_{ij} = C_{ijkl}\varepsilon_{kl} \tag{6.2}$$

$$\varepsilon_{kl} = \frac{1}{2}(u_{k,l} + u_{l,k}). \tag{6.3}$$

Notice that we use linear elasticity, the linearized strain-displacement relation, the infinitesimal strain, and the Cauchy stress, which we justify based on the perturbation solution being two successive linear solutions to the nonlinear problem. We choose to use Cauchy stress and the particle velocity (time derivative of the displacement) as the primary variables. Thus, we invert equation (6.2) using the compliance tensor $\mathbf{S}$ and then substitute equation (6.3) into it. Additionally, we write the body force in equation (6.1) as $f_i = \rho b_i$ and move it to the right-hand side. The two field equations in terms of stress and velocity are now

$$\sigma_{ij,j}^{[1]} = \rho \dot{v}_i^{[1]} - f_i^{[1]} \tag{6.4}$$

$$\frac{1}{2}\left(v_{i,j}^{[1]} + v_{j,i}^{[1]}\right) = S_{ijkl}\dot{\sigma}_{kl}^{[1]}, \tag{6.5}$$

where the superscript [1] denotes this to be field 1 in the plate. We wish to determine a relationship between this wavefield 1 and another wavefield 2 by manipulating the field equations (6.4) and (6.5). Therefore, multiply equation (6.4) by the complex conjugate (denoted with an *) of the wavefield 2 particle velocity $v_i^{[2]*}$, then multiply equation (6.5) by the complex conjugate of the wavefield 2 stress $\sigma_{ij}^{[2]*}$. Then add those two equations together to obtain

$$v_i^{[2]*}\sigma_{ij,j}^{[1]} + \sigma_{ij}^{[2]*}\frac{1}{2}\left(v_{i,j}^{[1]} + v_{j,i}^{[1]}\right) = v_i^{[2]*}\rho\dot{v}_i^{[1]} - v_i^{[2]*}f_i^{[1]} + \sigma_{ij}^{[2]*}S_{ijkl}\dot{\sigma}_{kl}^{[1]}. \tag{6.6}$$

The analogous operations are repeated by replacing wavefield 1 in equations (6.4) and (6.5) with the complex conjugate of wavefield 2 and then multiplying the first equation by $v_i^{[1]}$, multiplying the second equation by $\sigma_{ij}^{[1]}$, and then adding them together, which gives

$$v_i^{[1]}\sigma_{ij,j}^{[2]*} + \sigma_{ij}^{[1]}\frac{1}{2}\left(v_{i,j}^{[2]*} + v_{j,i}^{[2]*}\right) = v_i^{[1]}\rho\dot{v}_i^{[2]*} - v_i^{[1]}f_i^{[2]*} + \sigma_{ij}^{[1]}S_{ijkl}\dot{\sigma}_{kl}^{[2]*}. \tag{6.7}$$

Equations (6.6) and (6.7) are now added together

$$v_i^{[2]*}\sigma_{ij,j}^{[1]} + \sigma_{ij}^{[2]*}\frac{1}{2}\left(v_{i,j}^{[1]} + v_{j,i}^{[1]}\right) + v_i^{[1]}\sigma_{ij,j}^{[2]*} + \sigma_{ij}^{[1]}\frac{1}{2}\left(v_{i,j}^{[2]*} + v_{j,i}^{[2]*}\right)$$
$$= v_i^{[2]*}\rho\dot{v}_i^{[1]} - v_i^{[2]*}f_i^{[1]} + \sigma_{ij}^{[2]*}S_{ijkl}\dot{\sigma}_{kl}^{[1]} + v_i^{[1]}\rho\dot{v}_i^{[2]*} \tag{6.8}$$
$$- v_i^{[1]}f_i^{[2]*} + \sigma_{ij}^{[1]}S_{ijkl}\dot{\sigma}_{kl}^{[2]*}.$$

Notice that equation (6.8) can be simplified by using the chain rule to write

$$v_i\sigma_{ij,j} = [v_i\sigma_{ij}]_{,j} - v_{i,j}\sigma_{ij}, \tag{6.9}$$

the symmetry of the stress tensor to write,

$$\sigma_{ij}v_{j,i} = \sigma_{ij}v_{i,j} \tag{6.10}$$

and the major symmetry of the compliance tensor,

$$S_{ijkl} = S_{klij} \tag{6.11}$$

to obtain

$$\left[v_i^{[2]*}\sigma_{ij}^{[1]} + v_i^{[1]}\sigma_{ij}^{[2]*} \right]_{,j}$$

$$= \frac{\partial}{\partial t}\left[v_i^{[2]*}\rho v_i^{[1]} + \sigma_{ij}^{[2]*}S_{ijkl}\sigma_{kl}^{[1]} \right] - v_i^{[2]*}f_i^{[1]} - v_i^{[1]}f_i^{[2]*}, \tag{6.12}$$

which Auld [5] refers to as the *complex reciprocity relation* and Achenbach [6] would classify as a local reciprocity relation because it can be applied pointwise, whereas a global reciprocity relation involves an integral over the domain of interest. We will specialize and integrate the complex reciprocity relation in the next section.

6.2 Orthogonality

We now take the body forces to be zero and the wavefields 1 and 2 to be at the same harmonic frequency, thus time-dependence is through $e^{-i\omega t}$. Recall that if $z = e^{-i\omega t}$, then $z^*z = e^{i\omega t}e^{-i\omega t} = 1$. Thus, the entire right-hand side of equation (6.12) is zero (after ignoring the body forces) because it is time invariant, leaving just

$$\left[v_i^{[2]*}\sigma_{ij}^{[1]} + v_i^{[1]}\sigma_{ij}^{[2]*} \right]_{,j} = 0. \tag{6.13}$$

As a concrete example, let's evaluate the divergence for an infinite plate waveguide, as shown in figure 6.1, with waves propagating in the x_1-direction:

$$\frac{\partial}{\partial x_1}\left[v_i^{[2]*}\sigma_{i1}^{[1]} + v_i^{[1]}\sigma_{i1}^{[2]*} \right] + \frac{\partial}{\partial x_2}\left[v_i^{[2]*}\sigma_{i2}^{[1]} + v_i^{[1]}\sigma_{i2}^{[2]*} \right]$$

$$+ \frac{\partial}{\partial x_3}\left[v_i^{[2]*}\sigma_{i3}^{[1]} + v_i^{[1]}\sigma_{i3}^{[2]*} \right] = 0. \tag{6.14}$$

Limiting the analysis to planar waves (i.e. straight-crested Lamb or SH waves), $\frac{\partial[]}{\partial x_2} = 0$, yields

$$\frac{\partial}{\partial x_1}\left[v_i^{[2]*}\sigma_{i1}^{[1]} + v_i^{[1]}\sigma_{i1}^{[2]*} \right] + \frac{\partial}{\partial x_3}\left[v_i^{[2]*}\sigma_{i3}^{[1]} + v_i^{[1]}\sigma_{i3}^{[2]*} \right] = 0. \tag{6.15}$$

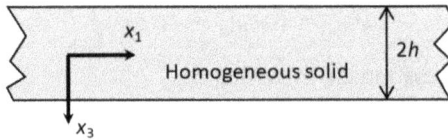

Figure 6.1. Infinite plate waveguide

Integrating over the cross-section of the plate (having unit width) gives

$$\int_{-h}^{h}\int_{0}^{1}\frac{\partial}{\partial x_1}\left[v_i^{[2]*}\sigma_{i1}^{[1]} + v_i^{[1]}\sigma_{i1}^{[2]*}\right]dx_2dx_3 + [v_i^{[2]*}\sigma_{i3}^{[1]} + v_i^{[1]}\sigma_{i3}^{[2]*}]\,\Big|_{-h}^{h} = 0. \quad (6.16)$$

The traction-free boundary conditions on the lateral surfaces of the plate make the boundary term zero, leaving just

$$\int_{-h}^{h}\frac{\partial}{\partial x_1}\left[v_i^{[2]*}\sigma_{i1}^{[1]} + v_i^{[1]}\sigma_{i1}^{[2]*}\right](1)dx_3 = 0. \quad (6.17)$$

Take wavefields 1 and 2 to be associated with guided wave modes m and n, respectively, such that

$$v_i^{[1]} = \frac{1}{2}V_i^{[m]}(x_3)e^{i(k_m x_1 - \omega t)} + \text{c. c.}$$

$$v_i^{[2]} = \frac{1}{2}V_i^{[n]}(x_3)e^{i(k_n x_1 - \omega t)} + \text{c. c.} \quad (6.18)$$

$$v_i^{[2]*} = \frac{1}{2}V_i^{[n]*}(x_3)e^{-i(k_n^* x_1 - \omega t)} + \text{c. c.}$$

and

$$\sigma_{i1}^{[1]} = \frac{1}{2}\Sigma_{i1}^{[m]}(x_3)e^{i(k_m x_1 - \omega t)} + \text{c. c.}$$

$$\sigma_{i1}^{[2]} = \frac{1}{2}\Sigma_{i1}^{[n]}(x_3)e^{i(k_n x_1 - \omega t)} + \text{c. c.} \quad (6.19)$$

$$\sigma_{i1}^{[2]*} = \frac{1}{2}\Sigma_{i1}^{[n]*}(x_3)e^{-i(k_n^* x_1 - \omega t)} + \text{c. c.} \,,$$

where $V_i^{[m]}(x_3)$ and $\Sigma_{i1}^{[m]}(x_3)$ are related directly to $U_i^{[m]}(x_3)$ and are given explicitly in example 6.3. The complex conjugates, denoted by c. c., are used to ensure that we are working with real-valued quantities. Upon substituting these wavefields into equation (6.17) and differentiating we obtain

$$\int_{-h}^{h}\left[\begin{array}{c}\frac{V_i^{[n]*}}{2}\frac{\Sigma_{i1}^{[m]}}{2}(-ik_n^* + ik_m)e^{-ik_n^* x_1}e^{ik_m x_1}e^{-i\omega t} \\ + \frac{V_i^{[m]}}{2}\frac{\Sigma_{i1}^{[n]*}}{2}(-ik_n^* + ik_m)e^{-ik_n^* x_1}e^{ik_m x_1}e^{-i\omega t}\end{array}\right]dx_3 = 0$$

$$-i(k_n^* - k_m)\int_{-h}^{h}\left[\frac{V_i^{[n]*}}{2}\frac{\Sigma_{i1}^{[m]}}{2} + \frac{V_i^{[m]}}{2}\frac{\Sigma_{i1}^{[n]*}}{2}\right]dx_3 = 0 \quad (6.20)$$

and defining the integral to be

$$P_{mn} \equiv -\int_{-h}^{h} \left[\frac{V_i^{[n]*}}{2} \frac{\Sigma_{i1}^{[m]}}{2} + \frac{V_i^{[m]}}{2} \frac{\Sigma_{i1}^{[n]*}}{2} \right] dx_3 \qquad (6.21)$$

we arrive at the relation that

$$i(k_n{}^* - k_m)P_{mn} = 0. \qquad (6.22)$$

Note that in the literature (e.g. Auld [5], de Lima and Hamilton [7], Lanza di Scalea *et al* [8], and Rose [4]) that P_{mn} is defined in such a way that there would be a 4 in equation (6.22). That convention is not followed here, although this approach follows that of de Lima and Hamilton [7]. Thus, an orthogonality relation is

$$P_{mn} = 0 \text{ for } k_n{}^* \neq k_m. \qquad (6.23)$$

Recall from chapter 5 that the wavenumbers can be real, imaginary, or complex. Real wavenumbers give propagating modes while imaginary and complex wavenumbers give evanescent modes. Interested readers are referred to Auld's section 10. J [5] for a more detailed explanation of the orthogonality of evanescent modes. For propagating modes, when $n = m$, we can use $z + z^* = 2 \, \text{Re}\,(z)$ to obtain

$$P_{mm} = -\text{Re} \left(\frac{1}{2} \int_{-h}^{h} V_i^{[m]*} \Sigma_{i1}^{[m]} dx_3 \right), \qquad (6.24)$$

which is interpreted as the average power flow in the x_1-direction per unit width of the plate and is often referred to as the x_1 component of the Poynting vector. For the modes to be orthonormal, that is $P_{mn} = \delta_{mn}$, requires that the wavestructures be normalized with respect to the square root of the average power flow.

Three example problems are worked out to demonstrate different aspects of orthogonality. Example 6.1 investigates the use of Auld's [5] *real reciprocity relation* to obtain an orthogonality relation. Then example 6.2 shows analytically that SH wave modes are orthogonal. Finally, example 6.3 takes a numerical approach to show the orthogonality of Lamb wave modes.

Example 6.1. Auld's real reciprocity relation.

We just developed the complex reciprocity relation and used it to derive an orthogonality criterion for plates. Auld [5] also derived a real reciprocity relation, which for purely acoustic fields reduces to Lamb's reciprocity relation [9]:

$$\left[v_i^{[1]} \sigma_{ij}^{[2]} - v_i^{[2]} \sigma_{ij}^{[1]} \right]_{,j} = v_i^{[2]} f_i^{[1]} - v_i^{[1]} f_i^{[2]}.$$

Can Lamb's reciprocity relation be used in the same way as the complex reciprocity relation to derive an orthogonality relation? Let's try it. Take the body forces to be zero and then evaluate the divergence as before,

$$\frac{\partial}{\partial x_1} \left[v_i^{[1]} \sigma_{i1}^{[2]} - v_i^{[2]} \sigma_{i1}^{[1]} \right] + 0 + \frac{\partial}{\partial x_3} \left[v_i^{[1]} \sigma_{i3}^{[2]} - v_i^{[2]} \sigma_{i3}^{[1]} \right] = 0.$$

Integrate over the cross-section of the plate

$$\int_{-h}^{h} \frac{\partial}{\partial x_1}\left[v_i^{[1]}\sigma_{i1}^{[2]} - v_i^{[2]}\sigma_{i1}^{[1]}\right]dx_3 + 0 + \left[v_i^{[1]}\sigma_{i3}^{[2]} - v_i^{[2]}\sigma_{i3}^{[1]}\right]_{-h}^{h} = 0.$$

The second term is zero because we have planar wavefields and the third term is zero due to the traction-free boundary conditions. Let the wavefields be given by (6.18) and (6.19) (ignoring the complex conjugates now). We now have

$$\int_{-h}^{h} \frac{\partial}{\partial x_1}\left[v_i^{[1]}\sigma_{i1}^{[2]} - v_i^{[2]}\sigma_{i1}^{[1]}\right]dx_3 = 0$$

$$\int_{-h}^{h} \frac{\partial}{\partial x_1}\left[\frac{V_i^m}{2}\frac{\Sigma_{i1}^n}{2}e^{i(k_m x_1 - \omega t)}e^{i(k_n x_1 - \omega t)} - \frac{V_i^n}{2}\frac{\Sigma_{i1}^m}{2}e^{i(k_n x_1 - \omega t)}e^{i(k_m x_1 - \omega t)}\right]dx_3 = 0$$

$$\int_{-h}^{h}\left[V_i^m\Sigma_{i1}^n(ik_m + ik_n)e^{i(k_m x_1 - \omega t)}e^{i(k_n x_1 - \omega t)} - V_i^n\Sigma_{i1}^m(ik_m + ik_n)e^{i(k_m x_1 - \omega t)}e^{i(k_n x_1 - \omega t)}\right]dx_3 = 0$$

$$i(k_m + k_n)\int_{-h}^{h}[V_i^m\Sigma_{i1}^n - V_i^n\Sigma_{i1}^m]dx_3 = 0.$$

We seek an orthogonality relation such that $\int_{-h}^{h}[V_i^m\Sigma_{i1}^n - V_i^n\Sigma_{i1}^m]dx_3 = \delta_{mn}$, but if we have the same mode, $m = n$, the integrand is identically zero. The leading term $(k_m + k_n)$ provides a clue that we might need to be more careful as to how we investigate the modes m and n; they could be co-directional or counter-propagating. In fact, Achenbach [6] (in their section 9.4) is able to derive orthogonality relations for antisymmetric Lamb waves based on whether they are co-directional or counter-propagating. Similar relations are given for symmetric Lamb waves. However, the orthogonality relation given by equation (6.23) is more commonly used in practice.

Example 6.2. Orthogonality of SH waves.

In this example we directly investigate the orthogonality of SH wave modes using equation (6.23). We select an fd product of $2c_T$ (see figure 5.3) such that we can show the orthogonality between the SH0, SH1, and SH2 modes. Returning to chapter 4, the particle displacement is

$$u_2 = f(x_3)e^{i(kx_1 - \omega t)},$$

the wavestructures are

$$f^{\mathrm{sym}}(x_3) = A_2 \cos\left(\frac{m\pi x_3}{2h}\right) \text{ for } m = 0, 2, 4, \ldots$$

$$f^{\mathrm{anti}}(x_3) = A_1 \sin\left(\frac{m\pi x_3}{2h}\right) \text{ for } m = 1, 3, 5, \ldots$$

for symmetric and antisymmetric modes, respectively. The wavenumber and phase velocity are

$$k^2 = \left(\frac{\omega}{c_T}\right)^2 - \left(\frac{n\pi}{2h}\right)^2$$

$$c_p = \frac{\omega}{k},$$

respectively. We need to specialize the integrand of equation (6.21) for SH wave modes, which only have particle motion in the x_2-direction,

$$V_i^{[n]*}\Sigma_{i1}^{[m]} + V_i^{[m]}\Sigma_{i1}^{[n]*} = V_2^{[n]*}\Sigma_{21}^{[m]} + V_2^{[m]}\Sigma_{21}^{[n]*},$$

where

$$v_2 = -i\omega u_2$$

$$v_2 = V_2(x_3)e^{i(kx_1-\omega t)} = -i\omega f(x_3)e^{i(kx_1-\omega t)}$$

$$\therefore V_2(x_3) = -i\omega f(x_3)$$

and

$$\sigma_{21} = \mu(u_{1,2} + u_{2,1}) = \mu u_{2,1}$$

$$\sigma_{21} = \Sigma_{21}(x_3)e^{i(kx_1-\omega t)} = ik\mu f(x_3)e^{i(kx_1-\omega t)}$$

$$\therefore \Sigma_{21}(x_3) = ik\mu f(x_3)$$

For:

SH0 ($m = 0$)	$f^{\text{sym}}(x_3) = A_2 \cos\left(\frac{0\pi x_3}{2h}\right) = A_2$
SH1 ($m = 1$)	$f^{\text{anti}}(x_3) = A_1 \sin\left(\frac{1\pi x_3}{2h}\right)$
SH2 ($m = 2$)	$f^{\text{sym}}(x_3) = A_2 \cos\left(\frac{2\pi x_3}{2h}\right).$

Considering SH0 and SH1 as modes $m = 0$ and $n = 1$, respectively, equation (6.21) becomes

$$P_{01} = -\frac{1}{4} \int_{-h}^{h} \left[V_2^{[1]*}\Sigma_{21}^{[0]} + V_2^{[0]}\Sigma_{21}^{[1]*} \right] dx_3$$

$$P_{01} = -\frac{1}{4} \int_{-h}^{h} \left[\left(-i\omega A_1 \sin\left(\frac{\pi x_3}{2h}\right)\right)^* ik\mu A_2 - i\omega A_2\left(ik\mu A_1 \sin\left(\frac{\pi x_3}{2h}\right)\right)^* \right] dx_3$$

$$P_{01} = -\frac{1}{4} \int_{-h}^{h} \left[\left(i\omega A_1 \sin\left(\frac{\pi x_3}{2h}\right)\right)ik\mu\, A_2 + i\omega A_2\left(ik\mu\, A_1 \sin\left(\frac{\pi x_3}{2h}\right)\right) \right] dx_3$$

$$P_{01} = \frac{\omega k\mu\, A_1 A_2}{4} \int_{-h}^{h} \left[\sin\left(\frac{\pi x_3}{2h}\right) + \sin\left(\frac{\pi x_3}{2h}\right) \right] dx_3 = 0.$$

Considering SH1 and SH2 as modes $m = 1$ and $n = 2$, respectively, equation (6.21) becomes

$$P_{12} = -\frac{1}{4} \int_{-h}^{h} \left[V_2^{[2]*} \Sigma_{21}^{[1]} + V_2^{[1]} \Sigma_{21}^{[2]*} \right] dx_3$$

$$P_{12} = -\frac{1}{4} \int_{-h}^{h} \left[\left(-i\omega A_2 \cos\left(\frac{2\pi x_3}{2h}\right) \right)^* ik\mu A_1 \sin\left(\frac{\pi x_3}{2h}\right) - i\omega A_1 \sin\left(\frac{\pi x_3}{2h}\right) \left(ik\mu A_2 \cos\left(\frac{2\pi x_3}{2h}\right) \right)^* \right] dx_3$$

$$P_{12} = -\frac{1}{4} \int_{-h}^{h} \left[\left(i\omega A_2 \cos\left(\frac{2\pi x_3}{2h}\right) \right) ik\mu A_1 \sin\left(\frac{\pi x_3}{2h}\right) + i\omega A_1 \sin\left(\frac{\pi x_3}{2h}\right) \left(ik\mu A_2 \cos\left(\frac{2\pi x_3}{2h}\right) \right) \right] dx_3$$

$$P_{12} = \frac{\omega k\mu \, A_1 A_2}{4} \int_{-h}^{h} 2 \cos\left(\frac{2\pi x_3}{2h}\right) \sin\left(\frac{\pi x_3}{2h}\right) dx_3 = 0.$$

The reader can verify that P_{02} for the SH0 and SH2 modes is also zero. Let us also consider P_{11} for the SH1 mode using equation (6.24),

$$P_{11} = -\text{Re}\left(\frac{1}{2} \int_{-h}^{h} \left[V_2^{[1]*} \Sigma_{21}^{[1]} \right] dx_3 \right)$$

$$P_{11} = -\text{Re}\left(\frac{1}{2} \int_{-h}^{h} \left[\left(-i\omega A_1 \sin\left(\frac{\pi x_3}{2h}\right) \right)^* ik\mu A_1 \sin\left(\frac{\pi x_3}{2h}\right) \right] dx_3 \right)$$

$$P_{11} = \text{Re}\left(\frac{1}{2} \int_{-h}^{h} \left[\left(\omega A_1 \sin\left(\frac{\pi x_3}{2h}\right) \right) k\mu A_1 \sin\left(\frac{\pi x_3}{2h}\right) \right] dx_3 \right)$$

$$P_{11} = \frac{\omega k\mu \, A_1^2}{2} \int_{-h}^{h} \sin^2\left(\frac{\pi x_3}{2h}\right) dx_3 = \frac{\omega k\mu \, A_1^2}{2} \left[\frac{x_3}{2} - \frac{2h}{4\pi} \sin\left(\frac{2\pi x_3}{2h}\right) \right]_{-h}^{h}$$

$$P_{11} = \frac{\omega k\mu \, A_1^2}{2} \left[\frac{h}{2} - \frac{2h}{4\pi} \sin(\pi) + \frac{h}{2} + \frac{2h}{4\pi} \sin(-\pi) \right] = \frac{\omega h k\mu \, A_1^2}{2}.$$

This is the normalization factor that makes the modes orthonormal, i.e. $P_{mn} = \delta_{mn}$.

Example 6.3. Orthogonality of Lamb waves.

The orthogonality of Lamb waves can also be demonstrated, but the equations have more terms. We start with the integrand of equation (6.21) as before:

$$V_i^{[n]*} \Sigma_{i1}^{[m]} + V_i^{[m]} \Sigma_{i1}^{[n]*} = V_1^{[n]*} \Sigma_{11}^{[m]} + V_3^{[n]*} \Sigma_{31}^{[m]} + V_1^{[m]} \Sigma_{11}^{[n]*} + V_3^{[m]} \Sigma_{31}^{[n]*}.$$

The velocity and stress components that we need come from the displacement components given in (5.5), which are

$$u_i(x_j, t) = U_i(x_3) e^{i(kx_1 - \omega t)}$$

and for Lamb waves we only need indices $i = 1$ and $i = 3$. The displacement wavestructures are given by equations (5.31)–(5.34):

Symmetric

$$U_1(x_3) = ikA \cos{(px_3)} - qD_2 \cos(qx_3)$$
$$U_3(x_3) = -pA \sin{(px_3)} + ikD_2 \sin(qx_3)$$

Antisymmetric

$$U_1(x_3) = ikB \sin{(px_3)} + qC_2 \sin(qx_3)$$
$$U_3(x_3) = pB \cos{(px_3)} + ikC_2 \cos(qx_3)$$

$$A = D_2 \frac{2i\mu kq \cos(qh)}{[(\lambda + 2\mu)p^2 + \lambda k^2]\cos(ph)}$$
$$B = -C_2 \frac{2i\mu kq \sin(qh)}{[(\lambda + 2\mu)p^2 + \lambda k^2]\sin(ph)}.$$

We can write the velocity and stress wavestructure components from equations (6.18) and (6.19), respectively, in terms of $U_i(x_3)$

$$V_1(x_3) = -i\omega U_1(x_3)$$

$$V_3(x_3) = -i\omega U_3(x_3)$$

$$\Sigma_{11}(x_3) = (\lambda + 2\mu)ik U_1(x_3) + \lambda U_{3,3}(x_3)$$

$$\Sigma_{31}(x_3) = \mu U_{1,3}(x_3) + \mu ik U_3(x_3),$$

where we used Hooke's law, equation (2.44), and the strain-displacement relations, equation (2.8), for the stress terms. The integrand of equation (6.21) can now be written as

$$V_i^{[n]*}\Sigma_{i1}^{[m]} + V_i^{[m]}\Sigma_{i1}^{[n]*}$$

$$= [-i\omega U_1(x_3)]^{[n]*}[(\lambda + 2\mu)ik U_1(x_3) + \lambda U_{3,3}(x_3)]^{[m]}$$

$$+ [-i\omega U_3(x_3)]^{[n]*}[\mu U_{1,3}(x_3) + \mu ik U_3(x_3)]^{[m]}$$

$$+ [-i\omega U_1(x_3)]^{[m]}[(\lambda + 2\mu)ik U_1(x_3) + \lambda U_{3,3}(x_3)]^{[n]*}$$

$$+ [-i\omega U_3(x_3)]^{[m]}[\mu U_{1,3}(x_3) + \mu ik U_3(x_3)]^{[n]*}$$

Table 6.1. Values of P_{mn} for 1 mm thick aluminum plate, with driving frequency of 3.1 MHz.

m	Mode type	Wavenumber (rad m^{-1})	n	Mode type	Wavenumber (rad m^{-1})	P_{mn} ($\times 10^{15}$ Pa s^{-1})
2	S0	~6082	2	S0	~6082	0.004 48
2	S0	~6082	4	S1	~2874	$-2.351 \times 10^{-13} \sim 0$
2	S0	~6082	3	A1	~3366	$-3.591 \times 10^{-19} \sim 0$
4	S1	~2874	3	A1	~3366	$-8.528 \times 10^{-19} \sim 0$
1	A0	~6979	3	A1	~3366	$-4.856 \times 10^{-13} \sim 0$
3	A1	~3366	3	A1	~3366	0.0732
1	A0	~6979	1	A0	~6979	0.0752

and the integral can be evaluated for various Lamb wave modes. Evaluation requires us to select the plate thickness and material as well as the frequency at which the wavenumbers (or equivalently phase velocities) will be tested. All of the other variables are computed from these inputs. The computations are difficult in the SI unit system (kg, m, s) because the values are quite large. However, using base units of kg, mm, and μs works well. The values for P_{mn} from equation (6.21) are provided in table 6.1 for a number of cases.

6.3 Completeness

If the eigenfunctions form a complete set, then the eigenfunction expansion can represent any piecewise smooth function and the orthogonality of the eigenfunctions can be used to determine the expansion coefficients. A simple example of a complete set is a Fourier sine series, which can be used to represent any piecewise smooth function. The temperature distribution $u(x, t)$ in a 1D rod, $0 \leqslant x \leqslant L$, having the boundary conditions $u(0, t) = u(L, t) = 0$, can be represented by

$$u(x, t) = \sum_{n=1}^{\infty} B_n \sin\left(\frac{n\pi x}{L}\right) e^{-(n\pi/L)\kappa t}$$

given that the initial conditions are described by a complete sine series

$$u(x, 0) = f(x) = \sum_{n=1}^{\infty} B_n \sin\left(\frac{n\pi x}{L}\right).$$

Kirrmann [10] investigated the completeness of Lamb waves, but his definition of Lamb waves is unconventional due to his interest in seismic waves. In his problem the top surface of the plate is traction-free, but the bottom surface is fixed. The fixed boundary conditions are justified by the increasing stiffness in strata with depth. Kirrmann [10] provides a rigorous mathematical definition of completeness and a completeness theorem for Lamb and SH waves (as he defined them). For our purposes, it should suffice to recognize that we are going to use only a subset of the modes anyway, and therefore we will not exactly satisfy completeness.

6.4 Normal mode expansion

For SH waves there exist a finite number of modes and for Lamb waves there exist an infinite number of modes. The modes have been proven to be orthogonal and can be normalized to make them orthonormal. The normal mode expansion [11]

$$u_i(x_j, t) = \frac{1}{2}\sum_m A_m(x_1) U_i^m(x_3) e^{-i\omega t} + \text{c. c.} \tag{6.25}$$

assumes that the set of m modes is complete, although for actual computations the summation may be truncated, presuming that the modes left out have minimal effect on the overall displacement field. The coefficients of the expansion, A_m, sometimes known as modal participation factors, are allowed to depend upon position, meaning that the expansion itself changes in the propagation direction. They are determined based upon some forced response (the nonlinearity in this case), as opposed to the free response from which the modal characteristics (wavenumber and wavestructure) were found.

6.5 Perturbation approach

The nonlinear boundary value problem (BVP) formulated in chapter 4 was general enough to apply to both self-interactions and mutual interactions. For clarity of presentation, let's tackle the second harmonic generation problem first. We are following the successive approximations approach pioneered by de Lima and Hamilton [7], although the notation and coordinate system are different, originating in [12]

$$\mathbf{u} = \mathbf{u}_a + \mathbf{u}_{aa} \tag{6.26}$$

$$\mathbf{H} = \mathbf{H}_a + \mathbf{H}_{aa} \tag{6.27}$$

$$\mathbf{S}(\mathbf{H}) = \mathbf{S}^L(\mathbf{H}_a) + \mathbf{S}^L(\mathbf{H}_{aa}) + \mathbf{S}^{NL}(\mathbf{H}_a, \mathbf{H}_a, 2) \tag{6.28}$$

with the nonlinear term given by equation (4.121). The sequential BVPs are given below.
Primary waves a (wavefield given by $\mathbf{u}_a$ or $\mathbf{H}_a$):

$$\nabla_X \cdot \mathbf{S}^L(\mathbf{H}_a) - \rho_0 \ddot{\mathbf{u}}_a = 0 \text{ for } |X_3| \leqslant h \tag{6.29}$$

$$\mathbf{N} \cdot \mathbf{S}_L(\mathbf{H}_a) = 0 \text{ for } X_3 = \pm h. \tag{6.30}$$

The first approximation is the linear problem that we solved in chapter 4 for Cauchy stress and infinitesimal strain. The solutions, given by the Lamb wave and SH wave dispersion curves in chapter 5, should be familiar by now.
Self-interaction of waves a give the secondary waves (wavefield given by $\mathbf{u}_{aa}$ or $\mathbf{H}_{aa}$):

$$\nabla_X \cdot \mathbf{S}^L(\mathbf{H}_{aa}) - \rho_0 \ddot{\mathbf{u}}_{aa} = -\nabla_X \cdot \mathbf{S}^{NL}(\mathbf{H}_a, \mathbf{H}_a, 2) \text{ for } |X_3| \leqslant h \tag{6.31}$$

$$\mathbf{N} \cdot \mathbf{S}^L(\mathbf{H}_{aa}) = -\mathbf{N} \cdot \mathbf{S}^{NL}(\mathbf{H}_a, \mathbf{H}_a, 2) \text{ for } X_3 = \pm h. \tag{6.32}$$

The second approximation is that the secondary waves are described by linear, but inhomogeneous, equations having the nonlinear stress as the driving force. The nonlinear stress is taken to depend only on the primary displacement gradient field, and thus is known.

Although we can use equation (4.121) to write the nonlinear driving forces in terms of the displacement gradient of the primary wavefield, we do not know which secondary modes are propagating. That is, we do not know $\ddot{\mathbf{u}}_{aa}$ and $\mathbf{H}_{aa}$ for the self-interaction BVP. Therefore, let's return to the complex reciprocity relation, equation (6.12), this time with body force and surface traction terms, and apply the normal mode expansion.

Let the wavefield 1 be a modal expansion of the secondary wavefield at the second harmonic frequency 2ω. The displacement field is expanded,

$$u_i^{[1]}(X_k, t) = \frac{1}{2}\sum_m A_m(X_1) U_i^m(X_3) e^{-i2\omega t} + \text{c. c.} , \tag{6.33}$$

where c. c. denotes the complex conjugate. Likewise, the particle velocity and stress fields can be written

$$v_i^{[1]}(X_k, t) = \frac{1}{2}\sum_m A_m(X_1) V_i^m(X_3) e^{-i2\omega t} + \text{c. c.} \tag{6.34}$$

$$\sigma_{ij}^{[1]}(X_k, t) = \frac{1}{2}\sum_m A_m(X_1) \Sigma_{ij}^m(X_3) e^{-i2\omega t} + \text{c. c.} , \tag{6.35}$$

where the profiles $V_i^m(X_3)$ and $\Sigma_{ij}^m(X_3)$ can be written in terms of the displacement wavestructure $U_i^m(X_3)$ using equations (5.8)–(5.10). The complex conjugate is used because the reciprocity relation uses wavefield products that we want to be real-valued functions [7]. Despite the nonlinear driving forces being from the first Piola–Kirchhoff stress $\mathbf{S}$, the stresses in equation (6.35) are Cauchy stresses. This is reasonable because the secondary wavefield is small ($\mathbf{u}_{aa} \ll \mathbf{u}_a$).

The 'forcing' used to determine the coefficients $A_m(X_1)$ includes both nonlinear body force and nonlinear surface traction and is included as part of wavefield 1. The nonlinear body force is denoted

$$f_i^{[1]} = f_i^{2\omega}(X_3) e^{2i(kX_1 - \omega t)} \text{ from } \mathbf{f}^{[1]} = \nabla_X \cdot \mathbf{S}^{NL}(\mathbf{H}_a, \mathbf{H}_a, 2), \tag{6.36}$$

where the $f_i^{2\omega}(X_3)$ comes from the divergence of the nonlinear stress, $\nabla_X \cdot \mathbf{S}^{NL}(\mathbf{H}_a, \mathbf{H}_a, 2)$, evaluated from the primary wavefield at the second harmonic frequency 2ω and wavenumber $2k$ (k being the wavenumber of the primary wavefield) based on $\mathbf{u}_{aa} \ll \mathbf{u}_a$. Likewise, the nonlinear surface traction applied at $X_3 = \pm h$ is denoted

$$\overline{S}_{ij}^{[1]} = \overline{S}_{ij}^{2\omega}(X_3) e^{2i(kX_1 - \omega t)} \text{ from } \overline{\mathbf{S}}^{[1]} = \mathbf{S}^{NL}(\mathbf{H}_a, \mathbf{H}_a, 2) \tag{6.37}$$

and is also evaluated from the primary wavefield at the second harmonic frequency 2ω and wavenumber $2k$. $\overline{S}_{ij}^{2\omega}(X_3)$ only needs to be evaluated at $X_3 = \pm h$.

Let the wavefield 2 be an arbitrary mode n at the second harmonic frequency 2ω:

$$v_i^{[2]}(X_k, t) = \frac{1}{2} V_i^n(X_3) e^{i(k^n X_1 - 2\omega t)} + \text{c. c.} \tag{6.38}$$

$$\sigma_{ij}^{[2]}(X_k, t) = \frac{1}{2} \Sigma_{ij}^n(X_3) e^{i(k^n X_1 - 2\omega t)} + \text{c. c.} \tag{6.39}$$

$$f_i^{[2]} = 0 \text{ from } \mathbf{f}^{[2]} = 0.$$

There is no 'forcing' of wavefield 2.

Substituting these fields into the complex reciprocity relation

$$\left[v_i^{[2]*}\sigma_{ij}^{[1]} + v_i^{[1]}\sigma_{ij}^{[2]*}\right]_{,j} = \frac{\partial}{\partial t}\left[v_i^{[2]*}\rho v_i^{[1]} + \sigma_{ij}^{[2]*}S_{ijkl}\sigma_{kl}^{[1]}\right] - v_i^{[2]*}f_i^{[1]} - v_i^{[1]}f_i^{[2]*}, \quad (6.12)$$

$$\left[v_i^{[2]*}\sigma_{ij}^{[1]} + v_i^{[1]}\sigma_{ij}^{[2]*}\right]_{,j} = 0 - v_i^{[2]*}f_i^{[1]} - 0.$$

Integrating over the cross-section of the plate we obtain

$$\int_{-h}^{h}\left[v_i^{[2]*}\sigma_{ij}^{[1]} + v_i^{[1]}\sigma_{ij}^{[2]*}\right]_{,j}(1)dX_3 = \int_{-h}^{h}-v_i^{[2]*}f_i^{[1]}(1)dX_3 \quad (6.40)$$

$$\int_{-h}^{h}\left\{\frac{\partial}{\partial X_1}\left[v_i^{[2]*}\sigma_{i1}^{[1]} + v_i^{[1]}\sigma_{i1}^{[2]*}\right] + 0 + \frac{\partial}{\partial X_3}\left[v_i^{[2]*}\sigma_{i3}^{[1]} + v_i^{[1]}\sigma_{i3}^{[2]*}\right]\right\}dX_3 = \int_{-h}^{h}-v_i^{[2]*}f_i^{[1]}dX_3,$$

where the second term is zero because the waves are planar. Integrating the third term gives

$$\int_{-h}^{h}\frac{\partial}{\partial X_1}\left[v_i^{[2]*}\sigma_{i1}^{[1]} + v_i^{[1]}\sigma_{i1}^{[2]*}\right]dX_3 + \left[v_i^{[2]*}\sigma_{i3}^{[1]} + v_i^{[1]}\sigma_{i3}^{[2]*}\right]_{-h}^{h}$$
$$= \int_{-h}^{h}-v_i^{[2]*}f_i^{[1]}dX_3. \quad (6.41)$$

When we substitute in for the velocity and stress the equation gets rather lengthy, so let's work on each of the three terms individually:

Term 1 $\quad \int_{-h}^{h}\frac{\partial}{\partial X_1}\left[v_i^{[2]*}\sigma_{i1}^{[1]} + v_i^{[1]}\sigma_{i1}^{[2]*}\right]dX_3$

$$= \int_{-h}^{h}\frac{\partial}{\partial X_1}\left[\left[\frac{V_i^{n*}(X_3)}{2}e^{i(-k^{n*}X_1-2\omega t)}\right]\left[\sum_m A_m(X_1)\frac{\Sigma_{i1}^{m}(X_3)}{2}e^{-i2\omega t}\right]\right.$$

$$\left. + \left[\sum_m A_m(X_1)\frac{V_i^{m}(X_3)}{2}e^{-i2\omega t}\right]\left[\frac{\Sigma_{i1}^{n*}(X_3)}{2}e^{i(-k^{n*}X_1-2\omega t)}\right]\right]dX_3$$

$$= \int_{-h}^{h}\frac{\partial}{\partial X_1}\left[e^{i(-k^{n*}X_1-4\omega t)}\left[\frac{V_i^{n*}(X_3)}{2}\sum_m A_m(X_1)\frac{\Sigma_{i1}^{m}(X_3)}{2} + \frac{\Sigma_{i1}^{n*}(X_3)}{2}\sum_m A_m(X_1)\frac{V_i^{m}(X_3)}{2}\right]\right]dX_3$$

Term 2 $\quad \left[v_i^{[2]*}\sigma_{i3}^{[1]} + v_i^{[1]}\sigma_{i3}^{[2]*}\right]_{-h}^{h} = \left[\left[\frac{V_i^{n*}(X_3)}{2}e^{i(-k^{n*}X_1-2\omega t)}\right]\sigma_{i3}^{[1]} + v_i^{[1]}\sigma_{i3}^{[2]*}\right]_{-h}^{h}$

The nonlinear surface traction in equation (6.32) is applied through the term $\sigma_{i3}^{[1]} = -\overline{S}_{i3}^{2\omega}(X_3)$ and wavefield 2 is traction-free at $X_3 = \pm h$. Thus, we obtain

Term 2

$$\left[\left[\frac{V_i^n{}^*(X_3)}{2}e^{i(-k^{n*}X_1-2\omega t)}\right]\left(-\overline{S}_{i3}^{2\omega}(X_3)e^{2i(kX_1-\omega t)}\right)\right]_{-h}^{h}$$

$$= \left[e^{i(-k^{n*}X_1-2\omega t)}e^{2i(kX_1-\omega t)}\frac{V_i^n{}^*(X_3)}{2}\left(-\overline{S}_{i3}^{2\omega}(X_3)\right)\right]_{-h}^{h}$$

$$= \left[e^{i(2kX_1-k^{n*}X_1-4\omega t)}\frac{V_i^n{}^*(X_3)}{2}\left(-\overline{S}_{i3}^{2\omega}(X_3)\right)\right]_{-h}^{h}$$

Term 3

$$\int_{-h}^{h}-v_i^{[2]*}f_i^{[1]}dX_3 = -\int_{-h}^{h}\left[\frac{V_i^n{}^*(X_3)}{2}e^{i(-k^{n*}X_1-2\omega t)}\right]f_i^{2\omega}(X_3)e^{2i(kX_1-\omega t)}dX_3$$

$$= -\int_{-h}^{h}e^{i(2kX_1-k^{n*}X_1-4\omega t)}\frac{V_i^n{}^*(X_3)}{2}f_i^{2\omega}(X_3)dX_3.$$

The time harmonic $e^{-4i\omega t}$ is common to all three terms and can be divided out. Reassembling equation (6.41) we have

$$\int_{-h}^{h}\frac{\partial}{\partial X_1}\left[e^{i(-k^{n*}X_1)}\left[\frac{V_i^n{}^*(X_3)}{2}\sum_m A_m(X_1)\frac{\Sigma_{i1}^m(X_3)}{2} + \frac{\Sigma_{i1}^n{}^*(X_3)}{2}\sum_m A_m(X_1)\frac{V_i^m(X_3)}{2}\right]\right]dX_3$$

$$+ \left[e^{i(2kX_1-k^{n*}X_1)}\frac{V_i^n{}^*(X_3)}{2}(-\overline{S}_{i3}^{2\omega}(X_3))\right]_{-h}^{h}$$

$$= -\int_{-h}^{h}e^{i(2kX_1-k^{n*}X_1)}\frac{V_i^n{}^*(X_3)}{2}f_i^{2\omega}(X_3)dX_3.$$

Use the chain rule to take the derivative in the first term,

$$\int_{-h}^{h}\left[e^{i(-k^{n*}X_1)}\left(-ikn*+\frac{\partial}{\partial X_1}\right)\left[\frac{V_i^n{}^*(X_3)}{2}\sum_m A_m(X_1)\frac{\Sigma_{i1}^m(X_3)}{2} + \frac{\Sigma_{i1}^n{}^*(X_3)}{2}\sum_m A_m(X_1)\frac{V_i^m(X_3)}{2}\right]\right]dX_3$$

$$+ \left[e^{i(2kX_1-k^{n*}X_1)}\frac{V_i^n{}^*(X_3)}{2}(-\overline{S}_{i3}^{2\omega}(X_3))\right]_{-h}^{h} = -\int_{-h}^{h}e^{i(2kX_1-k^{n*}X_1)}\frac{V_i^n{}^*(X_3)}{2}f_i^{2\omega}(X_3)dX_3.$$

Now we can divide out $e^{i(-k^n*X_1)}$ from all terms and switch the order of differentiation and integration:

$$\left(-ik^{n*}+\frac{\partial}{\partial X_1}\right)\int_{-h}^{h}\sum_m A_m(X_1)\left[\frac{V_i^n{}^*(X_3)}{2}\frac{\Sigma_{i1}^m(X_3)}{2} + \frac{V_i^m(X_3)}{2}\frac{\Sigma_{i1}^n{}^*(X_3)}{2}\right]dX_3$$

$$- e^{2ikX_1}\left[\frac{V_i^n{}^*(X_3)}{2}\overline{S}_{i3}^{2\omega}(X_3)\right]_{-h}^{h}$$

$$= - e^{2ikX_1}\int_{-h}^{h}\frac{V_i^n{}^*(X_3)}{2}f_i^{2\omega}(X_3)dX_3.$$

For any mode m in the expansion we must have

$$\left(-\mathrm{i}k^{n*}+\frac{\partial}{\partial X_1}\right)A_m(X_1)\int_{-h}^{h}\left[\frac{V_i^{n*}(X_3)}{2}\frac{\Sigma_{i1}^m(X_3)}{2}+\frac{V_i^m(X_3)}{2}\frac{\Sigma_{i1}^{n*}(X_3)}{2}\right]dX_3$$

$$-\mathrm{e}^{2\mathrm{i}kX_1}\left[\frac{V_i^{n*}(X_3)}{2}\overline{S}_{i3}^{2\omega}(X_3)\right]_{-h}^{h}=-\mathrm{e}^{2\mathrm{i}kX_1}\int_{-h}^{h}\frac{V_i^{n*}(X_3)}{2}f_i^{2\omega}(X_3)dX_3. \tag{6.42}$$

We can now define new variables to simplify writing the equation. The nonlinear surface traction and nonlinear body force are

$$F_n^{\mathrm{surf}}\equiv-\left[\frac{V_i^{n*}(X_3)}{2}\overline{S}_{i3}^{2\omega}(X_3)\right]_{-h}^{h} \tag{6.43}$$

$$F_n^{\mathrm{vol}}\equiv\int_{-h}^{h}\frac{V_i^{n*}(X_3)}{2}f_i^{2\omega}(X_3)dX_3, \tag{6.44}$$

respectively, where we use capital F to distinguish these force-like quantities having units of N s^{-1} from frequencies, and in equation (6.21) we already defined

$$P_{mn}\equiv-\int_{-h}^{h}\left[\frac{V_i^{n*}(X_3)}{2}\frac{\Sigma_{i1}^m(X_3)}{2}+\frac{V_i^m(X_3)}{2}\frac{\Sigma_{i1}^{n*}(X_3)}{2}\right]dX_3, \tag{6.21}$$

where $V_i^m(X_3)$ and $\Sigma_{i1}^m(X_3)$ are computed directly from $U_i^m(X_3)$, turning equation (6.42) into

$$\left(-\mathrm{i}k^{n*}+\frac{\partial}{\partial X_1}\right)A_m(X_1)(-P_{mn})+\mathrm{e}^{2\mathrm{i}kX_1}F_n^{\mathrm{surf}}=-\mathrm{e}^{2\mathrm{i}kX_1}F_n^{\mathrm{vol}}.$$

We let the orthogonality relation do its mode sifting work, i.e. the only time P_{mn} is not zero is when $m=n$. Therefore, letting $m=n$

$$\left(-\mathrm{i}k^{n*}+\frac{d}{dX_1}\right)A_n(X_1)=\mathrm{e}^{2\mathrm{i}kX_1}\frac{(F_n^{\mathrm{vol}}+F_n^{\mathrm{surf}})}{P_{nn}}. \tag{6.45}$$

Out of these lengthy algebraic manipulations comes the beautiful result,

$$\frac{d}{dX_1}A_n(X_1)-\mathrm{i}k^{n*}A_n(X_1)=F_n\mathrm{e}^{2\mathrm{i}kX_1} \tag{6.46}$$

$$F_n\equiv\frac{F_n^{\mathrm{vol}}+F_n^{\mathrm{surf}}}{P_{nn}}, \tag{6.47}$$

which holds for any mode n (not summed). Later, in chapter 8, we will see that F_n directly defines the normalized nonlinear 'driving force' of the second harmonic waves and also that it can be extended to mutual interactions, where it gives the mixing power of the interacting waves. Equation (6.46) is a nonhomogeneous first order ordinary differential equation that can be solved by an integrating factor,

$$A_n(X_1) = e^{ik^{n*}X_1} F_n \int_0^{X_1} e^{i(2k-k^{n*})x} dx. \qquad (6.48)$$

These expansion coefficients tell us the participation of each mode based on the driving forces encoded in the parameter F_n. It is now apparent that the orthogonality relation P_{mn} plays a leading role in determining $A_n(X_1)$. In closure, we have set up the boundary value problems and used a perturbation method (successive approximations) to formulate the solution. We did not actually solve the boundary value problems (although we could), but through the normal mode expansion and orthogonality we have a technique for determining which secondary modes are generated.

6.6 Internal resonance

Let us closely examine the modal expansion coefficients given by equation (6.48): if $2k - k^{n*} \neq 0$

$$A_n(X_1) = \frac{e^{ik^{n*}X_1}}{i(2k - k^{n*})} F_n[e^{i(2k-k^{n*})X_1} - 1]$$

or

$$A_n(X_1) = \frac{-i}{(2k - k^{n*})} F_n[e^{i2kX_1} - e^{ik^{n*}X_1}]. \qquad (6.49)$$

If $2k - k^{n*} = 0$

$$A_n(X_1) = F_n X_1 e^{i2kX_1}. \qquad (6.50)$$

In both cases, $A_n(X_1 = 0) = 0$, meaning that there are no secondary waves at the source point. Writing the two solutions together we have

$$A_n(X_1) = F_n \begin{cases} \dfrac{-i}{(2k - k^{n*})} [e^{i2kX_1} - e^{ik^{n*}X_1}] & \text{for } k^{n*} \neq 2k \\[2mm] X_1 e^{i2kX_1} & \text{for } k^{n*} = 2k. \end{cases} \qquad (6.51)$$

Equation (6.51) is the source of the *internal resonance* criteria and serves as the foundation for selecting primary guided wave modes that generate cumulative second harmonics. In a broader sense, variants of this equation apply to cumulative secondary waves whether they are due to self-interaction or mutual interactions, as well as whether they are second order or higher order. The term *resonance* implies that the driving force is at a natural frequency of the system, which we can interpret to be the frequency of a propagating mode. We can infer from the adjective *internal* that the driving force is not externally applied, but rather comes from the material (and finite strain-induced) nonlinearity. In a sense, the nonlinear driving force acts like a distributed source term. Chapters 7 and 8 will focus on selecting primary wave modes and frequencies where the internal resonance criteria are met. Here, we provide interpretation of equation (6.51).

Let's write the second harmonic wavefield determined by equation (6.51) to be

$$\mathbf{u}_{aa} = \sum_{m \neq n} [A_m(X_1)\mathbf{U}^m(X_3)e^{-i2\omega t}] + A_{n=m}(X_1)\mathbf{U}^{n=m}(X_3)e^{-i2\omega t} \tag{6.52}$$

$$\mathbf{u}_{aa} = \sum_n F_n \frac{-i}{(2k - k^{n*})}[e^{i2kX_1} - e^{ik^{n*}X_1}]\mathbf{U}^n(X_3)e^{-i2\omega t} \text{ for } k^{n*} \neq 2k \tag{6.53}$$

$$\mathbf{u}_{aa} = \sum_{m \neq n} \left[F_m \frac{-i}{(2k - k^{m*})}[e^{i2kX_1} - e^{ik^{m*}X_1}]\mathbf{U}^m(X_3)e^{-i2\omega t} \right]$$
$$+ F_n X_1 e^{i2kX_1}\mathbf{U}^n(X_3)e^{-i2\omega t} \text{ for } k^{n*} = 2k. \tag{6.54}$$

For clarity, in equation (6.53) the modal expansion is indexed by n, but in equation (6.54) it is indexed by m while n is fixed. In equations (6.53) and (6.54) the F_n parameter defined in equation (6.46) appears in every term. In the denominator of F_n is the average power flow, P_{nn}, equation (6.24), which is nonzero for propagating modes n. In the numerator of F_n are the nonlinear driving force terms that must be nonzero for $\mathbf{u}_{aa}$ to be nonzero in both equation (6.53) and equation (6.54). This is known as the nonzero power flow condition. Wave mode types that have no power flow can be identified by parity analysis, which is done in chapter 7. Furthermore, primary to secondary wave pairs can be quantitatively compared by the power flow, which is done in chapter 8.

The condition that $k^{n*} = 2k$ is known as phase matching or synchronism, meaning that the secondary mode is synchronized with the nonlinear driving force, while the alternative condition $k^{n*} \neq 2k$ is sometimes referred to as de-tuning. Whether the primary and secondary waves are phase-matched dictates if the secondary waves are cumulative, i.e. the amplitude increases linearly with propagation distance X_1 as in the second term of equation (6.54) or is oscillatory as in equation (6.53). In equation (6.54) the second term dominates the expansion after a finite propagation distance and the summation can be neglected away from the source as shown in example 6.4.

What about the units? P_{nn} (equation (6.24)) has units of power (N-m s^{-1} = W) and the nonlinear driving forces $F_n^{vol} + F_n^{surf}$ have units of force rate (N s^{-1}), thus the units on F_n are inverse length (m^{-1}). Of course, the secondary displacement field, $\mathbf{u}_{aa}$, has length units (m) because those are the units of $\mathbf{U}^m(X_3)$ and $\mathbf{U}^n(X_3)$.

Let's examine further the cumulative case first equation (6.54), keeping in mind the context for which it was formulated—straight-crested waves in a lossless traction-free plate. How does the amplitude of the second harmonic $\mathbf{u}_{aa}$ compare with second harmonics of longitudinal bulk waves? From equation (3.12) for longitudinal bulk waves, the amplitude of the second harmonic is

$$A_2 = \frac{1}{8}\beta k^2 A_1^2 X_1 \tag{3.12}$$

and for Lamb waves, (equation (6.54)) we just found that

$$A_2 = \max(\mathbf{u}_{aa}) = F_n \max(\mathbf{U}^n(X_3))X_1. \tag{6.55}$$

The term $F_n \max(\mathbf{U}^n(X_3))$ for Lamb waves is analogous to the $\frac{1}{8}\beta k^2 A_1^2$ term for bulk longitudinal waves and is extremely useful for estimating which internally resonant modes and frequencies to select for an application as will be investigated further in chapter 8. De Lima and Hamilton [7] provide an analogous result. Now consider the relative nonlinearity coefficient by dividing equation (6.55) by A_1^2,

$$\beta' = \frac{A_2}{A_1^2} = \frac{\max(\mathbf{u}_{aa})}{(\max(\mathbf{u}_a))^2} = F_n \frac{\max(\mathbf{U}^n(X_3))}{[\max(\mathbf{U}_a(X_3))]^2}X_1, \tag{6.56}$$

which inspires definition of a parameter useful for comparing the 'rate of non-linearity accumulation', or in a more general sense the 'mixing power':

$$M_p \equiv F_n \frac{\max(\mathbf{U}^n(X_3))}{[\max(\mathbf{U}_a(X_3))]^2}. \tag{6.57}$$

The denominator of the mixing power is slightly generalized for mutual wave interactions,

$$M_p \equiv F_n \frac{\max(\mathbf{U}^n(X_3))}{\max(\mathbf{U}_a(X_3))\max(\mathbf{U}_b(X_3))}. \tag{6.58}$$

Finally, we can simply write, $\beta' = M_p X_1$.

There has to be a limiting wave propagation distance based on the perturbation condition, $\mathbf{u}_{aa} \ll \mathbf{u}_a$. A first estimate can be obtained when the amplitude of $\mathbf{u}_{aa} = \mathbf{u}_a$ by taking

$$X_{\text{perturb}} = \frac{\max(\mathbf{U}_a)}{F_n \max(\mathbf{U}^n)}.$$

Clearly, this is a rough estimate, but any refinements should include the effects of attenuation and potentially diffraction, which are beyond the scope of this chapter.

When the primary and secondary waves are not synchronized, the secondary wavefield is given by equation (6.53). When the wavenumbers k^{n*} and $2k$ are in close proximity a beat phenomenon occurs having a beat length (also known as the dispersion length) of

$$L = \frac{2\pi}{|k^{n*} - 2k|}.$$

The resulting secondary waves are bounded oscillations having periodicity L and are studied by Matsuda and Biwa [13] and [14].

Example 6.4. Second harmonic generation of Lamb waves.

We examine the modal expansion solutions for Lamb waves, first equation (6.54) for S1 lamb waves at $fd = 3.59$ MHz-mm and then equation (6.53) for S0 lamb waves at $fd = 0.5$ MHz-mm. These are not random selections; the first is an internal resonance point that is discussed in detail in chapters 8 and 11, and the second is a nonresonant point in the nearly nondispersive low frequency region of the fundamental symmetric Lamb mode. The modal expansion is applied at the second harmonic frequency and the modal characteristics necessary to compute equation (6.54) are provided below for all propagating modes at the second harmonic frequency. We take $d = 1$ mm, thus $c_p = 6153$ m s^{-1}, $k = 3664$ rad m^{-1}, $fd = 3.588$ MHz, and $2 fd = 7.176$ MHz.

Mode	c_p (m s^{-1})	k (rad m^{-1})	F_n (m^{-1})	$U_1(X_3 = h)$ (nm)
S0	2894	15 577	0.1412	0.339i
S1	4048	11 138	0.1621	−0.0757
S2	6153	7327	0.099 65	−0.445
S3	7884	5719	0.070 56	−0.489
A0	2884	15 635	0	
A1	3326	13 555	0	
A2	5375	8389	0	
A3	6153	7327	0	
A4	10 633	4240	0	
SH0	3100	14 547	0	
SH1	3174	14 203	0	
SH2	3437	13 120	0	
SH3	4069	11 081	0	
SH4	6153	7327	0	

The first thing we notice is that the nonlinear driving force F_n is only nonzero for the symmetric Lamb wave modes, which will be explained in chapter 7. The column on the far-right is the wavestructure normalized with respect to the average power. Only modes having a nonzero F_n will contribute to the expansion. It is the S2 mode, having a wavenumber of 7327 rad m^{-1}, that is internally resonant because 7327 rad m^{-1} = $2k$. This is mode $m = n$. We choose to represent the second harmonic displacement field by the component at the top surface because it could be measured there. We choose the displacement component in the X_1 direction since it typically dominates the symmetric Lamb modes. Only propagating modes were used in the expansion because the evanescent modes will only contribute very near the source. The second harmonic displacement component in the X_1 direction at $X_3 = h$ is plotted as a function of propagation distance in figure 6.2. Clearly, it is linear (but remember the restrictions of the model). The inset plot shows a zoomed-in view of the contributions of the nonresonant terms from the S0, S1, and S3 modes.

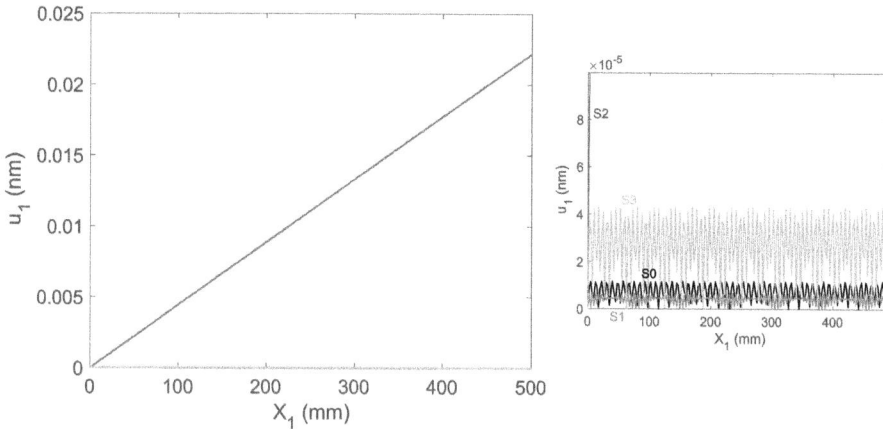

Figure 6.2. Displacement component u_1 at second harmonic frequency for S1 lamb waves at $fd = 3.59$ MHz-mm.

Considering now 0.5 MHz S0 lamb waves in a 1 mm thick plate ($c_p = 5336$ m s^{-1}, $k = 588.8$ rad m^{-1}), we tabulate the wave propagation characteristics needed to compute the second harmonic displacement field from equation (6.53). Again, it is only the symmetric Lamb wave mode with a nonlinear driving force, thus it is the only term in the expansion since the second harmonic frequency is below the cutoff of the S1 mode.

Mode	c_p (m s^{-1})	k (rad m^{-1})	F_n (m^{-1})	$U_l(X_3 = h)$ (nm)
S0	5269	1192	0.022 88	−1.76
A0	2314	2714	0	
SH0	3100	2027	0	

As before, we plot the second harmonic displacement as a function of propagation distance in figure 6.3. These primary and secondary waves are not synchronized, i.e. the wavenumber of the S0 mode at $2f$ (1192 rad m^{-1}) is not equal to $2k$ (1178 rad m^{-1}). The result of equation (6.53) is bounded oscillation as shown in figure 6.3.

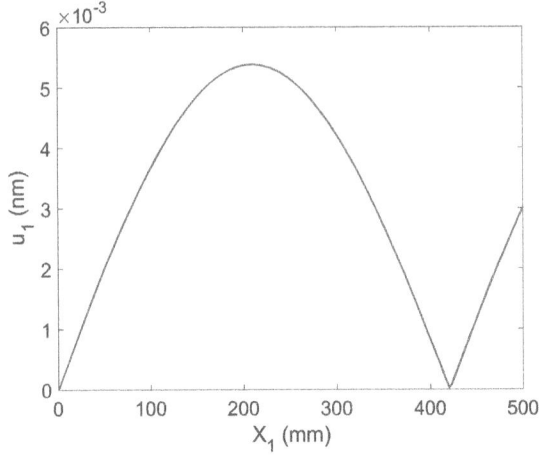

Figure 6.3. Displacement component u_1 at second harmonic frequency for S0 lamb waves at $fd = 0.5$ MHz-mm.

6.7 Wave mixing

The wave mixing problem follows from the second harmonic generation problem but has different nonlinear driving force terms. Recall from chapter 4 that the stress is an explicit function of the displacement gradient component and can be decomposed into

$$\mathbf{S}(\mathbf{H}) = \mathbf{S}^L(\mathbf{H}_a) + \mathbf{S}^L(\mathbf{H}_b) + \mathbf{S}^L(\mathbf{H}_{aa}) + \mathbf{S}^L(\mathbf{H}_{bb}) + \mathbf{S}^L(\mathbf{H}_{ab}) + \mathbf{S}^{NL}(\mathbf{H}_a + \mathbf{H}_b). \quad (4.118)$$

The nonlinear part can be subdivided based on the three types of wave interaction:

$$\mathbf{S}^{NL}(\mathbf{H}_a + \mathbf{H}_b) = \mathbf{S}^{NL}(\mathbf{H}_a, \mathbf{H}_a, 2) + \mathbf{S}^{NL}(\mathbf{H}_b, \mathbf{H}_b, 2) + \mathbf{S}^{NL}(\mathbf{H}_a, \mathbf{H}_b, 2). \quad (4.119)$$

The parallel BVPs are given below.
Primary waves a:

$$\nabla_X \cdot \mathbf{S}^L(\mathbf{H}_a) - \rho_o \ddot{\mathbf{u}}_a = 0 \text{ for } |X_3| \leqslant h \quad (4.123)$$

$$\mathbf{N} \cdot \mathbf{S}_L(\mathbf{H}_a) = 0 \text{ for } X_3 = \pm h. \quad (4.124)$$

Primary waves b:

$$\nabla_X \cdot \mathbf{S}^L(\mathbf{H}_b) - \rho_o \ddot{\mathbf{u}}_b = 0 \text{ for } |X_3| \leqslant h \quad (4.125)$$

$$\mathbf{N} \cdot \mathbf{S}^L(\mathbf{H}_b) = 0 \text{ for } X_3 = \pm h. \quad (4.126)$$

Self-interaction of waves a:

$$\nabla_X \cdot \mathbf{S}^L(\mathbf{H}_{aa}) - \rho_o \ddot{\mathbf{u}}_{aa} = -\nabla_X \cdot \mathbf{S}^{NL}(\mathbf{H}_a, \mathbf{H}_a, 2) \text{ for } |X_3| \leqslant h \quad (4.127)$$

$$\mathbf{N} \cdot \mathbf{S}^L(\mathbf{H}_{aa}) = -\mathbf{N} \cdot \mathbf{S}^{NL}(\mathbf{H}_a, \mathbf{H}_a, 2) \text{ for } X_3 = \pm h. \tag{4.128}$$

Self-interaction of waves b:

$$\nabla_X \cdot \mathbf{S}^L(\mathbf{H}_{bb}) - \rho_o \ddot{\mathbf{u}}_{bb} = -\nabla_X \cdot \mathbf{S}^{NL}(\mathbf{H}_b, \mathbf{H}_b, 2) \text{ for } |X_3| \leqslant h \tag{4.129}$$

$$\mathbf{N} \cdot \mathbf{S}^L(\mathbf{H}_{bb}) = -\mathbf{N} \cdot \mathbf{S}^{NL}(\mathbf{H}_b, \mathbf{H}_b, 2) \text{ for } X_3 = \pm h. \tag{4.130}$$

Mutual interaction of waves a and b:

$$\nabla_X \cdot \mathbf{S}^L(\mathbf{H}_{ab}) - \rho_o \ddot{\mathbf{u}}_{ab} = -\nabla_X \cdot \mathbf{S}^{NL}(\mathbf{H}_a, \mathbf{H}_b, 2) \text{ for } |X_3| \leqslant h \tag{4.131}$$

$$\mathbf{N} \cdot \mathbf{S}^L(\mathbf{H}_{ab}) = -\mathbf{N} \cdot \mathbf{S}^{NL}(\mathbf{H}_a, \mathbf{H}_b, 2) \text{ for } X_3 = \pm h, \tag{4.132}$$

where we have simply specified the domain and boundary of the domain to be that of a traction-free plate. We have just solved the self-interaction problems, equations (4.127)–(4.130), now we analyse the mutual interaction problem, equations (4.131)–(4.132).

We must reformulate the complex reciprocity relation based on the nonlinear driving force $\mathbf{S}^{NL}(\mathbf{H}_a, \mathbf{H}_b, 2)$ applied to the volume (by the body force) and the surface (by the traction). Let the wavefield 1 be a modal expansion of the secondary wavefield at the mixing frequencies $\omega_a \pm \omega_b$ (for $\omega_a > \omega_b$). We can evaluate either the sum frequency ($\omega_a + \omega_b$) or the difference frequency ($\omega_a - \omega_b$) from the $\pm$ sign. The particle velocity and stress fields are expanded:

$$v_i^{[1]}(X_k, t) = \frac{1}{2}\sum_m A_m(X_1) V_i^m(X_3) e^{-i(\omega_a \pm \omega_b)t} + \text{c. c.} \tag{6.59}$$

$$\sigma_{ij}^{[1]}(X_k, t) = \frac{1}{2}\sum_m A_m(X_1) \Sigma_{ij}^m(X_3) e^{-i(\omega_a \pm \omega_b)t} + \text{c. c.} \tag{6.60}$$

The exponentials describing the various waves are $e^{i[(k_a \pm k_b)X_1 - (\omega_a \pm \omega_b)t]}$ for waves propagating in the positive X_1 direction and $e^{-i[(k_a \pm k_b)X_1 - (\omega_a \pm \omega_b)t]}$ for waves propagating in the negative X_1 direction. Again, the $\pm$ sign means that we can either evaluate waves at the sum frequency or the difference frequency. The secondary waves at these sum and difference frequencies are second order (at third order they are at frequencies $|\pm 2\omega_a \pm \omega_b|$ and $|\pm \omega_a \pm 2\omega_b|$) and are sometimes known as combinational harmonics.

The 'forcing' used to determine the coefficients $A_m(X_1)$ includes both nonlinear body forces and nonlinear surface tractions and is included as part of wavefield 1 as before. The nonlinear body force is denoted

$$f_i^{[1]} = f_i^{\omega_a \pm \omega_b}(X_3) e^{i[(k_a \pm k_b)X_1 - (\omega_a \pm \omega_b)t]} \text{ from } \mathbf{f}^{[1]} = \nabla_X \cdot \mathbf{S}^{NL}(\mathbf{H}_a, \mathbf{H}_b, 2), \tag{6.61}$$

where the $f_i^{\omega_a \pm \omega_b}(X_3)$ comes from the divergence of the nonlinear stress, $\nabla_X \cdot \mathbf{S}^{NL}(\mathbf{H}_a, \mathbf{H}_b, 2)$, evaluated from the primary wavefield at the mixing frequencies $\omega_a \pm \omega_b$ and wavenumbers $k_a \pm k_b$ based on $\mathbf{u}_{ab} \ll \mathbf{u}_a, \mathbf{u}_b$. Likewise, the nonlinear surface traction applied at $X_3 = \pm h$ is denoted

$$\overline{S}_{ij}^{[1]} = \overline{S}_{ij}^{\omega_a \pm \omega_b}(X_3)e^{i[(k_a \pm k_b)X_1 - (\omega_a \pm \omega_b)t]} \text{ from } \overline{\mathbf{S}}^{[1]} = \mathbf{S}^{NL}(\mathbf{H}_a, \mathbf{H}_b, 2) \tag{6.62}$$

and is also evaluated from the primary wavefield at the mixing frequencies $\omega_a \pm \omega_b$ and wavenumbers $k_a \pm k_b$. $\overline{S}_{ij}^{\omega_a \pm \omega_b}(X_3)$ only needs to be evaluated at $X_3 = \pm h$.

Let the wavefield 2 be an arbitrary mode n at the mixing frequencies $\omega_a \pm \omega_b$:

$$v_i^{[2]}(X_k, t) = \frac{1}{2} V_i^n(X_3)e^{i[(k_a \pm k_b)X_1 - (\omega_a \pm \omega_b)t]} + \text{c. c.} \tag{6.63}$$

$$\sigma_{ij}^{[2]}(X_k, t) = \frac{1}{2}\Sigma_{ij}^n(X_3)e^{i[(k_a \pm k_b)X_1 - (\omega_a \pm \omega_b)t]} + \text{c. c.} \tag{6.64}$$

$$f_i^{[2]} = 0 \quad \text{from} \quad \mathbf{f}^{[2]} = 0.$$

Again, there is no 'forcing' of wavefield 2.

Substituting these fields into the complex reciprocity relation and performing similar manipulations as for the second harmonic we get for any mode m in the expansion:

$$\left(-ik^{n*} + \frac{\partial}{\partial X_1}\right)A_m(X_1) \int_{-h}^{h} \left[\frac{V_i^{n*}(X_3)}{2}\frac{\Sigma_{i1}^m(X_3)}{2} + \frac{V_i^m(X_3)}{2}\frac{\Sigma_{i1}^{n*}(X_3)}{2}\right]dX_3$$

$$- e^{i(k_a \pm k_b)X_1}\left[\frac{V_i^{n*}(X_3)}{2}\overline{S}_{i3}^{\omega_a \pm \omega_b}(X_3)\right]_{-h}^{h} \tag{6.65}$$

$$= - e^{i(k_a \pm k_b)X_1}\int_{-h}^{h}\frac{V_i^{n*}(X_3)}{2}f_i^{\omega_a \pm \omega_b}(X_3)dX_3.$$

We can now define new variables to simplify the equation. The nonlinear surface traction and nonlinear body force are

$$F_n^{\text{surf}} \equiv -\left[\frac{V_i^{n*}(X_3)}{2}\overline{S}_{i3}^{\omega_a \pm \omega_b}(X_3)\right]_{-h}^{h} \tag{6.66}$$

$$F_n^{\text{vol}} \equiv \int_{-h}^{h}\frac{V_i^{n*}(X_3)}{2}f_i^{\omega_a \pm \omega_b}(X_3)dX_3 \tag{6.67}$$

turning equation (6.65) into

$$\left(-ik^{n*} + \frac{\partial}{\partial X_1}\right)A_m(X_1)(-P_{mn}) + e^{i(k_a \pm k_b)X_1}F_n^{\text{surf}} = -e^{i(k_a \pm k_b)X_1}F_n^{\text{vol}}.$$

Let the orthogonality relation again do its mode sifting work,

$$\left(-ik^{n*} + \frac{d}{dX_1}\right)A_n(X_1) = e^{i(k_a \pm k_b)X_1}\frac{(F_n^{\text{vol}} + F_n^{\text{surf}})}{P_{nn}}. \tag{6.68}$$

Out of the algebraic manipulations emerges the result,

$$\frac{d}{dX_1}A_n(X_1) - ik^{n*}A_n(X_1) = F_n e^{i(k_a \pm k_b)X_1},\tag{6.69}$$

which holds for any mode n and can be solved by an integrating factor,

$$A_n(X_1) = e^{ik^{n*}X_1}F_n \int_0^{X_1} e^{i((k_a \pm k_b)-k^{n*})x}dx,\tag{6.70}$$

where n is not summed. The modal coefficients become

$$A_n(X_1) = F_n \begin{cases} \dfrac{-i}{((k_a \pm k_b)-k^{n*})}[e^{i(k_a \pm k_b)X_1} - e^{ik^{n*}X_1}] & \text{for } k^{n*} \neq k_a \pm k_b \\ X_1 e^{i(k_a \pm k_b)X_1} & \text{for } k^{n*} = k_a \pm k_b \end{cases}.\tag{6.71}$$

To solve the full BVP, we must also include the second harmonic generation at frequencies $2\omega_a$ and $2\omega_b$, which we know how to do.

The mixing power defined by equation (6.58) is used to account for the mutual interactions between primary modes.

Related analyses using the normal mode expansion are reported for third order harmonics [15], composite plates [16], wave mixing [17, 18] in plates and for axisymmetric waves [19] and flexural modes [20] in hollow cylinders.

6.8 Closure

The internal resonance criteria on power flow and phase matching for cumulative secondary waves were formulated using the normal mode expansion. The Lamb and SH wave modes are made orthonormal to enable use of orthogonality. Both nonzero power flow and phase matching will be investigated in chapter 7.

References

[1] Ditri J J and Rose J L 1992 Excitation of guided elastic wave modes in hollow cylinders by applied surface tractions *J. Appl. Phys.* **72** 2589–97

[2] Gao H and Rose J L 2010 Goodness dispersion curves for ultrasonic guided wave based SHM: a sample problem in corrosion monitoring *Aeronaut. J.* **114** 49–56

[3] Puthillath P, Galan J M, Ren B, Lissenden C J and Rose J L 2013 Ultrasonic guided wave propagation across waveguide transitions: energy transfer and mode conversion *J. Acoust. Soc. Am.* **133** 2624–33

[4] Rose J L 2014 *Ultrasonic Guided Waves in Solid Media* (Cambridge: Cambridge University Press)

[5] Auld B A 1973 *Acoustic Fields and Waves in Solids* (New York: Wiley)

[6] Achenbach J D 2003 *Reciprocity in Elastodynamics.* (Cambridge: Cambridge University Press)

[7] de Lima W J N and Hamilton M F 2003 Finite-amplitude waves in isotropic elastic plates *J. Sound Vib.* **265** 819–39

[8] Lanza di Scalea F, Srivastava A and Nucera C 2019 Nonlinear guided waves and thermal stresses *Nonlinear Ultrasonic and Vibro-Acoustical Techniques for Nondestructive Evaluation* (Cham: Springer Nature) pp 345–417

[9] Lamb H 1887 On reciprocal theorems in dynamics *Proc. Lond. Math. Soc.* **19** 144–51

[10] Kirrmann P 1995 On the completeness of Lamb modes *J. Elast.* **37** 39–69

[11] Auld B A and Kino G S 1971 Normal mode theory for acoustic waves and its application to the interdigital transducer *IEEE Trans. Electron Devices* **18** 898–908

[12] Chillara V K and Lissenden C J 2012 Interaction of guided wave modes in isotropic weakly nonlinear elastic plates: higher harmonic generation *J. Appl. Phys.* **111** 124909

[13] Matsuda N and Biwa S 2014 Frequency dependence of second-harmonic generation in Lamb waves *J. Nondestruct. Eval.* **33** 169–77

[14] Kanda K and Sugiura T 2018 Analysis of damped guided waves using the method of multiple scales *Wave Motion* **82** 86–95

[15] Liu Y, Chillara V K, Lissenden C J and Rose J L 2013 Third harmonic shear horizontal and Rayleigh Lamb waves in weakly nonlinear plates *J. Appl. Phys.* **114** 114908

[16] Zhao J, Chillara V K, Ren B, Cho H, Qiu J and Lissenden C J 2016 Second harmonic generation in composites: theoretical and numerical analyses *J. Appl. Phys.* **119** 064902

[17] Hasanian M and Lissenden C J 2017 Second order harmonic guided wave mutual interactions in plate: vector analysis, numerical simulation, and experimental results *J. Appl. Phys.* **122** 084901

[18] Hasanian M and Lissenden C J 2018 Second order ultrasonic guided wave mutual interactions in plate: arbitrary angles, internal resonance, and finite interaction region *J. Appl. Phys.* **124** 164904

[19] Liu Y, Khajeh E, Lissenden C J and Rose J L 2013 Interaction of torsional and longitudinal guided waves in weakly nonlinear circular cylinders *J. Acoust. Soc. Am.* **133** 2541–53

[20] Liu Y, Lissenden C J and Rose J L 2014 Higher order interaction of elastic waves in weakly nonlinear hollow circular cylinders. I. Analytical foundation *J. Appl. Phys.* **115** 214901

Chapter 7

Internal resonance in plates

What can we learn from the internal resonance criteria? We will see in this chapter that there is much to be gleaned without even specifying a frequency and mode. Two general types of analysis are presented; the first is a parity analysis based on the symmetry and antisymmetry of the wavestructures, the second is a detailed analysis of the Rayleigh–Lamb and SH dispersion relations.

This chapter takes the view that when monitoring material nonlinearity, it is important to utilize a secondary wave mode that is cumulative for ease of measurement. Although the phrase 'ease of measurement' may be misleading because measurements of weak nonlinearity are not easy. Measurements of cumulative secondary waves are generally less difficult for the reasons to be discussed in chapter 11. Thus, this chapter is devoted entirely to the internal resonance criteria of chapter 6. We start with a parity analysis of power flow to identify wave mode types that do *not* support power flow from primary waves to secondary waves. Both Lamb waves and shear-horizontal waves are analysed. For brevity we adopt RL as an abbreviation for Lamb waves because they come from the Rayleigh–Lamb dispersion relation and SH for shear-horizontal waves. To denote symmetric and antisymmetric modes, RL and SH will be preceded by an S or an A, respectively. Wavestructures for isotropic homogeneous plates, and hence particle velocity, strain, and stress, are sinusoidal functions through the thickness of the plate. Parity analysis is based upon whether the functions are even $f(x) = f(-x)$ or odd $f(x) = -f(-x)$, which dictates the parity of the second rank tensors and vectors. Parity analysis will only tell whether power flows, not how much power flows, unless it is exactly zero. In fact, we are most interested in the types of wave modes to which no power flows. We'll treat the properties of even and odd functions as common knowledge as they are readily available on the internet.

doi:10.1088/978-0-7503-4911-6ch7

Let's compile the information we need to analyse power flow. The displacement wavefields are:

- for symmetric RL waves (from equation (4.43))

$$u_1 = \left[ik\frac{A}{D_2}\cos(pX_3) - q\cos(qX_3) \right] e^{i(kX_1-\omega t)}$$

$$u_2 = 0 \tag{4.43}$$

$$u_3 = \left[-p\frac{A}{D_2}\sin(pX_3) + ik\sin(qX_3) \right] e^{i(kX_1-\omega t)}.$$

- for antisymmetric RL waves (from equation (4.46))

$$u_1 = \left[ik\frac{B}{C_2}\sin(pX_3) + q\sin(qX_3) \right] e^{i(kX_1-\omega t)}$$

$$u_2 = 0 \tag{4.46}$$

$$u_3 = \left[p\frac{B}{C_2}\cos(pX_3) + ik\cos(qX_3) \right] e^{i(kX_1-\omega t)}.$$

- for symmetric SH waves (from equation (4.62))

$$u_1 = 0$$
$$u_2 = A_2\cos(qX_3)e^{i(kX_1-\omega t)} \tag{4.62}$$
$$u_3 = 0.$$

- for antisymmetric SH waves (from equation (4.62))

$$u_1 = 0$$
$$u_2 = A_1\sin(qX_3)e^{i(kX_1-\omega t)} \tag{4.62}$$
$$u_3 = 0.$$

The coefficient ratios $\frac{A}{D_2}$ and $\frac{B}{C_2}$ are used for RL wave modes and $A_1 = A_2 = 1$ can be used for SH wave modes. The general form of the displacement gradient tensor (equation (2.2)) for straight-crested RL and SH waves is

$$\mathbf{H} = \begin{bmatrix} ikU_1(X_3) & 0 & U_{1,3}(X_3) \\ ikU_2(X_3) & 0 & U_{2,3}(X_3) \\ ikU_3(X_3) & 0 & U_{3,3}(X_3) \end{bmatrix} e^{i(kX_1-\omega t)}, \tag{7.1}$$

which is the foundation for the parity analysis.

7.1 Power flow for self-interaction

7.1.1 Second order

The parity of the RL and SH wavestructures and displacement gradients is summarized in table 7.1 based on their symmetry/antisymmetry in the X_3 direction. We will show how these results were obtained. The symbols S and A are used to denote even (symmetric) functions and odd (antisymmetric) functions of X_3. The prefixes S and A are used to denote symmetric and antisymmetric wavestructures, respectively, for both RL and SH wave modes. Strictly speaking, the symmetry of the first Piola–Kirchhoff stress cannot be assessed because it is a two-point tensor; here we refer simply to the matrix that represents its components.

The nonlinear first Piola–Kirchhoff stress for second order self-interaction from equation (4.121) is

$$\mathbf{S}^{NL}(\mathbf{H}_a, \mathbf{H}_a, 2) =$$

$$\frac{\lambda}{2}\mathrm{tr}\big(\mathbf{H}_a + \mathbf{H}_a^T\big)\mathbf{H}_a + \frac{\lambda}{2}\mathrm{tr}\big(\mathbf{H}_a^T\mathbf{H}_a\big)\mathbf{I}$$

$$+ \mu\big(\mathbf{H}_a\mathbf{H}_a + \mathbf{H}_a\mathbf{H}_a^T + \mathbf{H}_a^T\mathbf{H}_a\big)$$

$$+ \frac{\mathscr{A}}{4}\big(\mathbf{H}_a\mathbf{H}_a + \mathbf{H}_a\mathbf{H}_a^T + \mathbf{H}_a^T\mathbf{H}_a + \mathbf{H}_a^T\mathbf{H}_a^T\big)$$

$$+ \frac{\mathscr{B}}{4}\mathrm{tr}\big(\mathbf{H}_a\mathbf{H}_a + \mathbf{H}_a\mathbf{H}_a^T + \mathbf{H}_a^T\mathbf{H}_a + \mathbf{H}_a^T\mathbf{H}_a^T\big)\mathbf{I}$$

$$+ \mathscr{B}\mathrm{tr}\big(\mathbf{H}_a\big)\big(\mathbf{H}_a + \mathbf{H}_a^T\big) + \mathscr{C}(\mathrm{tr}(\mathbf{H}_a))^2\,\mathbf{I} \tag{4.121}$$

and the nonlinear surface and volumetric forces are

$$F_n^{\mathrm{surf}} \equiv -\frac{1}{2}[V_i^{n*}(X_3)\overline{S}_{i3}^{2\omega}(X_3)]_{-h}^{h} \tag{6.43}$$

Table 7.1. Parity of wave mode variables for second order self-interaction.

	SRL	ARL	SSH	ASH
Displacement, $\mathbf{u}_a$	$\begin{Bmatrix} S \\ 0 \\ A \end{Bmatrix}$	$\begin{Bmatrix} A \\ 0 \\ S \end{Bmatrix}$	$\begin{Bmatrix} 0 \\ S \\ 0 \end{Bmatrix}$	$\begin{Bmatrix} 0 \\ A \\ 0 \end{Bmatrix}$
Displacement gradient, $\mathbf{H}_a$	$\begin{bmatrix} S & 0 & A \\ 0 & 0 & 0 \\ A & 0 & S \end{bmatrix}$	$\begin{bmatrix} A & 0 & S \\ 0 & 0 & 0 \\ S & 0 & A \end{bmatrix}$	$\begin{bmatrix} 0 & 0 & 0 \\ S & 0 & A \\ 0 & 0 & 0 \end{bmatrix}$	$\begin{bmatrix} 0 & 0 & 0 \\ A & 0 & S \\ 0 & 0 & 0 \end{bmatrix}$
Nonlinear stress, $\mathbf{S}^{NL}(\mathbf{H}_a, \mathbf{H}_a, 2)$	$\begin{bmatrix} S & 0 & A \\ 0 & S & 0 \\ A & 0 & S \end{bmatrix}$	$\begin{bmatrix} S & 0 & A \\ 0 & S & 0 \\ A & 0 & S \end{bmatrix}$	$\begin{bmatrix} S & 0 & A \\ 0 & S & 0 \\ A & 0 & S \end{bmatrix}$	$\begin{bmatrix} S & 0 & A \\ 0 & S & 0 \\ A & 0 & S \end{bmatrix}$
Divergence of nonlinear stress, $\nabla_X \cdot \mathbf{S}^{NL}(\mathbf{H}_a, \mathbf{H}_a, 2)$	$\begin{Bmatrix} S \\ 0 \\ A \end{Bmatrix}$	$\begin{Bmatrix} S \\ 0 \\ A \end{Bmatrix}$	$\begin{Bmatrix} S \\ 0 \\ A \end{Bmatrix}$	$\begin{Bmatrix} S \\ 0 \\ A \end{Bmatrix}$

$$F_n^{\text{vol}} \equiv \frac{1}{2} \int_{-h}^{h} V_i^{n*}(X_3) f_i^{2\omega}(X_3) dX_3, \tag{6.44}$$

where

$$\nabla_X \cdot \mathbf{S}^{NL}(\mathbf{H}_a, \mathbf{H}_a, 2) \rightarrow f_i^{2\omega}(X_3) e^{2i(kX_1 - \omega t)}$$

$$\mathbf{S}^{NL}(\mathbf{H}_a, \mathbf{H}_a, 2) \rightarrow \overline{S}_{ij}^{2\omega}(X_3) e^{2i(kX_1 - \omega t)}.$$

Since $\mathbf{H}_a^T$ and $\mathbf{H}_a$ have the same parity for RL waves, equation (4.121) indicates that we need to evaluate the parity of four displacement gradient terms within $\mathbf{S}^{NL}(\mathbf{H}_a, \mathbf{H}_a, 2)$ for each possible wave mode: $\text{tr}(\mathbf{H}_a)\mathbf{H}_a$, $\text{tr}(\mathbf{H}_a\mathbf{H}_a)\mathbf{I}$, $\mathbf{H}_a\mathbf{H}_a$, and $[\text{tr}(\mathbf{H}_a)]^2\mathbf{I}$. We analyse these terms for each wave mode type and provide the results in table 7.1.

7.1.1.1 Symmetric RL modes
For symmetric RL modes:

$$\text{tr}(\mathbf{H}_a)\mathbf{H}_a \rightarrow S\begin{bmatrix} S & 0 & A \\ 0 & 0 & 0 \\ A & 0 & S \end{bmatrix} = \begin{bmatrix} S & 0 & A \\ 0 & 0 & 0 \\ A & 0 & S \end{bmatrix}$$

$$\mathbf{H}_a\mathbf{H}_a \rightarrow \begin{bmatrix} S & 0 & A \\ 0 & 0 & 0 \\ A & 0 & S \end{bmatrix}\begin{bmatrix} S & 0 & A \\ 0 & 0 & 0 \\ A & 0 & S \end{bmatrix} = \begin{bmatrix} S & 0 & A \\ 0 & 0 & 0 \\ A & 0 & S \end{bmatrix}$$

$$\text{tr}(\mathbf{H}_a\mathbf{H}_a)\mathbf{I} \rightarrow \text{tr}\begin{bmatrix} S & 0 & A \\ 0 & 0 & 0 \\ A & 0 & S \end{bmatrix}\mathbf{I} = S\mathbf{I} = \begin{bmatrix} S & 0 & 0 \\ 0 & S & 0 \\ 0 & 0 & S \end{bmatrix}$$

$$[\text{tr}(\mathbf{H}_a)]^2\mathbf{I} \rightarrow S\mathbf{I} = \begin{bmatrix} S & 0 & 0 \\ 0 & S & 0 \\ 0 & 0 & S \end{bmatrix},$$

which add together to give

$$\mathbf{S}^{NL}(\mathbf{H}_a, \mathbf{H}_a, 2) \rightarrow \begin{bmatrix} S & 0 & A \\ 0 & 0 & 0 \\ A & 0 & S \end{bmatrix} + \begin{bmatrix} S & 0 & A \\ 0 & 0 & 0 \\ A & 0 & S \end{bmatrix} + \begin{bmatrix} S & 0 & 0 \\ 0 & S & 0 \\ 0 & 0 & S \end{bmatrix} + \begin{bmatrix} S & 0 & 0 \\ 0 & S & 0 \\ 0 & 0 & S \end{bmatrix} = \begin{bmatrix} S & 0 & A \\ 0 & S & 0 \\ A & 0 & S \end{bmatrix}$$

and the divergence of the nonlinear stress is

$$\nabla_X \cdot \mathbf{S}^{NL}(\mathbf{H}_a, \mathbf{H}_a, 2) \rightarrow \begin{bmatrix} S & 0 & A \\ 0 & S & 0 \\ A & 0 & S \end{bmatrix}\begin{Bmatrix} \dfrac{\partial}{\partial X_1} \\ \dfrac{\partial}{\partial X_2} \\ \dfrac{\partial}{\partial X_3} \end{Bmatrix} = \begin{Bmatrix} S + S \\ 0 \\ A + A \end{Bmatrix} = \begin{Bmatrix} S \\ 0 \\ A \end{Bmatrix}.$$

7.1.1.2 Antisymmetric RL modes

For antisymmetric RL modes

$$\text{tr}(\mathbf{H}_a)\mathbf{H}_a \rightarrow A \begin{bmatrix} A & 0 & S \\ 0 & 0 & 0 \\ S & 0 & A \end{bmatrix} = \begin{bmatrix} S & 0 & A \\ 0 & 0 & 0 \\ A & 0 & S \end{bmatrix}$$

$$\mathbf{H}_a\mathbf{H}_a \rightarrow \begin{bmatrix} A & 0 & S \\ 0 & 0 & 0 \\ S & 0 & A \end{bmatrix}\begin{bmatrix} A & 0 & S \\ 0 & 0 & 0 \\ S & 0 & A \end{bmatrix} = \begin{bmatrix} S & 0 & A \\ 0 & 0 & 0 \\ A & 0 & S \end{bmatrix}$$

$$\text{tr}(\mathbf{H}_a\mathbf{H}_a)\mathbf{I} \rightarrow \text{tr}\begin{bmatrix} S & 0 & A \\ 0 & 0 & 0 \\ A & 0 & S \end{bmatrix}\mathbf{I} = S\mathbf{I} = \begin{bmatrix} S & 0 & 0 \\ 0 & S & 0 \\ 0 & 0 & S \end{bmatrix}$$

$$[\text{tr}(\mathbf{H}_a)]^2\mathbf{I} \rightarrow S\mathbf{I} = \begin{bmatrix} S & 0 & 0 \\ 0 & S & 0 \\ 0 & 0 & S \end{bmatrix},$$

which add together to give

$$\mathbf{S}^{NL}(\mathbf{H}_a, \mathbf{H}_a, 2) \rightarrow \begin{bmatrix} S & 0 & A \\ 0 & 0 & 0 \\ A & 0 & S \end{bmatrix} + \begin{bmatrix} S & 0 & A \\ 0 & 0 & 0 \\ A & 0 & S \end{bmatrix} + \begin{bmatrix} S & 0 & 0 \\ 0 & S & 0 \\ 0 & 0 & S \end{bmatrix} + \begin{bmatrix} S & 0 & 0 \\ 0 & S & 0 \\ 0 & 0 & S \end{bmatrix} = \begin{bmatrix} S & 0 & A \\ 0 & S & 0 \\ A & 0 & S \end{bmatrix}$$

and the divergence of the nonlinear stress is again

$$\nabla_X \cdot \mathbf{S}^{NL}(\mathbf{H}_a, \mathbf{H}_a, 2) \rightarrow \begin{bmatrix} S & 0 & A \\ 0 & S & 0 \\ A & 0 & S \end{bmatrix}\begin{Bmatrix} \dfrac{\partial}{\partial X_1} \\ \dfrac{\partial}{\partial X_2} \\ \dfrac{\partial}{\partial X_3} \end{Bmatrix} = \begin{Bmatrix} S + S \\ 0 \\ A + A \end{Bmatrix} = \begin{Bmatrix} S \\ 0 \\ A \end{Bmatrix}.$$

7.1.1.3 Symmetric SH modes

For symmetric SH modes $\mathbf{H}_a^T$ and $\mathbf{H}_a$ are different and both need to be analysed. Eliminating the terms that are zero,

$$\begin{aligned}
\mathbf{S}^{NL}(\mathbf{H}_a, \mathbf{H}_a, 2) \\
&= \frac{\lambda}{2}\text{tr}(\mathbf{H}_a^T\mathbf{H}_a)\mathbf{I} + \mu(\mathbf{H}_a\mathbf{H}_a^T + \mathbf{H}_a^T\mathbf{H}_a) + \frac{\mathscr{A}}{4}(\mathbf{H}_a\mathbf{H}_a^T + \mathbf{H}_a^T\mathbf{H}_a) \\
&+ \frac{\mathscr{B}}{4}\text{tr}(\mathbf{H}_a\mathbf{H}_a^T + \mathbf{H}_a^T\mathbf{H}_a)\,\mathbf{I}.
\end{aligned} \qquad (7.2)$$

The nonzero terms contain

$$\mathbf{H}_a^T\mathbf{H}_a \rightarrow \begin{bmatrix} 0 & S & 0 \\ 0 & 0 & 0 \\ 0 & A & 0 \end{bmatrix}\begin{bmatrix} 0 & 0 & 0 \\ S & 0 & A \\ 0 & 0 & 0 \end{bmatrix} = \begin{bmatrix} S & 0 & A \\ 0 & 0 & 0 \\ A & 0 & S \end{bmatrix}$$

$$\mathbf{H}_a\mathbf{H}_a^T \rightarrow \begin{bmatrix} 0 & 0 & 0 \\ S & 0 & A \\ 0 & 0 & 0 \end{bmatrix}\begin{bmatrix} 0 & S & 0 \\ 0 & 0 & 0 \\ 0 & A & 0 \end{bmatrix} = \begin{bmatrix} 0 & 0 & 0 \\ 0 & S & 0 \\ 0 & 0 & 0 \end{bmatrix}$$

$$\text{tr}(\mathbf{H}_a^T\mathbf{H}_a)\mathbf{I} \rightarrow \text{tr}\begin{bmatrix} S & 0 & A \\ 0 & 0 & 0 \\ A & 0 & S \end{bmatrix}\mathbf{I} = S\mathbf{I} = \begin{bmatrix} S & 0 & 0 \\ 0 & S & 0 \\ 0 & 0 & S \end{bmatrix}$$

$$\text{tr}(\mathbf{H}_a\mathbf{H}_a^T + \mathbf{H}_a^T\mathbf{H}_a)\mathbf{I} \rightarrow \text{tr}\begin{bmatrix} S & 0 & A \\ 0 & S & 0 \\ A & 0 & S \end{bmatrix}\mathbf{I} = \begin{bmatrix} S & 0 & 0 \\ 0 & S & 0 \\ 0 & 0 & S \end{bmatrix},$$

which add together to give

$$\mathbf{S}^{NL}(\mathbf{H}_a, \mathbf{H}_a, 2) \rightarrow \begin{bmatrix} S & 0 & 0 \\ 0 & S & 0 \\ 0 & 0 & S \end{bmatrix} + \begin{bmatrix} 0 & 0 & 0 \\ 0 & S & 0 \\ 0 & 0 & 0 \end{bmatrix} + \begin{bmatrix} S & 0 & A \\ 0 & 0 & 0 \\ A & 0 & S \end{bmatrix} + \begin{bmatrix} S & 0 & 0 \\ 0 & S & 0 \\ 0 & 0 & S \end{bmatrix} = \begin{bmatrix} S & 0 & A \\ 0 & S & 0 \\ A & 0 & S \end{bmatrix}$$

and the divergence of the nonlinear stress is

$$\nabla_X \cdot \mathbf{S}^{NL}(\mathbf{H}_a, \mathbf{H}_a, 2) \rightarrow \begin{bmatrix} S & 0 & A \\ 0 & S & 0 \\ A & 0 & S \end{bmatrix}\begin{Bmatrix} \dfrac{\partial}{\partial X_1} \\ \dfrac{\partial}{\partial X_2} \\ \dfrac{\partial}{\partial X_3} \end{Bmatrix} = \begin{Bmatrix} S \\ 0 \\ A \end{Bmatrix}.$$

7.1.1.4 Antisymmetric SH modes
For antisymmetric SH modes the nonzero terms contain

$$\mathbf{H}_a^T\mathbf{H}_a \rightarrow \begin{bmatrix} 0 & A & 0 \\ 0 & 0 & 0 \\ 0 & S & 0 \end{bmatrix}\begin{bmatrix} 0 & 0 & 0 \\ A & 0 & S \\ 0 & 0 & 0 \end{bmatrix} = \begin{bmatrix} S & 0 & A \\ 0 & 0 & 0 \\ A & 0 & S \end{bmatrix}$$

$$\mathbf{H}_a\mathbf{H}_a^T \rightarrow \begin{bmatrix} 0 & 0 & 0 \\ A & 0 & S \\ 0 & 0 & 0 \end{bmatrix}\begin{bmatrix} 0 & A & 0 \\ 0 & 0 & 0 \\ 0 & S & 0 \end{bmatrix} = \begin{bmatrix} 0 & 0 & 0 \\ 0 & S & 0 \\ 0 & 0 & 0 \end{bmatrix}$$

$$\text{tr}(\mathbf{H}_a^T\mathbf{H}_a)\mathbf{I} \rightarrow \text{tr}\begin{bmatrix} S & 0 & A \\ 0 & 0 & 0 \\ A & 0 & S \end{bmatrix}\mathbf{I} = S\mathbf{I} = \begin{bmatrix} S & 0 & 0 \\ 0 & S & 0 \\ 0 & 0 & S \end{bmatrix}$$

$$\text{tr}(\mathbf{H}_a\mathbf{H}_a^T + \mathbf{H}_a^T\mathbf{H}_a)\mathbf{I} \rightarrow \text{tr}\begin{bmatrix} S & 0 & A \\ 0 & S & 0 \\ A & 0 & S \end{bmatrix}\mathbf{I} = \begin{bmatrix} S & 0 & 0 \\ 0 & S & 0 \\ 0 & 0 & S \end{bmatrix},$$

which add together to give

$$\mathbf{S}^{NL}(\mathbf{H}_a, \mathbf{H}_a, 2) \rightarrow \begin{bmatrix} S & 0 & 0 \\ 0 & S & 0 \\ 0 & 0 & S \end{bmatrix} + \begin{bmatrix} 0 & 0 & 0 \\ 0 & S & 0 \\ 0 & 0 & 0 \end{bmatrix} + \begin{bmatrix} S & 0 & A \\ 0 & 0 & 0 \\ A & 0 & S \end{bmatrix} + \begin{bmatrix} S & 0 & 0 \\ 0 & S & 0 \\ 0 & 0 & S \end{bmatrix} = \begin{bmatrix} S & 0 & A \\ 0 & S & 0 \\ A & 0 & S \end{bmatrix}$$

and the divergence of the nonlinear stress is

$$\nabla_X \cdot \mathbf{S}^{NL}(\mathbf{H}_a, \mathbf{H}_a, 2) \rightarrow \begin{bmatrix} S & 0 & A \\ 0 & S & 0 \\ A & 0 & S \end{bmatrix}\begin{Bmatrix} \dfrac{\partial}{\partial X_1} \\[4pt] \dfrac{\partial}{\partial X_2} \\[4pt] \dfrac{\partial}{\partial X_3} \end{Bmatrix} = \begin{Bmatrix} S \\ 0 \\ A \end{Bmatrix}.$$

To state the obvious from table 7.1, the $\mathbf{S}^{NL}(\mathbf{H}_a, \mathbf{H}_a, 2)$ and $\nabla_X \cdot \mathbf{S}^{NL}(\mathbf{H}_a, \mathbf{H}_a, 2)$ terms have the same parity for all four types of modes, SRL, ARL, SSH, and ASH. Now we evaluate equations (6.43) and (6.44) by letting the secondary mode n be of each possible propagating mode type.

$$F_n^{\text{surf}} \equiv -\tfrac{1}{2}\left[V_i^{n*}(X_3)\overline{S}_{i3}^{2\omega}(X_3) \right]_{-h}^{h}$$

$$F_n^{\text{surf}} = -\tfrac{1}{2}[V_1^{n*}S_{13}^{NL} + V_2^{n*}S_{23}^{NL} + V_3^{n*}S_{33}^{NL}]_{-h}^{h}$$

SRL $\quad \rightarrow -\tfrac{1}{2}[S \cdot A + 0 \cdot 0 + A \cdot S]_{-h}^{h} = [A]_{-h}^{h} \neq 0$

ARL $\quad \rightarrow -\tfrac{1}{2}[A \cdot A + 0 \cdot 0 + S \cdot S]_{-h}^{h} = [S]_{-h}^{h} = 0$

SSH $\quad \rightarrow -\tfrac{1}{2}[0 \cdot S + S \cdot 0 + 0 \cdot A]_{-h}^{h} = [0]_{-h}^{h} = 0$

ASH $\quad \rightarrow -\tfrac{1}{2}[0 \cdot S + A \cdot 0 + 0 \cdot A]_{-h}^{h} = [0]_{-h}^{h} = 0$

$$F_n^{\text{vol}} \equiv \tfrac{1}{2}\int_{-h}^{h} V_i^{n*}(X_3)f_i^{2\omega}(X_3)dX_3$$

$$F_n^{\text{vol}} = \tfrac{1}{2}\int_{-h}^{h}\left[V_1^{n*}S_{11,1}^{NL} + V_2^{n*}S_{21,1}^{NL} + V_3^{n*}S_{31,1}^{NL} \right]dX_3$$

SRL $\quad \rightarrow \tfrac{1}{2}\int_{-h}^{h}[S \cdot S + 0 \cdot 0 + A \cdot A]dX_3 = \int_{-h}^{h} SdX_3 = [A]_{-h}^{h} \neq 0$

ARL $\quad \rightarrow \tfrac{1}{2}\int_{-h}^{h}[A \cdot S + 0 \cdot 0 + S \cdot A]dX_3 = \int_{-h}^{h} AdX_3 = [S]_{-h}^{h} = 0$

SSH $\quad \rightarrow \tfrac{1}{2}\int_{-h}^{h}[0 \cdot S + S \cdot 0 + 0 \cdot A]dX_3 = \int_{-h}^{h} 0dX_3 = 0$

ASH $\quad \rightarrow \tfrac{1}{2}\int_{-h}^{h}[0 \cdot S + A \cdot 0 + 0 \cdot A]dX_3 = \int_{-h}^{h} 0dX_3 = 0$

Both F_n^{surf} and F_n^{vol} have the same result; no power flows from any mode to ARL, SSH, or ASH modes. But power flows from any mode to the SRL modes. Thus, self-interaction only generates second harmonics that are symmetric Lamb waves. Antisymmetric Lamb waves have a distribution of power through the plate thickness, but it is an even function that cancels when integrated. SH waves, being shear waves, cannot generate second harmonics as was indicated by equation (3.25).

7.1.2 Third order

In the same fashion, we now consider third harmonics generated from self-interaction. Example 2.4 showed that we can decompose the stress components into linear, quadratic, and cubic terms for a strain energy function of the fourth order. Thus,

$$\mathbf{S}^{NL}(\mathbf{H}_a, \mathbf{H}_a, 3) = \mathbf{S}^{NL\_Q} + \mathbf{S}^{NL\_C}, \tag{7.3}$$

where the superscripts Q and C represent the second order (quadratic) and third order (cubic) terms respectively in terms of the displacement gradient. The quadratic terms are

$$\mathbf{S}^{NL\_Q} = \mathbf{S}^{NL}(\mathbf{H}_a, \mathbf{H}_a, 2), \tag{7.4}$$

which we just analysed. Hence, we need to analyse only (!) the cubic terms $\mathbf{S}^{NL\_C}$. From example 2.4, where the notation T^C_{PK1} was used, we have

$$\mathbf{S}^{NL\_C} = \frac{\lambda}{2}\mathrm{tr}(\mathbf{H}^T\mathbf{H})\,\mathbf{H} + \mu\mathbf{H}\mathbf{H}^T\mathbf{H} \tag{7.5}$$

$$+ \frac{\mathscr{A}}{4}\mathbf{H}(\mathbf{H}\mathbf{H} + \mathbf{H}\mathbf{H}^T + \mathbf{H}^T\mathbf{H} + \mathbf{H}^T\mathbf{H}^T)$$

$$+ \frac{\mathscr{A}}{4}(\mathbf{H}\mathbf{H}^T\mathbf{H} + \mathbf{H}^T\mathbf{H}^T\mathbf{H} + \mathbf{H}^T\mathbf{H}\mathbf{H} + \mathbf{H}^T\mathbf{H}\mathbf{H}^T)$$

$$+ \frac{\mathscr{B}}{4}\mathrm{tr}(\mathbf{H}\mathbf{H} + \mathbf{H}\mathbf{H}^T + \mathbf{H}^T\mathbf{H} + \mathbf{H}^T\mathbf{H}^T)\mathbf{H} + \mathscr{B}\mathrm{tr}(\mathbf{H})(\mathbf{H}\mathbf{H} + \mathbf{H}\mathbf{H}^T + \mathbf{H}^T\mathbf{H})$$

$$+ \frac{\mathscr{B}}{2}\mathrm{tr}(\mathbf{H}^T\mathbf{H})(\mathbf{H} + \mathbf{H}^T) + \frac{\mathscr{B}}{4}(\mathbf{H}\mathbf{H}^T\mathbf{H} + \mathbf{H}^T\mathbf{H}\mathbf{H} + \mathbf{H}^T\mathbf{H}^T\mathbf{H} + \mathbf{H}^T\mathbf{H}\mathbf{H}^T)\mathbf{I}$$

$$+ \mathscr{C}(\mathrm{tr}(\mathbf{H}))^2\mathbf{H} + \mathscr{C}\mathrm{tr}(\mathbf{H})\mathrm{tr}(\mathbf{H}^T\mathbf{H})\mathbf{I}$$

$$+ \frac{1}{8}\mathscr{E}\mathrm{tr}\begin{pmatrix}\mathbf{H}\mathbf{H}\mathbf{H} + \mathbf{H}\mathbf{H}\mathbf{H}^T + \mathbf{H}\mathbf{H}^T\mathbf{H} + \mathbf{H}\mathbf{H}^T\mathbf{H}^T \\ +\mathbf{H}^T\mathbf{H}\mathbf{H} + \mathbf{H}^T\mathbf{H}\mathbf{H}^T + \mathbf{H}^T\mathbf{H}^T\mathbf{H} + \mathbf{H}^T\mathbf{H}^T\mathbf{H}^T\end{pmatrix}\mathbf{I}$$

$$+ \frac{3}{4}\mathscr{E}\mathit{tr}(\mathbf{H})(\mathbf{H}\mathbf{H} + \mathbf{H}\mathbf{H}^T + \mathbf{H}^T\mathbf{H} + \mathbf{H}^T\mathbf{H}^T)$$

$$+ \frac{1}{2}\mathscr{F}\mathrm{tr}(\mathbf{H})\mathrm{tr}(\mathbf{H}\mathbf{H} + \mathbf{H}\mathbf{H}^T + \mathbf{H}^T\mathbf{H} + \mathbf{H}^T\mathbf{H}^T)\mathbf{I} + \mathscr{F}(\mathrm{tr}(\mathbf{H}))^2(\mathbf{H} + \mathbf{H}^T)$$

$$+ \frac{1}{2}\mathscr{G}\mathrm{tr}(\mathbf{H}\mathbf{H} + \mathbf{H}\mathbf{H}^T + \mathbf{H}^T\mathbf{H} + \mathbf{H}^T\mathbf{H}^T)(\mathbf{H} + \mathbf{H}^T) + 4\mathscr{H}(\mathrm{tr}(\mathbf{H}))^3\mathbf{I}.$$

It is implied that these displacement gradients are for mode a self-interactions. For RL waves, where $\mathbf{H}^T$ and $\mathbf{H}$ have the same parity, we must analyse seven types of terms with cubic nonlinearity: $\text{tr}(\mathbf{HH})\mathbf{H}$, $\mathbf{HHH}$, $\text{tr}(\mathbf{H})\mathbf{HH}$, $[\text{tr}(\mathbf{H})]^2\mathbf{H}$, $\text{tr}(\mathbf{H})\text{tr}(\mathbf{HH})\mathbf{I}$, $\text{tr}(\mathbf{HHH})\mathbf{I}$, $[\text{tr}(\mathbf{H})]^3\mathbf{I}$.

7.1.2.1 Symmetric RL modes
For symmetric RL waves

$$\mathbf{HH} \rightarrow \begin{bmatrix} S & 0 & A \\ 0 & 0 & 0 \\ A & 0 & S \end{bmatrix}\begin{bmatrix} S & 0 & A \\ 0 & 0 & 0 \\ A & 0 & S \end{bmatrix} = \begin{bmatrix} S & 0 & A \\ 0 & 0 & 0 \\ A & 0 & S \end{bmatrix}$$

$$\text{tr}(\mathbf{HH})\mathbf{H} \rightarrow \text{tr}\begin{bmatrix} S & 0 & A \\ 0 & 0 & 0 \\ A & 0 & S \end{bmatrix}\begin{bmatrix} S & 0 & A \\ 0 & 0 & 0 \\ A & 0 & S \end{bmatrix} = \begin{bmatrix} S & 0 & A \\ 0 & 0 & 0 \\ A & 0 & S \end{bmatrix}$$

$$\mathbf{HHH} \rightarrow \begin{bmatrix} S & 0 & A \\ 0 & 0 & 0 \\ A & 0 & S \end{bmatrix}\begin{bmatrix} S & 0 & A \\ 0 & 0 & 0 \\ A & 0 & S \end{bmatrix} = \begin{bmatrix} S & 0 & A \\ 0 & 0 & 0 \\ A & 0 & S \end{bmatrix}$$

$$\text{tr}(\mathbf{H})\mathbf{HH} \rightarrow \text{tr}\begin{bmatrix} S & 0 & A \\ 0 & 0 & 0 \\ A & 0 & S \end{bmatrix}\begin{bmatrix} S & 0 & A \\ 0 & 0 & 0 \\ A & 0 & S \end{bmatrix} = \begin{bmatrix} S & 0 & A \\ 0 & 0 & 0 \\ A & 0 & S \end{bmatrix}$$

$$[\text{tr}(\mathbf{H})]^2\mathbf{H} \rightarrow \left(\text{tr}\begin{bmatrix} S & 0 & A \\ 0 & 0 & 0 \\ A & 0 & S \end{bmatrix}\right)^2\begin{bmatrix} S & 0 & A \\ 0 & 0 & 0 \\ A & 0 & S \end{bmatrix} = \begin{bmatrix} S & 0 & A \\ 0 & 0 & 0 \\ A & 0 & S \end{bmatrix}$$

$$\text{tr}(\mathbf{H})\text{tr}(\mathbf{HH})\mathbf{I} \rightarrow \text{tr}\begin{bmatrix} S & 0 & A \\ 0 & 0 & 0 \\ A & 0 & S \end{bmatrix}\text{tr}\begin{bmatrix} S & 0 & A \\ 0 & 0 & 0 \\ A & 0 & S \end{bmatrix}\mathbf{I} = \begin{bmatrix} S & 0 & 0 \\ 0 & S & 0 \\ 0 & 0 & S \end{bmatrix}$$

$$\text{tr}(\mathbf{HHH})\mathbf{I} \rightarrow \text{tr}\begin{bmatrix} S & 0 & A \\ 0 & 0 & 0 \\ A & 0 & S \end{bmatrix}\mathbf{I} = \begin{bmatrix} S & 0 & 0 \\ 0 & S & 0 \\ 0 & 0 & S \end{bmatrix}$$

$$[\text{tr}(\mathbf{H})]^3\mathbf{I} \rightarrow \left(\text{tr}\begin{bmatrix} S & 0 & A \\ 0 & 0 & 0 \\ A & 0 & S \end{bmatrix}\right)^3\mathbf{I} = \begin{bmatrix} S & 0 & 0 \\ 0 & S & 0 \\ 0 & 0 & S \end{bmatrix},$$

which add together to give

$$\mathbf{S}^{NL\_C} \rightarrow \begin{bmatrix} S & 0 & A \\ 0 & S & 0 \\ A & 0 & S \end{bmatrix}$$

and the divergence of the cubic nonlinear stress is

$$\nabla_X \cdot \mathbf{S}^{NL_C} \rightarrow \begin{bmatrix} S & 0 & A \\ 0 & S & 0 \\ A & 0 & S \end{bmatrix} \begin{Bmatrix} \frac{\partial}{\partial X_1} \\ \frac{\partial}{\partial X_2} \\ \frac{\partial}{\partial X_3} \end{Bmatrix} = \begin{Bmatrix} S \\ 0 \\ A \end{Bmatrix}.$$

Now evaluate equations (6.43) and (6.44) by letting the secondary mode n be of each possible propagating mode type:

	$F_n^{\text{surf}} = -\frac{1}{2}[V_1^{n*}S_{13}^{NL} + V_2^{n*}S_{23}^{NL} + V_3^{n*}S_{33}^{NL}]_{-h}^{h}$
SRL	$\rightarrow -\frac{1}{2}[S \cdot A + 0 \cdot 0 + A \cdot S]_{-h}^{h} = [A]_{-h}^{h} \neq 0$
ARL	$\rightarrow -\frac{1}{2}[A \cdot A + 0 \cdot 0 + S \cdot S]_{-h}^{h} = [S]_{-h}^{h} = 0$
SSH	$\rightarrow -\frac{1}{2}[0 \cdot A + S \cdot 0 + 0 \cdot S]_{-h}^{h} = [0]_{-h}^{h} = 0$
ASH	$\rightarrow -\frac{1}{2}[0 \cdot A + A \cdot 0 + 0 \cdot S]_{-h}^{h} = [0]_{-h}^{h} = 0$
	$F_n^{\text{vol}} = \frac{1}{2}\int_{-h}^{h}\left[V_1^{n*}S_{11,1}^{NL} + V_2^{n*}S_{21,1}^{NL} + V_3^{n*}S_{31,1}^{NL}\right]dX_3$
SRL	$\rightarrow \frac{1}{2}\int_{-h}^{h}[S \cdot S + 0 \cdot 0 + A \cdot A]dX_3 = \int_{-h}^{h}SdX_3 = [A]_{-h}^{h} \neq 0$
ARL	$\rightarrow \frac{1}{2}\int_{-h}^{h}[A \cdot S + 0 \cdot 0 + S \cdot A]dX_3 = \int_{-h}^{h}AdX_3 = [S]_{-h}^{h} = 0$
SSH	$\rightarrow \frac{1}{2}\int_{-h}^{h}[0 \cdot S + S \cdot 0 + 0 \cdot A]dX_3 = \int_{-h}^{h}0dX_3 = 0$
ASH	$\rightarrow \frac{1}{2}\int_{-h}^{h}[0 \cdot S + A \cdot 0 + 0 \cdot A]dX_3 = \int_{-h}^{h}0dX_3 = 0.$

Therefore, power flows only from SRL modes to SRL modes.

7.1.2.2 Antisymmetric RL modes

For antisymmetric RL waves:

$$\mathbf{HH} \rightarrow \begin{bmatrix} A & 0 & S \\ 0 & 0 & 0 \\ S & 0 & A \end{bmatrix}\begin{bmatrix} A & 0 & S \\ 0 & 0 & 0 \\ S & 0 & A \end{bmatrix} = \begin{bmatrix} S & 0 & A \\ 0 & 0 & 0 \\ A & 0 & S \end{bmatrix}$$

$$\text{tr}(\mathbf{HH})\mathbf{H} \rightarrow \text{tr}\begin{bmatrix} S & 0 & A \\ 0 & 0 & 0 \\ A & 0 & S \end{bmatrix}\begin{bmatrix} A & 0 & S \\ 0 & 0 & 0 \\ S & 0 & A \end{bmatrix} = \begin{bmatrix} A & 0 & S \\ 0 & 0 & 0 \\ S & 0 & A \end{bmatrix}$$

$$\mathbf{HHH} \rightarrow \begin{bmatrix} S & 0 & A \\ 0 & 0 & 0 \\ A & 0 & S \end{bmatrix}\begin{bmatrix} A & 0 & S \\ 0 & 0 & 0 \\ S & 0 & A \end{bmatrix} = \begin{bmatrix} A & 0 & S \\ 0 & 0 & 0 \\ S & 0 & A \end{bmatrix}$$

$$\text{tr}(\mathbf{H})\mathbf{HH} \rightarrow \text{tr}\begin{bmatrix} A & 0 & S \\ 0 & 0 & 0 \\ S & 0 & A \end{bmatrix}\begin{bmatrix} S & 0 & A \\ 0 & 0 & 0 \\ A & 0 & S \end{bmatrix} = \begin{bmatrix} A & 0 & S \\ 0 & 0 & 0 \\ S & 0 & A \end{bmatrix}$$

$$[\mathrm{tr}(\mathbf{H})]^2\mathbf{H} \rightarrow \left(\mathrm{tr}\begin{bmatrix} A & 0 & S \\ 0 & 0 & 0 \\ S & 0 & A \end{bmatrix}\right)^2 \begin{bmatrix} A & 0 & S \\ 0 & 0 & 0 \\ S & 0 & A \end{bmatrix} = \begin{bmatrix} A & 0 & S \\ 0 & 0 & 0 \\ S & 0 & A \end{bmatrix}$$

$$\mathrm{tr}(\mathbf{H})\mathrm{tr}(\mathbf{HH})\mathbf{I} \rightarrow \mathrm{tr}\begin{bmatrix} A & 0 & S \\ 0 & 0 & 0 \\ S & 0 & A \end{bmatrix}\mathrm{tr}\begin{bmatrix} S & 0 & A \\ 0 & 0 & 0 \\ A & 0 & S \end{bmatrix}\mathbf{I} = \begin{bmatrix} A & 0 & 0 \\ 0 & A & 0 \\ 0 & 0 & A \end{bmatrix}$$

$$\mathrm{tr}(\mathbf{HHH})\mathbf{I} \rightarrow \mathrm{tr}\begin{bmatrix} A & 0 & S \\ 0 & 0 & 0 \\ S & 0 & A \end{bmatrix}\mathbf{I} = \begin{bmatrix} A & 0 & 0 \\ 0 & A & 0 \\ 0 & 0 & A \end{bmatrix}$$

$$[\mathrm{tr}(\mathbf{H})]^3\mathbf{I} \rightarrow \left(\mathrm{tr}\begin{bmatrix} A & 0 & S \\ 0 & 0 & 0 \\ S & 0 & A \end{bmatrix}\right)^3\mathbf{I} = \begin{bmatrix} A & 0 & 0 \\ 0 & A & 0 \\ 0 & 0 & A \end{bmatrix}.$$

The cubic terms add together to give

$$\mathbf{S}^{NL\_C} \rightarrow \begin{bmatrix} A & 0 & S \\ 0 & A & 0 \\ S & 0 & A \end{bmatrix}$$

and the divergence of the cubic nonlinear stress is

$$\nabla_X \cdot \mathbf{S}^{NL\_C} \rightarrow \begin{bmatrix} A & 0 & S \\ 0 & A & 0 \\ S & 0 & A \end{bmatrix}\begin{Bmatrix} \frac{\partial}{\partial X_1} \\ \frac{\partial}{\partial X_2} \\ \frac{\partial}{\partial X_3} \end{Bmatrix} = \begin{Bmatrix} A \\ 0 \\ S \end{Bmatrix}.$$

Now we evaluate equations (6.43) and (6.44) by letting the secondary mode n be of each propagating mode type:

$$F_n^{\mathrm{surf}} = -\tfrac{1}{2}[V_1^{n*}S_{13}^{NL} + V_2^{n*}S_{23}^{NL} + V_3^{n*}S_{33}^{NL}]_{-h}^h$$

SRL $\quad \rightarrow -\tfrac{1}{2}[S \cdot S + 0 \cdot 0 + A \cdot A]_{-h}^h = [S]_{-h}^h = 0$

ARL $\quad \rightarrow -\tfrac{1}{2}[A \cdot S + 0 \cdot 0 + S \cdot A]_{-h}^h = [A]_{-h}^h \neq 0$

SSH $\quad \rightarrow -\tfrac{1}{2}[0 \cdot S + S \cdot 0 + 0 \cdot A]_{-h}^h = [0]_{-h}^h = 0$

ASH $\quad \rightarrow -\tfrac{1}{2}[0 \cdot S + A \cdot 0 + 0 \cdot A]_{-h}^h = [0]_{-h}^h = 0$

$$F_n^{\mathrm{vol}} = \tfrac{1}{2}\int_{-h}^h \left[V_1^{n*}S_{11,1}^{NL} + V_2^{n*}S_{21,1}^{NL} + V_3^{n*}S_{31,1}^{NL}\right]dX_3$$

SRL $\quad \rightarrow \tfrac{1}{2}\int_{-h}^h [S \cdot A + 0 \cdot 0 + A \cdot S]dX_3 = \int_{-h}^h AdX_3 = [S]_{-h}^h = 0$

ARL $\quad \rightarrow \tfrac{1}{2}\int_{-h}^h [A \cdot A + 0 \cdot 0 + S \cdot S]dX_3 = \int_{-h}^h SdX_3 = [A]_{-h}^h \neq 0$

SSH $\quad \rightarrow \tfrac{1}{2}\int_{-h}^h [0 \cdot A + S \cdot 0 + 0 \cdot S]dX_3 = \int_{-h}^h 0dX_3 = 0$

ASH $\quad \rightarrow \tfrac{1}{2}\int_{-h}^h [0 \cdot A + A \cdot 0 + 0 \cdot S]dX_3 = \int_{-h}^h 0dX_3 = 0$

Therefore, power flows only from ARL modes to ARL modes.

7.1.2.3 SH mode nonlinear stress

We now consider SH waves, where $\text{tr}(\mathbf{H}) = 0$ and $\mathbf{HH} = \mathbf{H}^T\mathbf{H}^T = 0$, the nonzero cubic terms in $\mathbf{H}$ are given by

$$\mathbf{S}^{NL\_C} = \frac{\lambda}{2}\text{tr}(\mathbf{H}^T\mathbf{H})\mathbf{H} + \mu\mathbf{HH}^T\mathbf{H}$$

$$+\frac{\mathscr{A}}{4}\mathbf{HH}^T\mathbf{H} + \frac{\mathscr{A}}{4}(\mathbf{HH}^T\mathbf{H} + \mathbf{H}^T\mathbf{HH}^T)$$

$$+\frac{\mathscr{B}}{4}\text{tr}(\mathbf{HH}^T + \mathbf{H}^T\mathbf{H})\mathbf{H} + \frac{\mathscr{B}}{2}\text{tr}(\mathbf{H}^T\mathbf{H})(\mathbf{H} + \mathbf{H}^T) + \frac{\mathscr{B}}{4}\text{tr}(\mathbf{HH}^T\mathbf{H} + \mathbf{H}^T\mathbf{HH}^T)\mathbf{I}$$

$$+\frac{1}{8}\mathscr{E}\,\text{tr}(\mathbf{HH}^T\mathbf{H} + \mathbf{H}^T\mathbf{HH}^T)\mathbf{I}$$

$$+\frac{1}{2}\mathscr{G}\,\text{tr}(\mathbf{HH}^T + \mathbf{H}^T\mathbf{H})(\mathbf{H} + \mathbf{H}^T).$$

Combining like-terms because we are only interested in parity for now we obtain

$$\mathbf{S}^{NL\_C} = z_1\mathbf{HH}^T\mathbf{H} + z_2\mathbf{H}^T\mathbf{HH}^T + S(\mathbf{H} + \mathbf{H}^T)$$

giving

$$\mathbf{S}^{NL\_C} \rightarrow \begin{bmatrix} 0 & 0 & 0 \\ S & 0 & A \\ 0 & 0 & 0 \end{bmatrix} + \begin{bmatrix} 0 & S & 0 \\ 0 & 0 & 0 \\ 0 & A & 0 \end{bmatrix} + S\begin{bmatrix} 0 & 0 & 0 \\ S & 0 & A \\ 0 & 0 & 0 \end{bmatrix} + S\begin{bmatrix} 0 & S & 0 \\ 0 & 0 & 0 \\ 0 & A & 0 \end{bmatrix} = \begin{bmatrix} 0 & S & 0 \\ S & 0 & A \\ 0 & A & 0 \end{bmatrix}$$

for SSH waves and

$$\mathbf{S}^{NL\_C} \rightarrow \begin{bmatrix} 0 & 0 & 0 \\ A & 0 & S \\ 0 & 0 & 0 \end{bmatrix} + \begin{bmatrix} 0 & A & 0 \\ 0 & 0 & 0 \\ 0 & S & 0 \end{bmatrix} + S\begin{bmatrix} 0 & 0 & 0 \\ A & 0 & S \\ 0 & 0 & 0 \end{bmatrix} + S\begin{bmatrix} 0 & A & 0 \\ 0 & 0 & 0 \\ 0 & S & 0 \end{bmatrix} = \begin{bmatrix} 0 & A & 0 \\ A & 0 & S \\ 0 & S & 0 \end{bmatrix}$$

for ASH waves.

The divergence of the cubic nonlinear stress is

$$\nabla_X \cdot \mathbf{S}^{NL\_C} \rightarrow \begin{bmatrix} 0 & S & 0 \\ S & 0 & A \\ 0 & A & 0 \end{bmatrix} \left\{ \begin{matrix} \frac{\partial}{\partial X_1} \\ \frac{\partial}{\partial X_2} \\ \frac{\partial}{\partial X_3} \end{matrix} \right\} = \left\{ \begin{matrix} 0 \\ S \\ 0 \end{matrix} \right\}$$

for SSH waves and

$$\nabla_X \cdot \mathbf{S}^{NL\_C} \rightarrow \begin{bmatrix} 0 & A & 0 \\ A & 0 & S \\ 0 & S & 0 \end{bmatrix} \left\{ \begin{matrix} \frac{\partial}{\partial X_1} \\ \frac{\partial}{\partial X_2} \\ \frac{\partial}{\partial X_3} \end{matrix} \right\} = \left\{ \begin{matrix} 0 \\ A \\ 0 \end{matrix} \right\}$$

for ASH waves. Next, we will evaluate equations (6.43) and (6.44) by letting the secondary mode n be of each propagating mode type.

7.1.2.4 Symmetric SH modes
For SSH waves

$$F_n^{\text{surf}} = -\frac{1}{2}[V_1^{n*}S_{13}^{NL} + V_2^{n*}S_{23}^{NL} + V_3^{n*}S_{33}^{NL}]_{-h}^{h}$$

SRL $\quad\to-\frac{1}{2}[S \cdot 0 + 0 \cdot A + A \cdot 0]_{-h}^{h} = [0]_{-h}^{h} = 0$

ARL $\quad\to-\frac{1}{2}[A \cdot 0 + 0 \cdot A + S \cdot 0]_{-h}^{h} = [0]_{-h}^{h} = 0$

SSH $\quad\to-\frac{1}{2}[0 \cdot 0 + S \cdot A + 0 \cdot 0]_{-h}^{h} = [A]_{-h}^{h} \neq 0$

ASH $\quad\to-\frac{1}{2}[0 \cdot 0 + A \cdot A + 0 \cdot 0]_{-h}^{h} = [S]_{-h}^{h} = 0$

$$F_n^{\text{vol}} = \frac{1}{2}\int_{-h}^{h}\left[V_1^{n*}S_{11,1}^{NL} + V_2^{n*}S_{21,1}^{NL} + V_3^{n*}S_{31,1}^{NL}\right]dX_3$$

SRL $\quad\to\frac{1}{2}\int_{-h}^{h}[S \cdot 0 + 0 \cdot S + A \cdot 0]dX_3 = \int_{-h}^{h}0\,dX_3 = 0$

ARL $\quad\to\frac{1}{2}\int_{-h}^{h}[A \cdot 0 + 0 \cdot S + S \cdot 0]dX_3 = \int_{-h}^{h}0\,dX_3 = 0$

SSH $\quad\to\frac{1}{2}\int_{-h}^{h}[0 \cdot 0 + S \cdot S + 0 \cdot 0]dX_3 = \int_{-h}^{h}S\,dX_3 = [A]_{-h}^{h} \neq 0$

ASH $\quad\to\frac{1}{2}\int_{-h}^{h}[0 \cdot 0 + A \cdot S + 0 \cdot 0]dX_3 = \int_{-h}^{h}A\,dX_3 = [S]_{-h}^{h} = 0.$

Therefore, power flows only from SSH modes to SSH modes.

7.1.2.5 Antisymmetric SH modes
For ASH waves

$$F_n^{\text{surf}} = -\frac{1}{2}[V_1^{n*}S_{13}^{NL} + V_2^{n*}S_{23}^{NL} + V_3^{n*}S_{33}^{NL}]_{-h}^{h}$$

SRL $\quad\to-\frac{1}{2}[S \cdot 0 + 0 \cdot S + A \cdot 0]_{-h}^{h} = [0]_{-h}^{h} = 0$

ARL $\quad\to-\frac{1}{2}[A \cdot 0 + 0 \cdot S + S \cdot 0]_{-h}^{h} = [0]_{-h}^{h} = 0$

SSH $\quad\to-\frac{1}{2}[0 \cdot 0 + S \cdot S + 0 \cdot 0]_{-h}^{h} = [S]_{-h}^{h} = 0$

ASH $\quad\to-\frac{1}{2}[0 \cdot 0 + A \cdot S + 0 \cdot 0]_{-h}^{h} = [A]_{-h}^{h} \neq 0$

$$F_n^{\text{vol}} = \frac{1}{2}\int_{-h}^{h}\left[V_1^{n*}S_{11,1}^{NL} + V_2^{n*}S_{21,1}^{NL} + V_3^{n*}S_{31,1}^{NL}\right]dX_3$$

SRL $\quad\to\frac{1}{2}\int_{-h}^{h}[S \cdot 0 + 0 \cdot A + A \cdot 0]dX_3 = \int_{-h}^{h}0\,dX_3 = 0$

ARL $\quad\to\frac{1}{2}\int_{-h}^{h}[A \cdot 0 + 0 \cdot A + S \cdot 0]dX_3 = \int_{-h}^{h}0\,dX_3 = 0$

SSH $\quad\to\frac{1}{2}\int_{-h}^{h}[0 \cdot 0 + S \cdot A + 0 \cdot 0]dX_3 = \int_{-h}^{h}A\,dX_3 = [S]_{-h}^{h} = 0$

ASH $\quad\to\frac{1}{2}\int_{-h}^{h}[0 \cdot 0 + A \cdot A + 0 \cdot 0]dX_3 = \int_{-h}^{h}S\,dX_3 = [A]_{-h}^{h} \neq 0.$

Therefore, power flows only from ASH modes to ASH modes. The results of the parity analysis for third order self-interaction are summarized in table 7.2. The third

Table 7.2. Parity of wave mode variables for third order self-interaction.

	SRL	ARL	SSH	ASH
Displacement, $\mathbf{u}_a$	$\begin{Bmatrix} S \\ 0 \\ A \end{Bmatrix}$	$\begin{Bmatrix} A \\ 0 \\ S \end{Bmatrix}$	$\begin{Bmatrix} 0 \\ S \\ 0 \end{Bmatrix}$	$\begin{Bmatrix} 0 \\ A \\ 0 \end{Bmatrix}$
Displacement gradient, $\mathbf{H}_a$	$\begin{bmatrix} S & 0 & A \\ 0 & 0 & 0 \\ A & 0 & S \end{bmatrix}$	$\begin{bmatrix} A & 0 & S \\ 0 & 0 & 0 \\ S & 0 & A \end{bmatrix}$	$\begin{bmatrix} 0 & 0 & 0 \\ S & 0 & A \\ 0 & 0 & 0 \end{bmatrix}$	$\begin{bmatrix} 0 & 0 & 0 \\ A & 0 & S \\ 0 & 0 & 0 \end{bmatrix}$
Nonlinear stress, $\mathbf{S}^{NL}(\mathbf{H}_a, \mathbf{H}_a, 3)$	$\begin{bmatrix} S & 0 & A \\ 0 & S & 0 \\ A & 0 & S \end{bmatrix}$	$\begin{bmatrix} A & 0 & S \\ 0 & A & 0 \\ S & 0 & A \end{bmatrix}$	$\begin{bmatrix} 0 & S & 0 \\ S & 0 & A \\ 0 & A & 0 \end{bmatrix}$	$\begin{bmatrix} 0 & A & 0 \\ A & 0 & S \\ 0 & S & 0 \end{bmatrix}$
Divergence of nonlinear stress, $\nabla_X \cdot \mathbf{S}^{NL}(\mathbf{H}_a, \mathbf{H}_a, 3)$	$\begin{Bmatrix} S \\ 0 \\ A \end{Bmatrix}$	$\begin{Bmatrix} A \\ 0 \\ S \end{Bmatrix}$	$\begin{Bmatrix} 0 \\ S \\ 0 \end{Bmatrix}$	$\begin{Bmatrix} 0 \\ A \\ 0 \end{Bmatrix}$

order self-interactions can only generate secondary waves of the same type: SRL $\rightarrow$ SRL, ARL $\rightarrow$ ARL, SSH $\rightarrow$ SSH, and ASH $\rightarrow$ ASH. This contrasts with second order self-interaction, which can only generate SRL secondary waves.

7.2 Power flow for mutual interaction

7.2.1 Second order co-directional

If we mix like wave modes, e.g. SRL with SRL or ASH with ASH, then the parity of the nonlinear stress and the divergence of the nonlinear stress will be the same as what we just analysed for self-interaction in section 7.1. Thus, in this section we will analyse mutual interaction of different wave mode types. Altogether there are six new combinations of modes a and b to consider:

- SRL–ARL, SRL–SSH, SRL–ASH,
- ARL–SSH, ARL–ASH, and
- SSH–ASH.

The nonlinear stress as a function of the displacement gradient was given in (4.122):

$$
\begin{aligned}
\mathbf{S}^{NL}&(\mathbf{H}_a, \mathbf{H}_b, 2) \\
&= \frac{\lambda}{4}\mathrm{tr}(\mathbf{H}_a + \mathbf{H}_a^T)\mathbf{H}_b + \frac{\lambda}{4}\mathrm{tr}(\mathbf{H}_b + \mathbf{H}_b^T)\mathbf{H}_a + \frac{\lambda}{4}\mathrm{tr}(\mathbf{H}_a^T\mathbf{H}_b + \mathbf{H}_b^T\mathbf{H}_a)\mathbf{I} \\
&+ \frac{\mu}{2}\Big[\mathbf{H}_a\mathbf{H}_b + \mathbf{H}_b\mathbf{H}_a + \mathbf{H}_a\mathbf{H}_b^T + \mathbf{H}_b\mathbf{H}_a^T + \mathbf{H}_a^T\mathbf{H}_b + \mathbf{H}_b^T\mathbf{H}_a\Big] \\
&+ \frac{\mathscr{A}}{8}\Big[\mathbf{H}_a\mathbf{H}_b + \mathbf{H}_b\mathbf{H}_a + \mathbf{H}_a\mathbf{H}_b^T + \mathbf{H}_b\mathbf{H}_a^T + \mathbf{H}_a^T\mathbf{H}_b + \mathbf{H}_b^T\mathbf{H}_a + \mathbf{H}_a^T\mathbf{H}_b^T + \mathbf{H}_b^T\mathbf{H}_a^T\Big] \\
&+ \frac{\mathscr{B}}{8}\mathrm{tr}\Big(\mathbf{H}_a\mathbf{H}_b + \mathbf{H}_b\mathbf{H}_a + \mathbf{H}_a\mathbf{H}_b^T + \mathbf{H}_b\mathbf{H}_a^T + \mathbf{H}_a^T\mathbf{H}_b + \mathbf{H}_b^T\mathbf{H}_a + \mathbf{H}_a^T\mathbf{H}_b^T + \mathbf{H}_b^T\mathbf{H}_a^T\Big)\mathbf{I} \\
&+ \frac{\mathscr{B}}{2}\mathrm{tr}(\mathbf{H}_a)\big(\mathbf{H}_b + \mathbf{H}_b^T\big) + \frac{\mathscr{B}}{2}\mathrm{tr}(\mathbf{H}_b)(\mathbf{H}_a + \mathbf{H}_a^T) + \mathscr{C}\,\mathrm{tr}(\mathbf{H}_a)\mathrm{tr}(\mathbf{H}_b)\mathbf{I}.
\end{aligned} \tag{4.122}
$$

Table 7.3. Parity of wave mode variables for second order mutual interaction.

	SRL–ARL SSH–ASH	SRL–SSH ARL–ASH	SRL–ASH ARL–SSH
Nonlinear stress, $\mathbf{S}^{NL}(\mathbf{H}_a, \mathbf{H}_b, 2)$	$\begin{bmatrix} A & 0 & S \\ 0 & A & 0 \\ S & 0 & A \end{bmatrix}$	$\begin{bmatrix} 0 & S & 0 \\ S & 0 & A \\ 0 & A & 0 \end{bmatrix}$	$\begin{bmatrix} 0 & A & 0 \\ A & 0 & S \\ 0 & S & 0 \end{bmatrix}$
Divergence of nonlinear stress, $\boldsymbol{\nabla}_X \cdot \mathbf{S}^{NL}(\mathbf{H}_a, \mathbf{H}_b, 2)$	$\begin{Bmatrix} A \\ 0 \\ S \end{Bmatrix}$	$\begin{Bmatrix} 0 \\ S \\ 0 \end{Bmatrix}$	$\begin{Bmatrix} 0 \\ A \\ 0 \end{Bmatrix}$

Table 7.4. For mutual interaction of different collinear wave mode types, is there power flow to second order waves?

Interacting primary waves	Secondary wave mode type			
	SRL	ARL	SSH	ASH
SRL–ARL	No	**Yes**	No	No
SSH–ASH	No	**Yes**	No	No
SRL–SSH	No	No	**Yes**	No
ARL–ASH	No	No	**Yes**	No
SRL–ASH	No	No	No	**Yes**
ARL–SSH	No	No	No	**Yes**

We will analyse one combination as example 7.1 and then give the resulting parity of nonlinear stress and its divergence in table 7.3 amongst all the possible combinations. These parity results are then combined with all possible secondary wave modes (also in example 7.1) to find the cases where no power flow occurs, and results are given in table 7.4. The results in tables 7.3 and 7.4 are sourced from Liu's [1] tables 4.1 and 4.2, respectively.

For the purposes of discussion, define the terms 'wave type' as either RL or SH and 'wave nature' as either symmetric or antisymmetric. In summary, power flows only to secondary wave modes as follows.

- If the primary waves are of the same type and nature, then the secondary waves are SRL.
- If the primary waves are of the same type but different nature, then the secondary waves are ARL.
- If the primary waves are of different types but the same nature, then the secondary waves are SSH.
- If the primary waves are of different types and nature, then the secondary waves are ASH.

Example 7.1. Parity analysis of mutual interaction between SRL and ASH wavefields.

Given the interaction between collinear SRL and ASH wavefields, to which secondary mode types does no power flow? We will use S and A to denote even (symmetric) and odd (antisymmetric) functions of X_3, respectively.

For an SRL mode,

$$\mathbf{H}_a = \mathbf{H}_a^T = \begin{bmatrix} S & 0 & A \\ 0 & 0 & 0 \\ A & 0 & S \end{bmatrix}, \ \mathrm{tr}(\mathbf{H}_a) = S.$$

For an ASH mode,

$$\mathbf{H}_b = \begin{bmatrix} 0 & 0 & 0 \\ A & 0 & S \\ 0 & 0 & 0 \end{bmatrix}, \ \mathbf{H}_b^T = \begin{bmatrix} 0 & A & 0 \\ 0 & 0 & 0 \\ 0 & S & 0 \end{bmatrix}, \ \mathrm{tr}(\mathbf{H}_b) = \mathrm{tr}(\mathbf{H}_b^T) = 0.$$

To determine the parity of the nonlinear stress we will carefully substitute

$$\mathbf{H}_a\mathbf{H}_b = \mathbf{H}_a^T\mathbf{H}_b = \mathbf{H}_b^T\mathbf{H}_a = \mathbf{H}_b^T\mathbf{H}_a^T = 0$$

$$\mathbf{H}_b\mathbf{H}_a = \mathbf{H}_b\mathbf{H}_a^T = \begin{bmatrix} 0 & 0 & 0 \\ A & 0 & S \\ 0 & 0 & 0 \end{bmatrix}$$

$$\mathbf{H}_a\mathbf{H}_b^T = \mathbf{H}_a^T\mathbf{H}_b^T = \begin{bmatrix} 0 & A & 0 \\ 0 & 0 & 0 \\ 0 & S & 0 \end{bmatrix}$$

into equation (4.122)

$$\mathbf{S}^{NL}(\mathbf{H}_a, \mathbf{H}_b, 2) = S\begin{bmatrix} 0 & 0 & 0 \\ A & 0 & S \\ 0 & 0 & 0 \end{bmatrix} + \begin{bmatrix} 0 & 0 & 0 \\ A & 0 & S \\ 0 & 0 & 0 \end{bmatrix} + \begin{bmatrix} 0 & A & 0 \\ 0 & 0 & 0 \\ 0 & S & 0 \end{bmatrix} + S\begin{bmatrix} 0 & A & 0 \\ 0 & 0 & 0 \\ 0 & S & 0 \end{bmatrix} = \begin{bmatrix} 0 & A & 0 \\ A & 0 & S \\ 0 & S & 0 \end{bmatrix},$$

then take the divergence

$$\nabla_X \cdot \mathbf{S}^{NL} = S_{i1,1}^{NL} + S_{i3,3}^{NL} = \begin{Bmatrix} 0 \\ A + A \\ 0 \end{Bmatrix} = \begin{Bmatrix} 0 \\ A \\ 0 \end{Bmatrix}.$$

Now we can compute the parity of the nonlinear surface and volumetric forces

$$F_n^{\mathrm{surf}} = -\frac{1}{2}[V_1^{n*}S_{13}^{NL} + V_2^{n*}S_{23}^{NL} + V_3^{n*}S_{33}^{NL}]_{-h}^{h}$$

$$F_n^{\mathrm{vol}} = \frac{1}{2}\int_{-h}^{h}\left[V_1^{n*}S_{11,1}^{NL} + V_2^{n*}S_{21,1}^{NL} + V_3^{n*}S_{31,1}^{NL}\right]dX_3.$$

For SRL

$$F_n^{\mathrm{surf}} = -\frac{1}{2}[S \cdot 0 + 0 \cdot S + A \cdot 0]_{-h}^{h} = 0$$

$$F_n^{\mathrm{vol}} = \frac{1}{2}\int_{-h}^{h}[S \cdot 0 + 0 \cdot A + A \cdot 0]dX_3 = 0.$$

For ARL

$$F_n^{\text{surf}} = -\frac{1}{2}[A \cdot 0 + 0 \cdot S + S \cdot 0]_{-h}^{h} = 0$$

$$F_n^{\text{vol}} = \frac{1}{2} \int_{-h}^{h} [A \cdot 0 + 0 \cdot A + S \cdot 0] dX_3 = 0.$$

For SSH

$$F_n^{\text{surf}} = -\frac{1}{2}[0 \cdot 0 + S \cdot S + 0 \cdot 0]_{-h}^{h} = [S]_{-h}^{h} = 0$$

$$F_n^{\text{vol}} = \frac{1}{2} \int_{-h}^{h} [0 \cdot 0 + S \cdot A + 0 \cdot 0] dX_3 = \int_{-h}^{h} [A] dX_3 = 0.$$

For ASH

$$F_n^{\text{surf}} = -\frac{1}{2}[0 \cdot 0 + A \cdot S + 0 \cdot 0]_{-h}^{h} = [A]_{-h}^{h} \neq 0$$

$$F_n^{\text{vol}} = \frac{1}{2} \int_{-h}^{h} [0 \cdot 0 + A \cdot A + 0 \cdot 0] dX_3 = \int_{-h}^{h} [S] dX_3 \neq 0.$$

Thus, for mutual interaction of collinear SRL and ASH wavefields power only flows to ASH secondary wavefields.

7.2.2 Third order co-directional

As demonstrated in example 4.1, third order interactions require expanded nomenclature to unambiguously define the interaction. That is, the notation $\mathbf{S}^{NL}(\mathbf{H}_a, \mathbf{H}_b, 3)$ does not indicate how the waves a and b are interacting, thus we generalize to notation suitable for any order interaction, $\mathbf{S}^{NL}(\mathbf{H}_a, \mathbf{H}_b, N_a, N_b)$, where N_a and N_b denote the number of times waves a and b, respectively, participate in the interaction. $N_a + N_b$ must be equal to the order of the interaction, thus we can write the interaction between wavefields a and b to third order as

$$\begin{aligned}
\mathbf{S}^{NL}(\mathbf{H}_a &+ \mathbf{H}_b) \\
&= \mathbf{S}^Q(\mathbf{H}_a, \mathbf{H}_b, 2, 0) + \mathbf{S}^Q(\mathbf{H}_a, \mathbf{H}_b, 0, 2) + \mathbf{S}^Q(\mathbf{H}_a, \mathbf{H}_b, 1, 1) \\
&+ \mathbf{S}^C(\mathbf{H}_a, \mathbf{H}_b, 3, 0) + \mathbf{S}^C(\mathbf{H}_a, \mathbf{H}_b, 0, 3) \\
&+ \mathbf{S}^C(\mathbf{H}_a, \mathbf{H}_b, 2, 1) + \mathbf{S}^C(\mathbf{H}_a, \mathbf{H}_b, 1, 2),
\end{aligned} \tag{7.6}$$

where superscripts Q and C denote quadratic and cubic terms of the interaction. To fully describe the third order interaction between wavefields a and b, we must compute all seven terms in equation (7.6). An analogous parity analysis can be carried out for each term based on the type (RL or SH) and nature (symmetric or antisymmetric) of wavefields a and b. The first five terms have already been analysed, leaving just $\mathbf{S}^C(\mathbf{H}_a, \mathbf{H}_b, 2, 1)$ and $\mathbf{S}^C(\mathbf{H}_a, \mathbf{H}_b, 1, 2)$, which can be done as one. But it is lengthy and tedious. Here we simply provide the results from Liu *et al* [2] in table 7.5.

Table 7.5. For the interaction of different collinear wave mode types, is there power flow to third order secondary waves?

Mode a	Mode b	N_a	N_b	Secondary wave mode type			
				SRL	ARL	SSH	ASH
SRL	SRL	3	0	**Yes**	No	No	No
SRL	ARL	1	2				
SRL	SSH	1	2				
SRL	ASH	1	2				
ARL	SRL	1	2	No	**Yes**	No	No
ARL	ARL	3	0				
ARL	SSH	1	2				
ARL	ASH	1	2				
SSH	SRL	1	2	No	No	**Yes**	No
SSH	ARL	1	2				
SSH	SSH	3	0				
SSH	ASH	1	2				
ASH	SRL	1	2	No	No	No	**Yes**
ASH	ARL	1	2				
ASH	SSH	1	2				
ASH	ASH	3	0				

7.3 Effect of directionality

Thus far, we have considered mutual interaction between plane waves propagating in the same direction, we will call that *co-directional* wave mixing. The harmonic function $e^{i(kX_1-\omega t)}$ was used to represent waves in the X_1 direction. If the wavenumber k is real, then they are propagating waves. If k is imaginary, then the waves are evanescent and decay rapidly. Positive real values of k mean the waves are propagating in the positive X_1 direction, and vice versa. Another way that collinear waves can interact is by propagating in opposite directions; these are *counter-propagating* waves. The third general descriptor for mutual guided wave interactions in a plate is simply *non-collinear*, and it includes the mixing of waves at any arbitrary angle, except at 0 and 180 degrees, which have already been defined. Unlike non-collinear bulk waves, which are free to propagate in 3D space, non-collinear waves in a plate are constrained to the plane of the plate. Since we are now interested in analysing counter-propagating and non-collinear wave mixing we need to extend the modeling from 1D to 2D by using the wavevector in place of the scalar wavenumber.

The relevant variables for describing non-collinear wave mixing of planar wavefields a and b in a plate are shown in figure 7.1. To be clear, the wavefields represent straight-crested guided waves and can be written

$$\mathbf{u}_a(\mathbf{X}, t) = \frac{1}{2}[U_a(X_3)e^{i(\mathbf{K}_a \cdot \mathbf{p}(X_1, X_2) - \omega_a t)} + \text{c. c.}]$$

$$\mathbf{u}_b(\mathbf{X},\, t) = \frac{1}{2}[U_b(X_3)e^{i(\mathbf{K}_b \cdot \mathbf{p}(X_1,\, X_2) - \omega_b t)} + \text{c. c.}], \tag{7.7}$$

where $\mathbf{p}(X_1,\, X_2)$ is the position vector for a point in the plane of the plate and $\mathbf{K}_a$ and $\mathbf{K}_b$ are the wavevectors. Following Hasanian and Lissenden [3], it is convenient to write the wavevectors in magnitude-direction format, where the magnitude is the wavenumber:

$$\mathbf{K}_a = k_a \mathbf{r}_a$$

$$\mathbf{K}_b = k_b \mathbf{r}_b \tag{7.8}$$

and $\mathbf{r}$ is the unit vector in the propagation direction. In the wave mixing zone the total displacement field can be decomposed into the primary and secondary wavefields, as we have done before:

$$\mathbf{u} = \mathbf{u}_a + \mathbf{u}_b + \mathbf{u}_{aa} + \mathbf{u}_{bb} + \mathbf{u}_{ab} \tag{4.116}$$

to second order. Here, we will only analyse the mutual interaction term $\mathbf{u}_{ab}$ since the self-interaction terms have already been analysed. Taking the same approach as in section 6.7 we find that the secondary harmonic terms occur at the sum and difference frequencies,

$$e^{i[(\mathbf{K}_a \pm \mathbf{K}_b)\cdot \mathbf{p} - (\omega_a \pm \omega_b)t]} \quad \text{and} \quad e^{-i[(\mathbf{K}_a \pm \mathbf{K}_b)\cdot \mathbf{p} - (\omega_a \pm \omega_b)t]}$$

and that the direction of secondary wave propagation is given by the sum and difference wavevectors. Figure 7.1 illustrates the wavevectors and angles involved for the sum $\mathbf{K}_a + \mathbf{K}_b$. In the figure $\mathbf{K}_m$ represents the wavevector of the secondary wave mode m. The wave interaction angle is θ and the angle at which the secondary waves propagate is γ. To provide a simple reference for θ, we take $\mathbf{K}_a$ to be in the X_1 direction. Naturally, the vector analysis that we will use also applies to the special cases of counter-propagating ($\theta = 180°$) and co-directional ($\theta = 0°$) wave mixing.

As before, a perturbation method is used to solve the boundary value problem. As before, the solution relies upon the complex reciprocity theorem, orthogonality, and the normal mode expansion. Thus, we write the secondary linear stress and velocity fields as

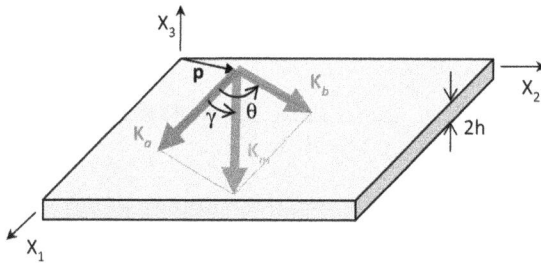

Figure 7.1. Non-collinear wave mixing in a plate.

$$\mathbf{S}^{L}(\mathbf{H}_{ab}) = \frac{1}{2}\left[\sum_{m} A_m(X_1, X_2)\mathbf{S}_m(X_3)e^{i(\omega_a \pm \omega_b)t} + \text{c. c.}\right]$$

$$\dot{\mathbf{u}}_{ab} = \frac{1}{2}\left[\sum_{m} A_m(X_1, X_2)\mathbf{V}_m(X_3)e^{i(\omega_a \pm \omega_b)t} + \text{c. c.}\right], \quad (7.9)$$

where $\mathbf{S}_m(X_3)$ and $\mathbf{V}_m(X_3)$ are stress and velocity fields for mode m and the expansion coefficients $A_m(X_1, X_2)$ are associated with the displacement field $\mathbf{U}_m(X_3)$. The modal displacement, displacement gradient, velocity, strain, and stress fields $\mathbf{U}_m$, $\mathbf{H}_m$, $\mathbf{V}_m$, $\mathbf{e}_m$, $\mathbf{S}_m$ are connected by the usual relations, including

$$\mathbf{e}_m = \frac{1}{2}(\mathbf{H}_m + \mathbf{H}_m^T)$$

$$\mathbf{S}_m = \lambda \text{tr}(\mathbf{e}_m)\mathbf{I} + 2\mu\mathbf{e}_m. \quad (7.10)$$

Auld's [4] complex reciprocity theorem (6.12) can be used to derive the partial differential equation

$$\mathbf{P}'_{mn} \cdot \mathbf{n}_1\left(\frac{\partial}{\partial X_1} - i\mathbf{K}_n^* \cdot \mathbf{n}_1\right)A_m(X_1, X_2) + \mathbf{P}'_{mn} \cdot \mathbf{n}_2\left(\frac{\partial}{\partial X_2} - i\mathbf{K}_n^* \cdot \mathbf{n}_2\right)A_m(X_1, X_2)$$
$$= (F_n^{\text{surf}} + F_n^{\text{vol}})e^{i(\mathbf{K}_a \pm \mathbf{K}_b)\cdot\mathbf{p}}, \quad (7.11)$$

whose counterpart, equation (6.68), was an ordinary differential equation for mixing collimated waves. The power flow density vector integrated through the thickness of the plate is

$$\mathbf{P}'_{mn} = -\frac{1}{4}\int_{-h}^{h}\left[\mathbf{S}_m\mathbf{V}_n^* + \mathbf{S}_n^*\mathbf{V}_m\right]dX_3$$

$$P_{mn} = \mathbf{P}'_{mn} \cdot \mathbf{r}_m, \quad (7.12)$$

where the indices m and n refer to modes m and n and not Einstein's index notation. The unit vector prescribing the direction of the wavevector for mode m is $\mathbf{r}_m$. Note that F_n^{surf} and F_n^{vol} are given in equations (6.43) and (6.44), respectively. For the special case of co-directional wave mixing, where $\theta = 0°$, equation (6.68) is recovered. For the general case, we choose mode n to be the mode not orthogonal to mode m such that $P_{mn} \neq 0$, therefore $m = n$. The two solutions to equation (7.12) are:

$$A_n(X_1, X_2) = \frac{-i}{\mid \mathbf{K}_n^* - (\mathbf{K}_a \pm \mathbf{K}_b)\mid}F_n[e^{i\mathbf{K}_n^*\cdot\mathbf{p}} - e^{i(\mathbf{K}_a \pm \mathbf{K}_b)\cdot\mathbf{p}}] \quad (7.13)$$

if $\mathbf{K}_n^* \neq \mathbf{K}_a \pm \mathbf{K}_b$, and

Table 7.6. For mutual interaction of non-collinear wave mode types, is there power flow to second order waves?

Interacting primary wave modes	Secondary wave mode type			
	SRL	ARL	SSH	ASH
SRL–SRL	Yes	No	Yes	No
ARL–ARL	Yes	No	Yes	No
SSH–SSH	Yes	No	Yes	No
ASH–ASH	Yes	No	Yes	No
SRL–ARL	No	Yes	No	Yes
SSH–ASH	No	Yes	No	Yes
SRL–SSH	Yes	No	Yes	No
ARL–ASH	Yes	No	Yes	No
SRL–ASH	No	Yes	No	Yes
ARL–SSH	No	Yes	No	Yes

$$A_n(X_1, X_2) = F_n \frac{(\mathbf{K}_a \pm \mathbf{K}_b) \cdot \mathbf{p}}{|\mathbf{K}_a \pm \mathbf{K}_b|} e^{i(\mathbf{K}_a \pm \mathbf{K}_b) \cdot \mathbf{p}} \tag{7.14}$$

if $\mathbf{K}_n^* = \mathbf{K}_a \pm \mathbf{K}_b$, where F_n is the same as in chapter 6, but can now been written more generally as

$$F_n = \frac{F_n^{\text{vol}} + F_n^{\text{surf}}}{P_{nn}} \tag{7.15}$$

$$\mathbf{P}'_{nn} = -\frac{1}{2} \operatorname{Re} \int_{-h}^{h} \mathbf{S}_n \mathbf{V}_n^* dX_3, \quad P_{nn} = \mathbf{P}'_{nn} \cdot \mathbf{r}_n. \tag{7.16}$$

Thus, for equation (7.14) there is cumulative growth in the direction of $\mathbf{p}$.

By repeating the parity analysis of wave mixing we find that the results shown in table 7.4 apply to both types of collinear waves: co-directional and counter-propagating. However, non-collinear mixing is less restrictive, as shown in table 7.6. Instead of just having one secondary wave classification (i.e. type and nature) for each type of wave mixing, there are two. For example, nonzero power flows from interacting waves of the same type and nature to both SRL and SSH waves.

7.4 Synchronism

7.4.1 Second order self-interaction

Synchronism is the second of the internal resonance criteria. The requirement for a cumulative second harmonic in equation (6.50), that $k^{n*}=2k$, can be visualized on the dispersion curves as shown in figure 7.2 for Lamb waves. Let's assume that we are interested in propagating modes, thus k^{n*} is real. We seek secondary waves having $(2k, 2\omega)$ where the primary waves have (k, ω). We showed earlier that power

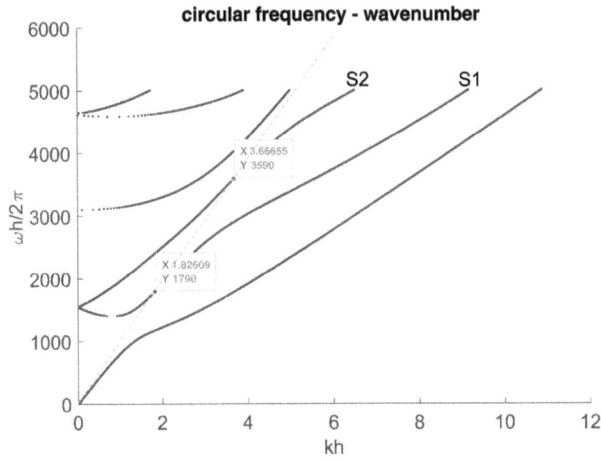

Figure 7.2. A synchronism point on the dispersion curves for symmetric Lamb waves.

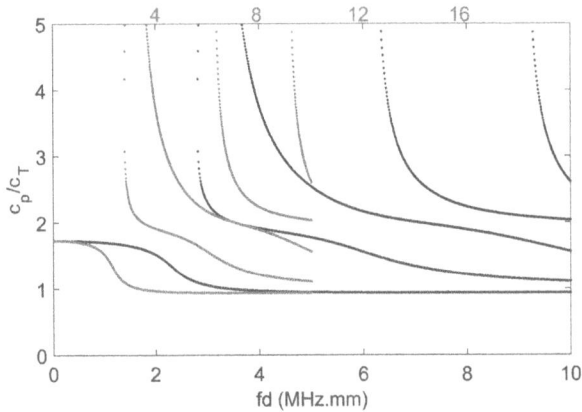

Figure 7.3. Phase velocity dispersion curves for symmetric Lamb waves at primary frequency (blue) and second harmonic frequency (red).

will flow only to symmetric Lamb waves for second harmonic generation, thus figure 7.2 shows only symmetric Lamb waves, although self-interaction of ARL, SSH, and ASH waves can also generate second harmonic SRL waves. Visual examination of synchronism is actually easier using phase velocity dispersion curves. As we know, phase velocity is inversely related to wavenumber through $c_p = \omega/k$. Figure 7.3 shows dispersion curves for symmetric Lamb waves plotted on two different fd axes, one for the primary waves (blue at bottom) and one for the secondary waves (red at top). Synchronism occurs at points where these primary and secondary dispersion curves intersect.

The terms synchronized and phase-matched are used interchangeably and imply the phase velocities are the same even though the actual criterion is based on the wavenumbers. It is also true that the wavelengths of the primary and secondary waves differ by a factor of two. Thus, the nodes of the primary waves correspond to nodes of the secondary waves.

7.4.1.1 Symmetric Lamb waves

A more rigorous mathematical approach entails analysing the Rayleigh–Lamb dispersion relations, equations (4.72) and (4.75), as shown by Müller *et al* [5] and Matsuda and Biwa [6] for Lamb waves and extended to SH waves by Liu *et al* [7]. We will see that synchronism occurs: (a) at the mode cutoffs where $k = 0$, (b) when the phase velocity is equal to the longitudinal wave speed $c_p = c_L$, (c) when the phase velocity is equal to the shear wave speed $c_p = c_T$, (d) at the Rayleigh wave speed $c_p = c_R$, and (e) at mode crossings. Since we know that cumulative second harmonics in a plate have to be symmetric Lamb waves, we will start with symmetric Lamb waves, whose dispersion relations are

$$\frac{\tan(qh)}{\tan(ph)} = \frac{-4k^2pq}{(q^2 - k^2)^2}. \tag{4.72}$$

We seek to determine from equation (4.72) at which frequencies ω (or fd products) do both (k, ω) and $(2k, 2\omega)$ lie on propagating modes. Recall that

$$p^2 \equiv \left(\frac{\omega}{c_L}\right)^2 - k^2, \quad q^2 \equiv \left(\frac{\omega}{c_T}\right)^2 - k^2. \tag{4.9}$$

Evaluating the right-hand side of equation (4.72) for $(2k, 2\omega)$ we find that

$$\frac{-4(2k)^2(2p)(2q)}{((2q)^2 - (2k)^2)^2} = \frac{-4k^2pq}{(q^2 - k^2)^2},$$

i.e. the right-hand side of equation (4.72) is the same for (k, ω) and $(2k, 2\omega)$. Thus,

$$\frac{\tan(qh)}{\tan(ph)} = \frac{\tan(2qh)}{\tan(2ph)} \tag{7.17}$$

provides the synchronism conditions for second harmonic generation of symmetric Lamb waves. Some manipulations lead to

$$2\sin(ph)\sin(qh)[\cos(2ph) - \cos(2qh)] = 0, \tag{7.18}$$

which has three distinct general solutions

$$qh = n\pi$$

$$ph = n\pi$$

$$qh \pm ph = n\pi, \tag{7.19}$$

where n is an arbitrary whole number. We can interpret these solutions by substituting them back into equation (4.72). Let's analyse equation (7.21a) first,

$$\frac{\tan(n\pi)}{\tan(ph)} = 0 = \frac{-4k^2pq}{(q^2-k^2)^2},$$

which has three possibilities.

Case 1. $k = 0$, occurs at mode cutoffs where $c_p \to \infty$. Note that these second harmonics do not propagate because $c_g \to 0$ at mode cutoffs (i.e. these are standing waves that resonate through the plate thickness).

Case 2. $p = 0$, gives $\omega = c_L k$ and therefore $c_p = c_L$. This is a nonleaky point analysed in detail by Rose [8] in his section 6.7.1. Each symmetric mode contains one of these points at

$$(fd)_n = \frac{nc_T}{\sqrt{1 - \left(\frac{c_T}{c_L}\right)^2}} \quad n = 1, 2, 3, \dots$$

and we note that $\omega h = \pi fd$.

Case 3. $q = 0$, requires a more complicated limit analysis of the dispersion relations and leads to $c_p = c_T$, which only occurs for the higher order modes at high frequency.

Moving on, we substitute equation (7.21b) into equation (4.72),

$$\frac{\tan(qh)}{\tan(n\pi)} = \frac{\tan(qh)}{0} = \frac{-4k^2pq}{(q^2-k^2)^2} \to q^2 - k^2 = 0 \to q^2 = k^2.$$

Using equation (4.9b) we obtain

$$c_p = \frac{\omega}{k} = \sqrt{2}\,c_T$$

under the stipulation that $\sqrt{2}\,c_T > c_L$ since p is a real number. The fd product for this case ($p = n\pi$) is

$$fd = \frac{\sqrt{2}\,nc_Lc_T}{\sqrt{2c_T^2 - c_L^2}}.$$

Finally, we recognize from equation (7.21c) that

$$qh - ph = n\pi \to \tan(qh) = \tan(ph)$$

and then substitute into equation (4.72),

$$1 = \frac{-4k^2pq}{(q^2-k^2)^2}.$$

We can also substitute into equation (4.75) for antisymmetric Lamb waves

$$1 = \frac{-(q^2-k^2)^2}{4k^2pq}.$$

These reciprocals are the same equation,

$$4k^2pq = -(q^2 - k^2)^2,$$

and we conclude that the conditions for both symmetric and antisymmetric Lamb waves are simultaneously satisfied, and thus these are mode crossing points.

The other solution from equation (7.21c) that

$$qh + ph = n\pi \rightarrow \tan(qh) = -\tan(ph)$$

leads to the same conclusion. Therefore, when primary waves exist at a mode crossing point on the symmetric and antisymmetric Lamb wave dispersion curves, there exist second harmonic waves at another mode crossing point.

The final possibility is only satisfied in the high frequency limit, where the fundamental symmetric and antisymmetric modes approach the Rayleigh wave speed. These waves are known as quasi-Rayleigh waves because as the wavelength decreases with frequency the plate appears to the waves more like a half-space. Only in the asymptotic limit is synchronism formally satisfied, but from a practical standpoint the de-tuning is minimal and both phase velocity matching and group velocity matching are very nearly satisfied in this nondispersive region. Quasi-Rayleigh wave synchronism is discussed in more detail by Müller *et al* [5] and Matsuda and Biwa [6].

7.4.1.2 Antisymmetric Lamb waves

The analysis above for symmetric Lamb waves can be applied to antisymmetric Lamb waves by simply changing the Rayleigh–Lamb dispersion relation from equation (4.72) (SRL) to equation (4.75) (ARL) for the primary waves. The secondary waves remain symmetric Lamb waves based on the power flow analysis. The dispersion relations for primary and secondary waves are

$$\frac{\tan(qh)}{\tan(ph)} = \frac{-(q^2 - k^2)^2}{4k^2pq} \tag{4.75}$$

$$\frac{\tan(2qh)}{\tan(2ph)} = \frac{-4k^2pq}{(q^2 - k^2)^2},$$

respectively. Combining them yields

$$\frac{\tan(qh)\tan(2qh)}{\tan(ph)\tan(2ph)} = 1, \tag{7.20}$$

or, after a little manipulation,

$$2\cos(qh)\cos(ph)[\cos(2ph) - \cos(2qh)] = 0. \tag{7.21}$$

As for the symmetric Lamb wave problem there are three distinct general solutions:

$$qh = \frac{\pi}{2}(2n + 1)$$

$$ph = \frac{\pi}{2}(2n + 1)$$

$$qh \pm ph = n\pi, \tag{7.22}$$

where n is an arbitrary whole number. Let's substitute equation (7.22a) into equation (4.75) first,

$$\frac{\tan\left(\frac{\pi}{2}(2n + 1)\right)}{\tan(ph)} = \frac{-(q^2 - k^2)^2}{4k^2pq} \rightarrow 4k^2pq = 0,$$

which has three possibilities:
    $k = 0$, occurs at mode cutoffs,
    $p = 0$, occurs where $c_p = c_L$,
    $q = 0$, occurs where $c_p = c_T$.

Substituting equation (7.22b) into equation (4.75)

$$\frac{\tan(qh)}{\tan\left(\frac{\pi}{2}(2n + 1)\right)} = \frac{-(q^2 - k^2)^2}{4k^2pq} \rightarrow -(q^2 - k^2)^2 = 0 \rightarrow q^2 = k^2,$$

using equation (4.9b) we obtain

$$c_p = \frac{\omega}{k} = \sqrt{2}\,c_T$$

under the stipulation that $\sqrt{2}\,c_T > c_L$ since p is a real number. The fd product for this case ($ph = \frac{\pi}{2}(2n + 1)$) is

$$fd = \frac{\sqrt{2}(2n + 1)c_L c_T}{2\sqrt{2c_T^2 - c_L^2}}.$$

The third possibility in equation (7.22c) is identical to the analysis for symmetric Lamb waves. Thus, we conclude that primary antisymmetric Lamb waves at a mode crossing point will generate second harmonic symmetric Lamb waves at mode crossing points.

7.4.1.3 SH waves
The dispersion relations for SH waves are equations (4.60) and (4.61),

$$q = \frac{n\pi}{2h},$$

where n is even for symmetric modes and odd for antisymmetric modes. The Rayleigh–Lamb dispersion relation for the second harmonic symmetric Lamb waves is

$$\frac{\tan(2qh)}{\tan(2ph)} = \frac{-4k^2pq}{(q^2 - k^2)^2}.$$

But substituting equation (4.60) gives

$$\frac{\tan{(n\pi)}}{\tan{(2ph)}} = 0 = \frac{-4k^2pq}{(q^2 - k^2)^2}.$$

Thus, $k = 0$ at the mode cutoffs is a solution, as is $p = 0$, where $c_p = c_L$, and $q = 0$, where $c_p = c_T$. The $q = 0$ case must be analysed in the limit and the $c_p = c_T$ solutions only occur at high frequencies. Another possible solution arises for $\tan{(2ph)} = 0$, i.e. $2ph = m\pi$. In this case the second harmonics occur at mode crossing points because both $\tan{(2ph)}$ and $\tan{(2qh)}$ are zero.

In summary, synchronism occurs for second order self-interaction in the following instances for RL and SH waves:

- cutoff frequencies of modes,
- phase velocity equal to c_L,
- phase velocity equal to $\sqrt{2}\,c_T$,
- phase velocity equal to c_T at high frequency, and
- crossing points between symmetric and antisymmetric modes,

when the secondary waves are always symmetric Lamb waves. In addition, it is essentially satisfied by quasi-Rayleigh waves.

7.4.2 Third order self-interaction

For brevity and to avoid monotony we will only consider third order self-interaction of SH waves. Interested readers can read the comparable analysis for Lamb waves in Liu *et al* [2]. The parity analysis of power flow showed that a primary wave generates a third harmonic of the same type (SRL, ARL, SSH, or ASH). Consider the SH dispersion relations for primary waves and third harmonic waves,

$$q = \frac{n\pi}{2h}$$

$$3q = \frac{m\pi}{2h},$$

where n and m are arbitrary whole numbers. It is necessary to have $m = 3n$ for synchronism to occur. If $n = 0$ then $m = 0$ and SH0 generates an SH0 third harmonic. If $n = 1$ then $m = 3$ and SH1 generates an SH3 third harmonic, and so on. Any frequency of any mode will generate a synchronized third harmonic. This has been called holo-internal resonance because the entire set of SH waves (symmetric and antisymmetric) exhibit internally resonant SH third harmonics at any driving frequency.

7.5 Group velocity matching

As evident from equation (6.50), group velocity matching between primary and secondary waves is not part of the internal resonance criteria—and yet it is. It turns out that group velocity matching is implicitly linked to the internal resonance criteria with one notable exception—continuous waves. The analyses in chapter 6 and up to this point in chapter 7 presume that the primary waves are continuous waves at the frequencies ω_a and ω_b (or just ω_a for the case of self-interaction). However, it is common in nondestructive evaluation to use discrete wave packets, often generated by a toneburst source. Both self-interactions and mutual interactions rely on the primary and secondary wavefields physically overlapping in what we can call a *mixing zone* for them to interact. Interaction in the mixing zone acts like a distributed source for the secondary waves; but if the group velocities are different, then the waves will eventually cease to interact as they travel forward. The cumulative nature of internally resonant secondary waves is quite affected by the mixing zone, which is determined by the wave directions, the duration of the primary wave packets, and the group velocities.

Since the mixing zone is where wave interactions occur, whether they are self-interactions or mutual interactions, the size of the mixing zone is important. Given an application to nondestructively evaluate the material state in a structural component that also acts as a waveguide, the waves interact with themselves (or each other) and the material. The material degradation of interest could be localized or uniformly distributed, and this can affect the nondestructive evaluation strategy. If degradation is localized it may be advantageous to employ a small mixing zone to accurately locate it. On the other hand, if the degradation is uniform, then sampling the material over a larger mixing zone can leverage the cumulative effect associated with internal resonance. In the remainder of this section, we discuss some advantages and disadvantages of the three types of wave mixing presuming that finite-length wave packets are used.

7.5.1 Co-directional wave mixing

Suppose we desire a large mixing zone. Then co-directional wave mixing with group velocity matching is an excellent option. Primary wave modes and frequencies having internal resonance and group velocity matching will be identified in chapter 8. Suppose on the other hand that we desire a small mixing zone. Then we could mix co-directional primary waves that have very different group velocities. Time delays could be employed to move the wave mixing zone to different positions along the wave propagation path. In addition, the duration of one of the wave packages can be minimized. In both cases, a through-transmission measurement system is typically appropriate because usually the secondary waves propagate in the same direction as the primary waves. A simplified model proposed by Hasanian and Lissenden [3] describes the effect that the mixing zone has on the amplitude of the secondary waves without having to do full-scale finite element modeling.

7.5.2 Counter-propagating wave mixing

Since the wave mixing zone of counter-propagating waves is explicitly limited by the durations of the wave packets, this mixing mode is better suited for creating small wave mixing zones. Obviously, sources at two different positions are required and the wavevectors dictate in which collinear direction the secondary waves propagate, unless the secondary waves are standing waves that do not propagate.

7.5.3 Non-collinear wave mixing

The size of the mixing zone for non-collinear waves is also limited by the durations of the wave packets. The unique advantage of non-collinear mixing is that the secondary waves can propagate in a direction distinct from the primary waves and the measurement system nonlinearities that they might carry with them. The price of this advantage is that the transducer layout is more complicated and potentially larger.

7.6 Comments on hollow cylinders

Although this chapter focusses on plates, there are some aspects of internal resonance in hollow cylinders that warrant discussion, but we will keep it brief. Wave mixing is limited to the co-directional and counter-propagating types. Wave modes in hollow cylinders are not defined as symmetric or antisymmetric, thus the parity analysis for plates does not translate to hollow cylinders. However, as shown by Liu *et al* [9] for axisymmetric longitudinal $L(0,n)$ and torsional $T(0,n)$ wave modes, the structure of the nonlinear stress components, which dictates the structure of the nonlinear driving forces (both on the surface and through the volume) enables us to make similar conclusions about to which types of secondary waves power does and does not flow. The results are that the self-interaction of both $L(0,n)$ and $T(0,n)$ waves results in nonzero power flow to secondary $L(0,n)$ waves, but no power flows to secondary $T(0,n)$ waves. These results are confirmed by a completely different type of analysis [10]. The same results hold for mutual interaction of the same wave type; $L(0,n)$–$L(0,n)$ and $T(0,n)$–$T(0,n)$. However, for mutual interaction of different wave types, $L(0,n)$–$T(0,n)$, nonzero power flows to both secondary $L(0,n)$ and $T(0,n)$ waves.

The dispersion relations for hollow cylinders are quite complicated, too complicated for mathematical analysis of synchronism. However, we can lean on overlaid dispersion curves as in figures 7.2 and 7.3 and the knowledge that the dispersion curves for axisymmetric waves in hollow cylinders are quite similar to Lamb waves and SH waves in plates. In chapter 8 we will investigate selecting primary waves in hollow cylinders that generate internally resonant secondary waves after our analysis of plates.

7.7 Closure

The internal resonance criteria for plates were analysed in detail. For self-interactions only symmetric Lamb waves receive power flow at the second harmonic frequency. Synchronism occurs for second harmonics at five different spots on the dispersion curves. Third harmonics were also analysed and the results tabulated. Mutual interactions were defined as co-directional, counter-propagating, and non-collinear and secondary wave modes to which no power flows were identified. The issue of group velocity matching was explained in terms of wave interactions within a mixing zone. With this knowledge we are ready to select wave modes and frequencies for detecting material nonlinearity.

References

[1] Liu Y 2014 Characterization of global and localized material degradation in plates and cylinders via nonlinear interaction of ultrasonic guided waves *PhD Thesis* The Pennsylvania State University, University Park, PA https://etda.libraries.psu.edu/catalog/22432

[2] Liu Y, Chillara V K, Lissenden C J and Rose J L 2013 Third harmonic shear horizontal and Rayleigh Lamb waves in weakly nonlinear plates *J. Appl. Phys.* **114** 114908

[3] Hasanian M and Lissenden C J 2018 Second order ultrasonic guided wave mutual interactions in plate: arbitrary angles, internal resonance, and finite interaction region *J. Appl. Phys.* **124** 164904

[4] Auld B A 1973 *Acoustic Fields and Waves in Solids* (Hoboken, NJ: Wiley)

[5] Müller M F, Kim J-Y, Qu J and Jacobs L J 2010 Characteristics of second harmonic generation of Lamb waves in nonlinear elastic plates *J. Acoust. Soc. Am.* **127** 2141–52

[6] Matsuda N and Biwa S 2011 Phase and group velocity matching for cumulative harmonic generation in Lamb waves *J. Appl. Phys.* **109** 094903

[7] Liu Y, Chillara V K and Lissenden C J 2013 On selection of primary modes for generation of strong internally resonant second harmonics in plate *J. Sound Vib.* **332** 4517–28

[8] Rose J L 2014 *Ultrasonic Guided Waves in Solid Media* (Cambridge: Cambridge University Press)

[9] Liu Y, Khajeh E, Lissenden C J and Rose J L 2013 Interaction of torsional and longitudinal guided waves in weakly nonlinear circular cylinders *J. Acoust. Soc. Am.* **133** 2541–53

[10] Chillara V K and Lissenden C J 2013 Analysis of second harmonic guided waves in pipes using a large-radius asymptotic approximation for axis-symmetric longitudinal modes *Ultrasonics* **53** 862–9

IOP Publishing

Nonlinear Ultrasonic Guided Waves

Cliff J Lissenden

Chapter 8

Selecting primary waves

History provides perspective. Between 1996 and 1999 Mingxi Deng published a sequence of articles on second harmonic generation of SH and Lamb waves in plates [1–3]. However, the partial waves method used makes the analysis cumbersome. In 2003 De Lima and Hamilton [4] applied the perturbation solution based on the normal mode expansion that enhanced understanding of the mechanics. These pioneering efforts got researchers interested in which wave modes and frequencies satisfy the internal resonance conditions. The next question was which primary waves transfer the most power to the secondary waves. This chapter addresses the internal resonance and the power flow questions. This is not the end of the story however, but rather the beginning, because after internal resonance we need to consider issues associated with making actual measurements with transducers and instruments.

After formulating the boundary value problem in chapter 4, solving the linear problem in chapter 5, applying a perturbation method to solve the nonlinear problem in chapter 6, and then analysing the internal resonance conditions in chapter 7, we are finally ready to select wave modes and frequencies for making nonlinear ultrasonic guided wave measurements. We will focus on satisfying the internal resonance conditions such that the secondary waves are cumulative. We start with plates and break the discussion into self-interactions and mutual interactions. Then we progress to hollow cylinders followed by arbitrary cross-sections. The chapter concludes by acknowledging that satisfying the internal resonance conditions is not an absolute requirement and looks at when non-cumulative secondary waves may be advantageous.

We seek a quantitative comparison of the power flow from the primary waves to the secondary waves at the various internal resonance points on the dispersion curves. The power flow analysis stems from the eigen-functions. The eigenvectors are the wavestructures (i.e. displacement profiles) and should have units of length regardless of how they are scaled. We will scale them to force $P_{mn} = \delta_{mn}$ to make them orthonormal. From the displacement fields we get the velocity fields and stress fields. P_{nn} is the average power flow through the thickness of the plate per unit width given by equation (6.24):

$$P_{nn} = -\text{Re}\left(\frac{1}{2}\int_{-h}^{h} V_i^{[n]*}\Sigma_{i1}^{[n]}dX_3\right), \tag{6.24}$$

where the velocity $V_i^{[n]}$ and stress $\Sigma_{i1}^{[n]}$ distributions for mode n are computed from the displacement wavestructure $U_i^{[n]}$ using equations (5.8) and (5.9). Once the wavestructure is normalized with respect to the average power flow then $P_{nn} = 1$. Likewise, the nonlinearity—i.e. the nonlinear surface force and nonlinear body force given by equations (6.43) and (6.44), respectively, (repeated here for convenience),

$$F_n^{\text{surf}} \equiv -\frac{1}{2}[V_i^{n*}(X_3)S_{i3}^{NL}(X_3)]_{-h}^{h} \tag{6.43}$$

$$F_n^{\text{vol}} \equiv \frac{1}{2}\int_{-h}^{h} V_i^{n*}(X_3)S_{ij,j}^{NL}(X_3)dX_3 \tag{6.44}$$

are force rates for a unit width of plate. S_{ij}^{NL} is the nonlinear part of the first Piola–Kirchhoff stress computed from the primary wavefield as per the perturbation method (equation (4.121)). F_n^{surf} and F_n^{vol} are combined into the nonlinear driving force F_n in equation (6.47),

$$F_n \equiv \frac{F_n^{\text{vol}} + F_n^{\text{surf}}}{P_{nn}}. \tag{6.47}$$

Finally, for the sake of comparisons, the mixing power (sometimes referred to as the rate of accumulation [5]) has been defined in equation (6.58):

$$M_p \equiv F_n\frac{\max\left(\mathbf{U}^n(X_3)\right)}{\max\left(\mathbf{U}_a(X_3)\right)\max\left(\mathbf{U}_b(X_3)\right)}, \tag{6.58}$$

where $a = b$ applies for self-interaction.

Results are provided for an aluminum plate represented by the following properties (here and elsewhere in this book),

$$\rho = 2700 \text{ kg m}^{-3}$$

$$E = 69 \text{ GPa}, \nu = 0.33$$

$$\mathscr{A} = -350 \text{ GPa}, \mathscr{B} = -155 \text{ GPa}, \mathscr{C} = -95 \text{ GPa}$$

from which other properties are computed,

$$\lambda = 50.35 \text{ GPa}, \mu = 25.94 \text{ GPa}$$

$$c_L = 6153 \text{ ms}^{-1}, c_T = 3100 \text{ ms}^{-1},$$

although we will study the material nonlinearity by parametrically changing the third order elastic constants.

8.1 Self-interaction in plates

We now dig into the problem of selecting primary modes for cumulative second harmonic generation in plates considering Lamb and SH waves.

8.1.1 Seond harmonic generation

In chapter 7, when studying the internal resonance condition, we found that power only flows to second harmonic waves that are symmetric Lamb modes. In addition, we showed that phase matching occurs for SRL waves when:

- $k = 0$ (mode cutoffs),
- $p = 0$ (where $c_p = c_L$),
- $q = 0$ (where $c_p \to c_T$ for higher order modes),
- $qh + ph = n\pi$ at mode crossing points,
- $p = n\pi$ ($c_p = \sqrt{2}\, c_T$, the Lamé modes),
- where $c_p \to c_R$ for quasi-Rayleigh waves.

Phase matching for ARL waves is similar (see section 7.4.1.2). Likewise, phase matching for SH waves was detailed in section 7.4.1.3.

Let's call points on the dispersion curves that satisfy both internal resonance criteria internal resonance (IR) points, but keep it in the back of our minds that due to the source influence [6] transducers do not actually activate a single point, but rather a zone. Thus, later we will discuss detuning, where the phase matching condition is not identically satisfied. But for now, we think of activating IR points. The phase velocity dispersion curves are shown in figure 8.1 with eight IR points marked. The blue-green dispersion curves are for the primary SRL, ARL, SSH, and ASH wave modes. The red dispersion curves are plotted at the second harmonic frequency along the top axis. Phase matching occurs when the primary and secondary dispersion curves intersect. There are eight IR points identified below the fd-product of 8 MHz-mm.

The equations for the coordinates (fd-products and phase velocities) of these points for isotropic homogeneous traction-free plates are given in table 8.1. The equations for IR point 1 and IR point 2 can be found in many sources, and Müller *et al* [7] and Matsuda and Biwa [8] provide the equations for IR point 3. The driving fd products for IR points 5–7 are obtained by substituting $c_p = c_L$ into equation (4.64), where n is the SH mode number. The primary waves for IR points 1–3 are Lamb modes and the primary waves for IR points 4–8 are SH modes. The second harmonic waves are always symmetric Lamb modes to satisfy the nonzero power flow criterion. Except for IR point 4, these IR points repeat with increasing fd-product. IR point 1 and IR point 6 have the same coordinates because the S1 and SH2 modes cross at this point. IR point 3 and IR point 8 also have the same coordinates because the S2, A2, and SH3 modes all cross at this point. The other IR points (at the mode cutoffs and at high fd-products) are thought to be difficult to use in practice and hence are not marked.

Considering a 1 mm thick aluminum plate, the details for IR points 1–8 are provided in tables 8.2 and 8.3. Table 8.2 provides all the intermediate variables used

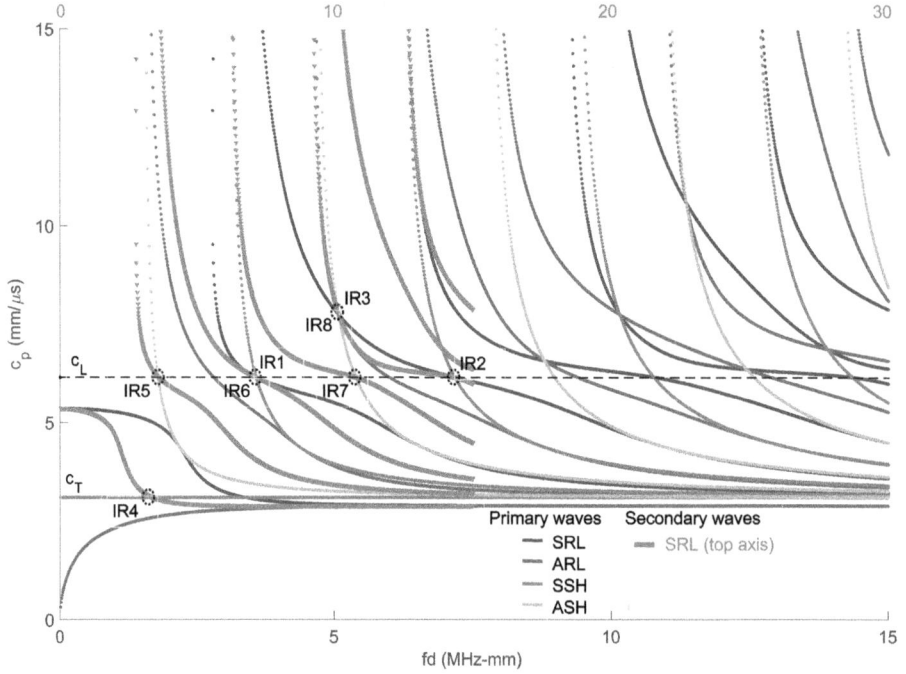

Figure 8.1. Dispersion curves for RL and SH waves in plates with SRL secondary waves plotted in red along the top axis such that intersection points between primary and secondary waves are internal resonance points. Eight internal resonance (IR) points are labeled.

Table 8.1. Internal resonance points 1–8, general formulae for the driving frequency and phase velocity, where $\xi = c_T/c_L$, $\eta = 1/3$.

IR point	Description	Driving fd product	Phase velocity c_p
1	Nonleaky SRL modes, S1–S2	$\dfrac{c_T}{\sqrt{1-\xi^2}}$	c_L
2	Nonleaky SRL modes, S2–S4	$\dfrac{2c_T}{\sqrt{1-\xi^2}}$	c_L
3	SRL/ARL mode crossing point, S2–S4 and A2–S4	$\dfrac{3c_T}{2}\sqrt{\dfrac{1-\eta^2}{1-\xi^2}}$	$c_T\sqrt{\dfrac{1-\eta^2}{\xi^2-\eta^2}}$
4	SH0–S0 crossing point	3.3913 MHz-mm	c_T
5	SH1–S1 mode crossing point	$\dfrac{1c_Tc_L}{2\sqrt{c_L^2-c_T^2}}$	c_L
6	SH2–S2 mode crossing point	$\dfrac{2c_Tc_L}{2\sqrt{c_L^2-c_T^2}}$	c_L
7	SH3–S3 mode crossing point	$\dfrac{3c_Tc_L}{2\sqrt{c_L^2-c_T^2}}$	c_L
8	SH3–S4 mode crossing point	$\dfrac{3c_T}{2}\sqrt{\dfrac{1-\eta^2}{1-\xi^2}}$	$c_T\sqrt{\dfrac{1-\eta^2}{\xi^2-\eta^2}}$

Table 8.2. Internal resonance points 1–8, specific details for 1 mm thick aluminum plate.

IR Point	Primary/secondary modes	Frequency (MHz)	Phase speed (m s⁻¹)	Wavenumber (rad m⁻¹)	Coef (m²)	Norm (√W)	Umax (nm)	Group Speed (m s⁻¹)
1	S1	3.588	6153	3664	−1.767i	1.43E13	0.8904	4376
	S2	7.176	6153	7327	+1.767i	5.73E13	0.4452	4376
2	S2	7.176	6153	7327	+1.767i	5.73E13	0.4452	4376
	S4	14.352	6153	14 654	+1.767i	2.29E14	0.2226	4376
3	S2	5.074	7738	4120	−1.989i	2.11E13	0.6532	3640
	S4	10.148	7738	8240	+1.081i	5.59E13	0.4964	2056
3	A2	5.074	7738	4120	+1.081i	1.40E13	0.9927	2094
	S4	10.148	7738	8240	+1.081i	5.59E13	0.4964	2056
4	SH0	1.695	3100	3436	1	6.89E8	1.452	3100
	S0	3.390	3100	6872	−0.003135i	2.38E11	0.8847	2389
5	SH1	1.794	6153	1831	1	3.66E8	2.733	1561
	S1	3.588	6153	3664	−1.767i	1.43E13	0.890	4378
6	SH2	3.588	6153	3664	1	7.33E8	1.366	1561
	S2	7.176	6153	7327	+1.767i	5.73E13	0.4452	4375
7	SH3	5.382	6153	5496	1	1.10E9	1.822	1561
	S3	10.76	6153	10 990	−1.767i	1.29E14	0.5936	4381
8	SH3	5.074	7738	4120	1	9.23E8	1.084	1242
	S4	10.15	7738	8240	+1.081i	5.59E13	0.496	2056

Table 8.3. Internal resonance points 1–8, nonlinear forces and mixing power.

IR point	F_n^{surf} (N s^{-1})	F_n^{vol} (N s^{-1})	F_n (m^{-1})	M_p (m^{-2})
L-waves, 3.588 MHz	0	0.0967	0.0967	99.34×10^6
L-waves, 7.176 MHz	0	0.1934	0.1934	397.4×10^6
1, S1–S2	0	0.0996	0.0996	55.96×10^6
2, S2–S4	0	0.1993	0.1993	223.8×10^6
3, S2–S4	0	0.0583	0.0583	67.81×10^6
3, A2–S4	0	0.0808	0.0808	40.69×10^6
4, SH0–S0	0.0211	0.0615	0.0825	34.66×10^6
5, SH1–S1	0	0.0094	0.0094	1.115×10^6
6, SH2–S2	0	0.0187	0.0187	4.461×10^6
7, SH3–S3	0	0.0281	0.0281	10.04×10^6
8, SH3–S4	0	0	0	0

to compute the variables in table 8.3 that we are really interested in comparing and studying further. In table 8.2 there are two rows for each IR point, the variables for the primary waves are given in the first row and the variables for the secondary waves are given in the second row. Readers interested in recreating these tables need to know that:

- The wavenumbers come from the eigenvalues of the dispersion relations and the phase speeds are computed from $2\pi f/k$.
- *Coef* is the coefficient for the displacement wavestructure prior to normalization, i.e. from equations (4.43a) and (4.46a), using A/D_2 for SRL waves and B/C_2 for ARL waves.
- *Norm* is the normalization factor $\sqrt{P_{mm}}$ used for the displacement wavestructure, i.e. it is the square root of the average power flow computed based on *Coef* for displacement, velocity, and stress computed from equations (5.9).
- *Umax* is the maximum value of the displacement wavestructure after it has been normalized with respect to the average power flow, and this is $\max(\mathbf{U}^n(X_3))$ used to compute the mixing power (this wavestructure parameter should not be confused with the wave amplitude).
- The group speed is computed as described in section 5.1.3.

Table 8.2 provides the input for computing the variables in table 8.3. We used Matlab (Mathworks.com) to develop an algorithm to compute the mixing power at the IR points. In fact, there are two algorithms, one for self-interactions that is then generalized for mutual interactions. We start by solving for the wavenumber at the frequency of interest, which provides the phase speed, group speed, and the displacement wavestructure, from which the particle velocity and stress are computed to enable use of equation (6.24) for the average power flow, which is in turn used to normalize the wavestructure for the primary mode. This procedure is repeated for the secondary mode. We confirm that the normalized $P_{mm} = 1$ and store the *Umax* values. The nonlinear driving force can now be determined by using the displacement gradient $\mathbf{H}_a$ from the primary waves to compute the nonlinear part of the first Piola–Kirchhoff stress $\mathbf{S}^{NL}(\mathbf{H}_a, \mathbf{H}_a, 2)$ and its divergence $\nabla_X \cdot \mathbf{S}^{NL}(\mathbf{H}_a, \mathbf{H}_a, 2)$. Symbolic variables are used because displacements and $\mathbf{H}_a$ depend on X_3. Equations (6.43), (6.44), (6.47), and (6.58) are used to compute F_n^{surf}, F_n^{vol}, F_n, and M_p in table 8.3. Note that due to the ambiguity in the wavestructure, even after normalization, that the sign of the nonlinear surface traction and/or nonlinear body force may be negative. This occurs because the wavestructure can always be multiplied by −1, which is the solution to obtaining a positive force.

Before discussing the results in table 8.3, we want to use the nonlinear bulk longitudinal waves that were analysed in section 3.1 as a reference. Specifically, we want to confirm that

$$A_2 = \max(\mathbf{u}_{aa}) = F_n X_1 \max(\mathbf{U}^n(X_3)) \tag{6.55}$$

reduces to

$$A_2 = \frac{1}{8}\beta k^2 A_1^2 X_1 \tag{3.12}$$

for 1D longitudinal waves. Consider the primary displacement field

$$u_a = \frac{1}{2}A_1 e^{i(k_a X_1 - \omega_a t)} + \text{c. c.} \tag{8.1}$$

Since there are no boundaries $F_n^{\text{surf}} = 0$. The nonlinear body force is

$$F_n^{\text{vol}} = \frac{1}{2}\int_{-h}^{h} V_1^{n*} S_{11,1}^{NL} dX_3 = \frac{2h}{2}(-i\omega_n U_n)^* S_{11,1}^{NL}.$$

The nonlinear stress comes from the displacement gradient:

$$\mathbf{u}_a = \left\{\begin{matrix} u_a \\ 0 \\ 0 \end{matrix}\right\} \rightarrow \mathbf{H}_a = \begin{bmatrix} u_a' & 0 & 0 \\ 0 & 0 & 0 \\ 0 & 0 & 0 \end{bmatrix},$$

which has only one nonzero component. From (4.121) we obtain

$$\mathbf{S}^{NL} = \lambda u_a' \mathbf{H}_a + \frac{\lambda}{2}(u_a')^2 \mathbf{I} + 3\mu\mathbf{H}_a\mathbf{H}_a + \mathscr{A}\mathbf{H}_a\mathbf{H}_a + \mathscr{B}(u_a')^2\mathbf{I} + \mathscr{B}u_a'2\mathbf{H}_a + \mathscr{C}(u_a')^2\mathbf{I},$$

$$S_{11}^{NL} = \lambda u_a' u_a' + \frac{\lambda}{2}(u_a')^2 + 3\mu u_a' u_a' + \mathscr{A}u_a' u_a' + \mathscr{B}(u_a')^2 + 2\mathscr{B}u_a' u_a' + \mathscr{C}(u_a')^2$$

$$S_{11,1}^{NL} = (2\lambda + \lambda + 6\mu + 2\mathscr{A} + 2\mathscr{B} + 4\mathscr{B} + 2\mathscr{C})u_a' u_a''$$

$$S_{11,1}^{NL} = \left[3(\lambda + 2\mu) + 2\mathscr{A} + 6\mathscr{B} + 2\mathscr{C}\right]u_a' u_a'', \tag{8.2}$$

where

$$u_a' = ik_a u_a = \frac{1}{2}ik_a A_1$$

$$u_a'' = -k_a^2 u_a = -\frac{1}{2}k_a^2 A_1.$$

The average power flow is

$$P_{nn} = -\text{Re}\left[\frac{1}{2}\int_{-h}^{h} V_1^{n*}\Sigma_{11}^n(1)dX_3\right] = -\text{Re}\left[\frac{2h}{2}(-i\omega_n U_n)^*(\lambda + 2\mu)ik_n U_n\right]$$

$$P_{nn} = \frac{2h}{2}(\lambda + 2\mu)\omega_n k_n U_n^2.$$

(8.3)

Substituting equations (8.1) and (8.2) into equation (6.47) we obtain

$$F_n = \frac{\frac{2h}{2}(-i\omega_n U_n)^*[3(\lambda + 2\mu) + 2\mathscr{A} + 6\mathscr{B} + 2\mathscr{C}]u_a'u_a'' + 0}{\frac{2h}{2}(\lambda + 2\mu)\omega_n k_n U_n^2}$$

$$F_n = \frac{\frac{2h}{2}(i\omega_n U_n)[3(\lambda + 2\mu) + 2\mathscr{A} + 6\mathscr{B} + 2\mathscr{C}]\left(\frac{1}{2}ik_a A_1\right)\left(-\frac{1}{2}k_a^2 A_1\right) + 0}{\frac{2h}{2}(\lambda + 2\mu)\omega_n k_n U_n^2}.$$

Recall from example 3.4 that

$$-\beta = \frac{3(\lambda + 2\mu) + 2\mathscr{A} + 6\mathscr{B} + 2\mathscr{C}}{(\lambda + 2\mu)}$$

in the absence of residual stress, and therefore

$$F_n = \frac{(i\omega_n U_n)[-\beta]\left(\frac{1}{2}ik_a A_1\right)\left(-\frac{1}{2}k_a^2 A_1\right)}{\omega_n k_n U_n^2}.$$

We use $k_n = 2k_a$ to help simplify further:

$$F_n = \frac{-\beta\left(\frac{1}{4}k_a^3 A_1^2\right)}{2k_a U_n} = \frac{-\beta k_a^2 A_1^2}{8U_n}.$$

(8.4)

Returning to equation (6.55) we now have

$$A_2 = \frac{-\beta k_a^2 A_1^2}{8U_n}\max(\mathbf{U}^n(X_3))X_1 = \frac{-\beta k_a^2 A_1^2}{8}X_1.$$

(8.5)

The negative sign is a consequence of the definition of β in equation (3.9) and there is no problem taking the absolute value of the $-\beta$ term. Thus, we have agreement with equation (3.12).

Finally, the mixing power is

$$M_p = F_n\frac{\max(\mathbf{U}^n(X_3))}{\max(\mathbf{U}_a(X_3))^2} = \frac{1}{8}|-\beta|k_a^2\frac{u_a^2}{u_n}\frac{\max(\mathbf{U}^n(X_3))}{\max(\mathbf{U}_a(X_3))^2} = \frac{1}{8}|-\beta|k_a^2$$

(8.6)

showing that the mixing power M_p reduces to the longitudinal wave result in section 3.1 (see equation (3.12)) as one would expect. Another way to contextualize the mixing power starts by analysing equation (6.54) at a finite propagation distance such that the modes in the expansion other than $m = n$ can be ignored. Considering only the maximum value of the real part of the secondary wavefield

$$\max(\mathrm{Re}\,(\mathbf{u}_{aa})) = F_n \max(\mathbf{U}^n(X_3))X_1,$$

which enables us to now write the relative nonlinearity as

$$\frac{\max(\mathbf{U}^n(X_3))}{\max(\mathbf{U}_a(X_3))^2} = \frac{F_n\max(\mathbf{U}^n(X_3))}{\max(\mathbf{U}_a(X_3))^2}X_1 = M_p X_1. \tag{8.7}$$

The mixing powers for bulk longitudinal waves at the driving frequencies of IR point 1 and IR point 2 are given in table 8.3. As for IR points 1 and 2, the difference in M_p values is 4 since the wavenumbers differ by a factor of 2. Notice that although the F_n values for longitudinal waves and IR points 1 and 2 correspond well, the M_p values are significantly higher for longitudinal waves than for IR points 1 and 2 (in fact they are 1.775 times higher).

We wonder why all but one F_n^{surf} value in table 8.3 is zero. Let's find out by investigating the defining equation,

$$F_n^{\mathrm{surf}} \equiv -\frac{1}{2}[V_i^{n*}(X_3)S_{i3}^{NL}(X_3)]_{-h}^{h}. \tag{6.43}$$

In all cases the secondary wavefields are SRL modes, thus $U_2^{n*}=V_2^{n*}=0$ and we need only examine U_1^{n*} and U_3^{n*}. The wavestructures for the SRL modes involved with IR points 1–3 are shown in figure 8.2. In all three cases $U_3^{n*}(X_3 = \pm h) = 0$, leaving only $U_1^{n*}(X_3 = \pm h) \neq 0$. Thus, we must assess the $S_{13}^{NL}(X_3 = \pm h)$ term. It will be shown later that $\mathbf{S}^{NL}(X_3 = \pm h)$ is diagonal for second order interactions, and hence $S_{13}^{NL} = 0$, which makes $F_n^{\mathrm{surf}} = 0$. Investigation of IR points 5–8 leads to an identical result because the secondary waves are SRL modes at either $c_p = c_L$ or a mode crossing point. In both cases $U_3^{n*}(X_3 = \pm h) = 0$.

The mixing power values in table 8.3 differ from previous publications [9, 10] for three reasons: (i) the TOECs are slightly different, (ii) the wavestructures have been normalized with respect to the average power flow making the modes orthonormal, and (iii) the definition of P_{mn} in equation (6.21) is different, in that the usual multiplier '4' for P_{nn} does not appear in equation (6.47).

The wavestructures for IR points 4–8 that are not shown in figure 8.2 are provided in figure 8.3. Each of these IR points has SH primary waves that generate SRL secondary waves with a completely different polarity. The source of the coupling between SH waves polarized in the X_1–X_2 plane and SRL waves polarized in the X_1–X_3 plane can be discovered by considering the wavestructure of the SH primary waves and tracing its effect on the nonlinear stress. For all SH waves

$$\mathbf{u} = \begin{Bmatrix} 0 \\ f(X_3) \\ 0 \end{Bmatrix} e^{i(kX_1-\omega t)} \rightarrow \mathbf{H} = \begin{bmatrix} 0 & 0 & 0 \\ ikf(X_3) & 0 & f'(X_3) \\ 0 & 0 & 0 \end{bmatrix} e^{i(kX_1-\omega t)},$$

where $f(X_3)$ is either $\cos(qX_3)$ or $\sin(qX_3)$. For second harmonic generation, the nonlinear stress is quadratic in $\mathbf{H}$, and comprises many combinations.

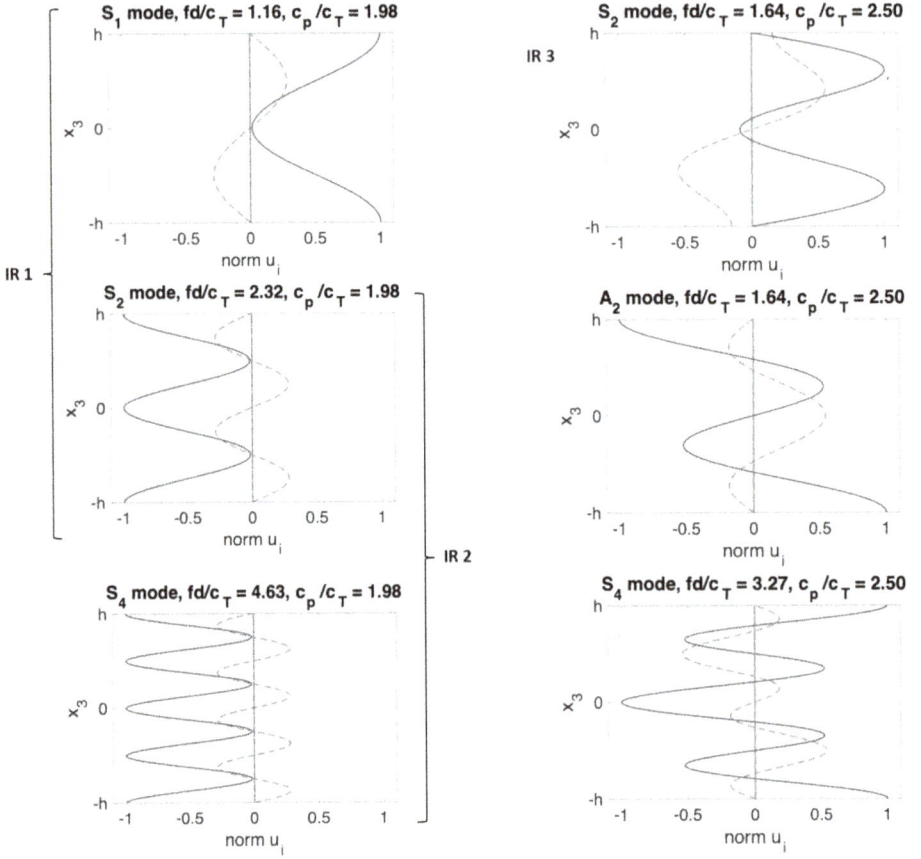

Figure 8.2. Wavestructures for the primary and secondary waves of IR points 1–3. Legend: blue = U_1, red = U_3, solid = real part, dashed = imaginary part.

Clearly $\mathrm{tr}(\mathbf{H}) = 0$, so we can eliminate terms in $\mathbf{S}^{NL}$ containing $\mathrm{tr}(\mathbf{H})$ from further consideration. We need to consider four quadratic variations of $\mathbf{H}$:

$$\mathbf{HH} = \begin{bmatrix} 0 & 0 & 0 \\ ikf(X_3) & 0 & f'(X_3) \\ 0 & 0 & 0 \end{bmatrix} \begin{bmatrix} 0 & 0 & 0 \\ ikf(X_3) & 0 & f'(X_3) \\ 0 & 0 & 0 \end{bmatrix} = \begin{bmatrix} 0 & 0 & 0 \\ 0 & 0 & 0 \\ 0 & 0 & 0 \end{bmatrix} \tag{8.8}$$

$$\mathbf{HH}^T = \begin{bmatrix} 0 & 0 & 0 \\ ikf(X_3) & 0 & f'(X_3) \\ 0 & 0 & 0 \end{bmatrix} \begin{bmatrix} 0 & ikf(X_3) & 0 \\ 0 & 0 & 0 \\ 0 & f'(X_3) & 0 \end{bmatrix} = \begin{bmatrix} 0 & 0 & 0 \\ 0 & -k^2 f^2(X_3) + f'^2(X_3) & 0 \\ 0 & 0 & 0 \end{bmatrix} \tag{8.9}$$

$$\mathbf{H}^T\mathbf{H} = \begin{bmatrix} 0 & ikf(X_3) & 0 \\ 0 & 0 & 0 \\ 0 & f'(X_3) & 0 \end{bmatrix} \begin{bmatrix} 0 & 0 & 0 \\ ikf(X_3) & 0 & f'(X_3) \\ 0 & 0 & 0 \end{bmatrix} = \begin{bmatrix} -k^2 f^2(X_3) & 0 & ikf(X_3)f'(X_3) \\ 0 & 0 & 0 \\ ikf(X_3)f'(X_3) & 0 & f'^2(X_3) \end{bmatrix} \tag{8.10}$$

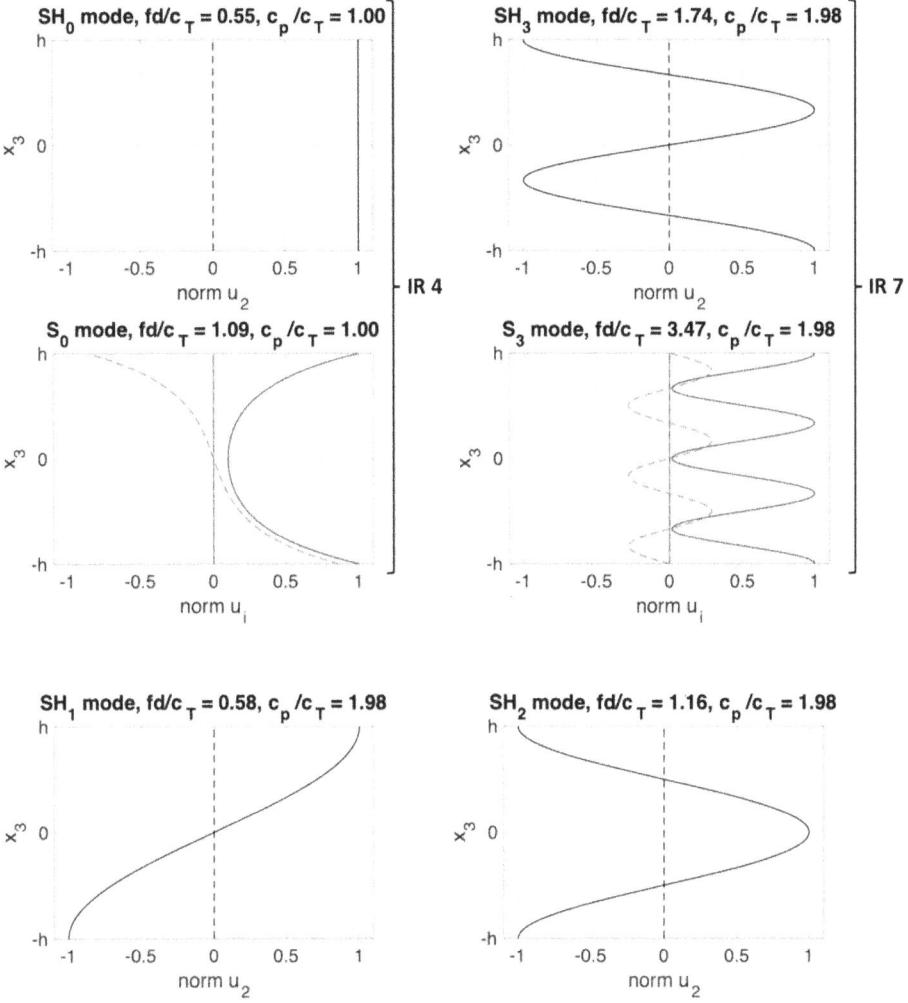

Figure 8.3. Wavestructures for the primary and secondary waves of IR points 4–8. Legend: blue = U_1, black = U_2, red = U_3, solid = real part, dashed = imaginary part.

$$\mathbf{H}^T\mathbf{H}^T = \begin{bmatrix} 0 & ikf(X_3) & 0 \\ 0 & 0 & 0 \\ 0 & f'(X_3) & 0 \end{bmatrix} \begin{bmatrix} 0 & ikf(X_3) & 0 \\ 0 & 0 & 0 \\ 0 & f'(X_3) & 0 \end{bmatrix} = \begin{bmatrix} 0 & 0 & 0 \\ 0 & 0 & 0 \\ 0 & 0 & 0 \end{bmatrix}. \tag{8.11}$$

Now we see that for SH primary waves

$$\mathbf{S}^{NL} = \frac{\lambda}{2}\mathrm{tr}(\mathbf{H}_a^T\mathbf{H}_a)\mathbf{I} + \mu(\mathbf{H}_a\mathbf{H}_a^T + \mathbf{H}_a^T\mathbf{H}_a)$$

$$+ \frac{\mathscr{A}}{4}(\mathbf{H}_a\mathbf{H}_a^T + \mathbf{H}_a^T\mathbf{H}_a) + \frac{\mathscr{B}}{4}\mathrm{tr}(\mathbf{H}_a\mathbf{H}_a^T + \mathbf{H}_a^T\mathbf{H}_a)\mathbf{I}$$

$$\mathbf{S}^{NL} = \left(\frac{\lambda}{2} + \frac{\mathcal{B}}{2}\right)\left(-k^2 f^2(X_3) + f'^2(X_3)\right)\mathbf{I}$$

$$+ \left(\mu + \frac{\mathcal{A}}{4}\right)\begin{bmatrix} -k^2 f^2(X_3) & 0 & ikf(X_3)f'(X_3) \\ 0 & -k^2 f^2(X_3) + f'^2(X_3) & 0 \\ ikf(X_3)f'(X_3) & 0 & f'^2(X_3) \end{bmatrix}. \qquad (8.12)$$

Thus, the off-diagonal terms are imaginary, and the diagonal terms are real. It is the terms on the diagonal that cause the coupling between primary SH waves and secondary SRL waves. The nonlinearity comes from both SOECs (λ and μ) and TOECs ($\mathcal{A}$ and $\mathcal{B}$, but not $\mathcal{C}$). Given that

$$f(X_3) = \begin{cases} \cos\left(\dfrac{n\pi}{2h}X_3\right) & n = 0,\, 2,\, 4\ldots \\[2ex] \sin\left(\dfrac{n\pi}{2h}X_3\right) & n = 1,\, 3,\, 5\ldots \end{cases}$$

we see that the $\mathbf{S}^{NL}{}_{13}(X_3 = \pm h) = \mathbf{S}^{NL}{}_{31}(X_3 = \pm h) = 0$ resulting in $\mathbf{S}^{NL}(X_3 = \pm h)$ being diagonal, and therefore $F_n^{\text{surf}} = 0$ as discussed earlier.

Turning attention now to F_n^{vol}, we assess

$$\nabla_X \cdot \mathbf{S}^{NL}(\mathbf{H}_a, \mathbf{H}_a, 2) = ik\mathbf{S}^{NL} + \frac{\partial}{\partial X_3}\mathbf{S}^{NL}.$$

For the power flow to the secondary waves to be nonzero the secondary waves must be SRL, therefore the nonlinear body force becomes

$$F_n^{\text{vol}} = \frac{1}{2}\int_{-h}^{h}\left[V_1^{n*}(X_3)S_{11,1}^{NL}(X_3) + V_3^{n*}(X_3)S_{33,3}^{NL}(X_3)\right]dX_3,$$

where (8.12) is used to determine the divergences of the nonlinear stress components

$$S_{11,1}^{NL} = ik\left[\left(\frac{\lambda}{2} + \frac{\mathcal{B}}{2}\right)(-k^2 f^2(X_3) + f'^2(X_3)) + \left(\mu + \frac{\mathcal{A}}{4}\right)(-k^2 f^2(X_3))\right]$$

$$S_{33,3}^{NL} = \frac{d}{dX_3}\left[\left(\frac{\lambda}{2} + \frac{\mathcal{B}}{2}\right)(-k^2 f^2(X_3) + f'^2(X_3)) + \left(\mu + \frac{\mathcal{A}}{4}\right)(f'^2(X_3))\right]$$

$$= (\lambda + \mathcal{B})(-k^2 f(X_3)f'(X_3) + 2 f'(X_3)f''(X_3)) + \left(2\mu + \frac{\mathcal{A}}{2}\right)(f'(X_3)f''(X_3)).$$

The presence of both material and geometric (i.e. finite strain) nonlinearity leads us to consider their relative strengths. Let's look at bulk longitudinal waves first because then (in the absence of residual stresses) we need only consider

$$\beta = -\frac{3(\lambda + 2\mu) + 2\mathcal{A} + 6\mathcal{B} + 2\mathcal{C}}{(\lambda + 2\mu)}.$$

If there is no material nonlinearity, then finite amplitude waves result in $\beta = -3$. With the representative TOECs for aluminum, $\beta = +14.8$, where the sign change is emphasized. We do the same analysis for IR points 1 and 4. If the TOECs are set to zero then the mixing power for IR points 1 and 4 are 8.44×10^6 and 8.665×10^6 m^{-2}, respectively, which are compared to 13.99×10^6 and 8.665×10^6 m^{-2} in table 8.3. The fraction of the mixing power associated with finite amplitude waves is:

bulk longitudinal waves: $3/14.8 = 0.203$,
S1–S2, IR point 1: $8.44/55.96 = 0.151$,
SH0–S0, IR point 4: $15.84/34.66 = 0.457$,

from which we can conclude that the material nonlinearity plays the largest role ($1 - 0.151 = 0.849$) for IR point 1 and the smallest role ($1 - 0.457 = 0.543$) for IR4, with the bulk longitudinal waves being in the middle (0.797). This is another important consideration when selecting wave modes and frequencies for nonlinear ultrasonic testing aimed at characterizing the material nonlinearity with finite amplitude waves that themselves have nonlinearity not associated with the material. We point out that this geometric nonlinearity due to finite amplitude waves (mathematically it stems from the nonlinear strain–displacement relation, equation (2.7)) is different than the measurement system nonlinearity in that it is cumulative, while measurement system nonlinearity is not. Thus, it can be even more difficult to distinguish material nonlinearity from geometric nonlinearity.

Returning for one last point on table 8.3, we examine the results for IR point 3 (and 8), which is the mode crossing point. Both S2 and A2 primary waves can generate S4 second harmonic waves, with the A2 mode displaying a lower mixing power than the S2 mode (40.69 to 67.81 mm^{-2}).

In summary, we have seen that the mixing power increases with the square of the driving frequency through equation (8.6) for bulk longitudinal waves since wavenumber and frequency are proportional. However, we also know that preferential excitation of high frequency Lamb or SH waves is difficult due to the presence of many modes. Perhaps a compromise can be struck by assessing the mixing power-driving frequency ratio (M_p/f_a in m^{-2} Hz^{-1}) below:

IR1	IR2	IR3 (S2)	IR3 (A2)	IR4	IR5	IR6	IR7	IR8
15.60	31.2	13.4	8.02	20.4	0.62	1.24	1.86	0.0

Based on the mixing power-driving frequency ratio, the top three IR points are IR2, IR4, and IR1. Whether this is a useful comparison remains to be determined. However, as previously discussed IR4 has a significant portion of the mixing power associated with the geometric nonlinearity. In addition, the group speeds for IR4 are mismatched. Without considering how to actuate the primary waves, IR points 1 and 2 appear to be the best choices. We will discuss actuation of IR points 1 and 2 in chapter 12.

IR point 8 is curious because there are no nonlinear forces and no mixing power (see table 8.3) even though there is synchronism and parity analysis indicates nonzero power flow. To answer why there is no power flow we plot $V_i^{n*}(X_3)S_{ij,j}^{NL}(X_3)$, which is the integrand for the nonlinear body force in equation (6.44), through the thickness of the plate in figure 8.4. The area under the curve (colored blue) is the nonlinear body force—and it is zero because the positive and negative parts are balanced. The nonlinear surface force has already been shown to be zero.

We now investigate how larger, or smaller, TOEC values affect the power flow from primary to secondary waves. We begin this analysis by assessing the stress–strain curve for a hyperelastic material described by Landau and Lifshitz's TOECs. The stress–strain curves for tension/compression and pure shear loading are shown in figure 8.5. The nonlinear tensile stress–strain curve has a decreasing slope, but the compressive stress–strain curve has an increasing slope; meaning that it stiffens. The shear stress–strain response is linear since the nonlinearity occurs on the diagonal of the matrix representing the first Piola–Kirchhoff stress tensor (for a third order strain energy function). There are nonzero shear stress components in the X_1–X_3 plane, but not in the X_1–X_2 plane. Let's examine the components of the stress tensor in terms of the TOECs and FOECs that were derived in example 2.4 for the case of shear loading, i.e. SH waves. For the SH wave displacement gradient,

$$\mathbf{H}_{SH} = \begin{bmatrix} 0 & 0 & 0 \\ H_{21} & 0 & H_{23} \\ 0 & 0 & 0 \end{bmatrix}.$$

The displacement gradient combinations we need are

$$\mathbf{HH}^T = \begin{bmatrix} 0 & 0 & 0 \\ 0 & H_{21}^2 + H_{23}^2 & 0 \\ 0 & 0 & 0 \end{bmatrix} \tag{8.13}$$

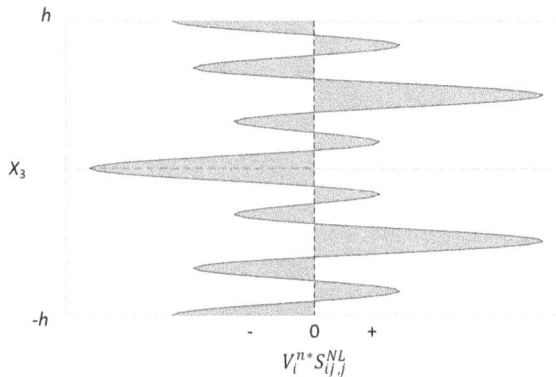

Figure 8.4. Plotting the integrand of the nonlinear body force (equation (6.44)) through the thickness of the plate. The shaded area is zero for IR point 8.

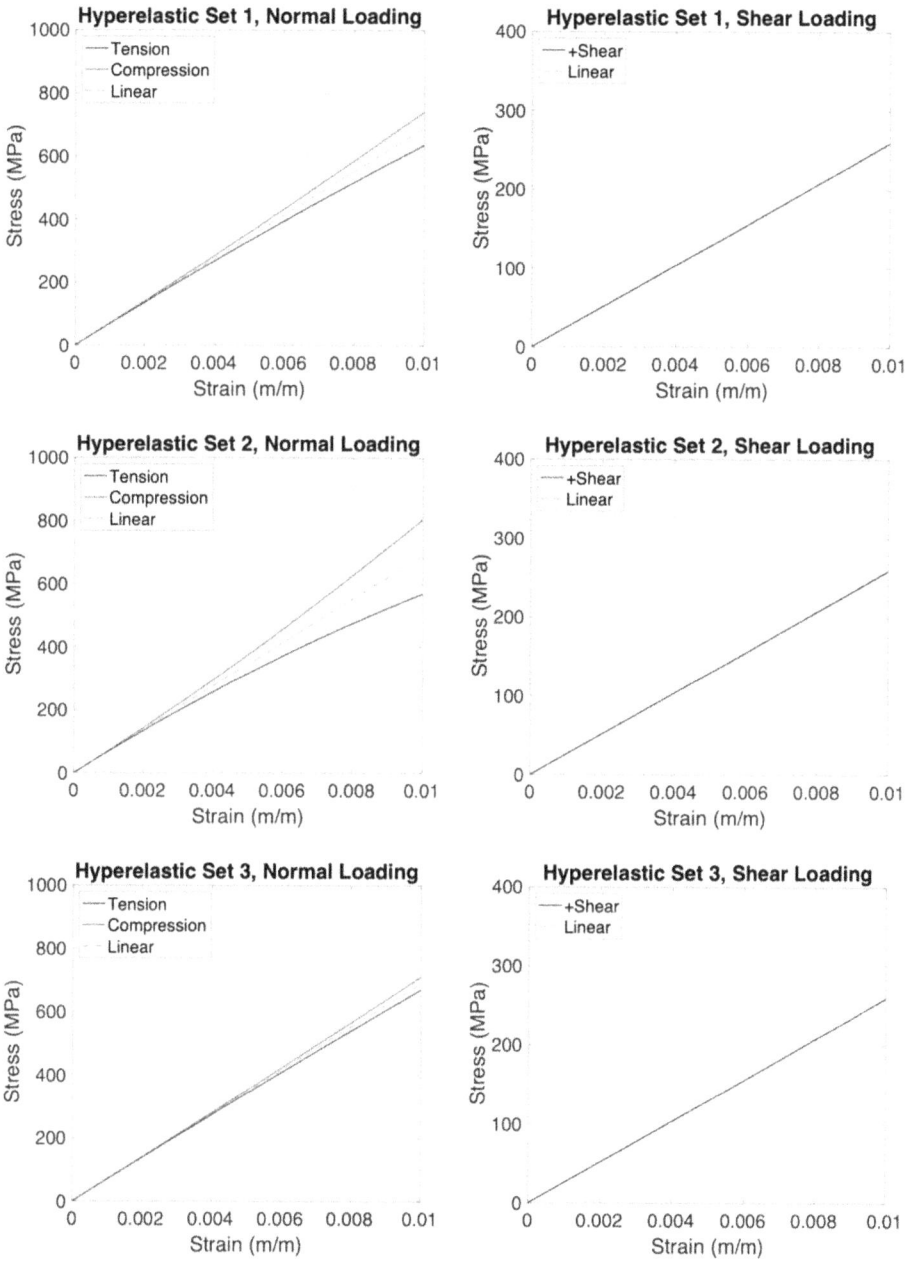

Figure 8.5. Normal and shear stress–strain curves for hyperelastic material having different TOECs. Set 1: TOEC × 1; Set 2: TOECs × 2; Set 3: TOECs × ½.

$$\mathbf{H}^T\mathbf{H} = \begin{bmatrix} H_{21}^2 & 0 & H_{21}H_{23} \\ 0 & 0 & 0 \\ H_{21}H_{23} & 0 & H_{23}^2 \end{bmatrix} \tag{8.14}$$

$$\mathbf{H}\mathbf{H}^T\mathbf{H} = \begin{bmatrix} 0 & 0 & 0 \\ 0 & H_{21}^2 + H_{23}^2 & 0 \\ 0 & 0 & 0 \end{bmatrix}\begin{bmatrix} 0 & 0 & 0 \\ H_{21} & 0 & H_{23} \\ 0 & 0 & 0 \end{bmatrix} = \begin{bmatrix} 0 & 0 & 0 \\ H_{21}\left(H_{21}^2 + H_{23}^2\right) & 0 & H_{23}\left(H_{21}^2 + H_{23}^2\right) \\ 0 & 0 & 0 \end{bmatrix} \tag{8.15}$$

$$\mathbf{H}^T\mathbf{H}\mathbf{H}^T = \begin{bmatrix} 0 & H_{21} & 0 \\ 0 & 0 & 0 \\ 0 & H_{23} & 0 \end{bmatrix}\begin{bmatrix} 0 & 0 & 0 \\ 0 & H_{21}^2 + H_{23}^2 & 0 \\ 0 & 0 & 0 \end{bmatrix} = \begin{bmatrix} 0 & H_{21}\left(H_{21}^2 + H_{23}^2\right) & 0 \\ 0 & 0 & 0 \\ 0 & H_{23}\left(H_{21}^2 + H_{23}^2\right) & 0 \end{bmatrix}, \tag{8.16}$$

all the other quadratic and cubic terms are identically zero. The quadratic nonlinear stress tensor comes from adding the tensors in equations (8.13) and (8.14). It is useful for power flow analyses to know that since H_{21} is always imaginary and H_{23} is always real, the diagonal of $\mathbf{S}^Q$ contains real terms while the nonzero off-diagonal terms are imaginary.

We can now identify how the SOECs (λ and μ), TOECs ($\mathscr{A}$, $\mathscr{B}$, $\mathscr{C}$), and FOECs ($\mathscr{E}$, $\mathscr{F}$, $\mathscr{G}$, $\mathscr{H}$) affect the first Piola–Kirchhoff stress tensor from the equations in example 2.4. The superscripts on the nonzero stress components denote which elastic constant affects that stress component. The linear, quadratic, and cubic parts of the stress tensor are, respectively,

$$\mathbf{S}^L = \begin{bmatrix} 0 & S_{21}^\mu & 0 \\ S_{21}^\mu & 0 & S_{23}^\mu \\ 0 & S_{23}^\mu & 0 \end{bmatrix} \tag{8.17}$$

$$\mathbf{S}^Q = \begin{bmatrix} S_{11}^{\lambda+\mu+\mathscr{A}+\mathscr{B}} & 0 & S_{13}^{\mu+\mathscr{A}} \\ 0 & S_{22}^{\lambda+\mu+\mathscr{A}+\mathscr{B}} & 0 \\ S_{13}^{\mu+\mathscr{A}} & 0 & S_{33}^{\lambda+\mu+\mathscr{A}+\mathscr{B}} \end{bmatrix} \tag{8.18}$$

$$\mathbf{S}^C = \begin{bmatrix} S_{11}^{\mathscr{B}+\mathscr{E}} & S_{12}^{\mathscr{A}+\mathscr{B}+\mathscr{G}} & 0 \\ S_{21}^{\lambda+\mu+\mathscr{A}+\mathscr{B}+\mathscr{G}} & S_{22}^{\mathscr{B}+\mathscr{E}} & S_{23}^{\lambda+\mu+\mathscr{A}+\mathscr{B}+\mathscr{G}} \\ 0 & S_{32}^{\mathscr{A}+\mathscr{B}+\mathscr{G}} & S_{33}^{\mathscr{B}+\mathscr{E}} \end{bmatrix}. \tag{8.19}$$

The absence of TOEC $\mathscr{C}$ and FOECs $\mathscr{F}$ and $\mathscr{H}$ in equations (8.18) and (8.19) demonstrates that none of these constants play a role in the nonlinearity generated by SH waves. The fact that $S_{21} = 0$ in equation (8.18) is responsible for the linear shear stress–strain curve in figure 8.5 and no second harmonic generation of SH waves. However, if cubic nonlinearity is included then the nonzero S_{21} term depends on λ, μ, $\mathscr{A}$, $\mathscr{B}$, and $\mathscr{G}$. It is fascinating that finite strain cubic nonlinearity affects the S_{21} and S_{23} terms, but not the S_{12} and S_{32} terms. This happens because the cubic nonlinearity from finite deformation comes only from the transformation from the second Piola–Kirchhoff stress to the first Piola–Kirchhoff stress. The nonlinear stress components comprised of the Kronecker delta times the trace of a function of the displacement gradient drive coupling between SH and Lamb waves at both quadratic and cubic nonlinearity, and the quadratic coupling is supplemented by the $\mathbf{HH}^T$ and $\mathbf{H}^T\mathbf{H}$ terms.

It's worth pointing out that the shear-normal coupling is a one-way action; shear loading causes normal stresses, but the mathematical form of the strain energy function does not permit normal loading to couple to shear stresses. The displacement gradient for Lamb waves is

$$\mathbf{H}_{RL} = \begin{bmatrix} H_{11} & 0 & H_{13} \\ 0 & 0 & 0 \\ H_{31} & 0 & H_{33} \end{bmatrix}.$$

Quadratic and cubic terms of $\mathbf{H}_{RL}$ have this same form with an additional H_{22} term that comes from terms having a trace multiplied by the Kronecker delta. But no combinations yield nonzero H_{21} and H_{23} terms, hence the Lamb waves do not generate SH waves.

Figure 8.5 only shows three TOEC sets:

Set 1: TOECs $\times$ 1
Set 2: TOECs $\times$ 2
Set 3: TOECs $\times$ ½.

We can now assess the effect of each TOEC on the mixing power. FOECs will be considered in the next section. The material nonlinearity associated with the TOECs affects the power flow from primary waves to secondary waves. The effect is quite straightforward for bulk longitudinal waves, as equation (8.6) indicates that the mixing power M_p increases proportionally to β, which contains the term $2(\mathscr{A} + 3\mathscr{B} + \mathscr{C})$. Thus, M_p is three times more sensitive to $\mathscr{B}$ than $\mathscr{A}$ and $\mathscr{C}$. Let's assess how the TOECs affect power flow at IR points 1 and 4. We increase each TOEC individually by a factor of two and then decrease each TOEC individually by

Table 8.4. Sensitivity of mixing power, M_p (m^{-2}) $\times 10^6$, to TOECs for self-interaction at IR points 1 and 4.

TOECs	IR point 1 (S1–S2)	IR point 4 (SH0–S0)
$\mathscr{A}$, $\mathscr{B}$, $\mathscr{C}$	55.96	34.66
$(\mathscr{A}, \mathscr{B}, \mathscr{C}) \times 2$	120.4	85.16
$(\mathscr{A}, \mathscr{B}, \mathscr{C}) \times$ ½	23.76	9.41
$\mathscr{A} \times 2$, $\mathscr{B}$, $\mathscr{C}$	82.49	54.49
$\mathscr{A}/2$, $\mathscr{B}$, $\mathscr{C}$	42.69	24.74
$\mathscr{A}$, $\mathscr{B} \times 2$, $\mathscr{C}$	88.36	65.33
$\mathscr{A}$, $\mathscr{B}/2$, $\mathscr{C}$	39.76	19.32
$\mathscr{A}$, $\mathscr{B}$, $\mathscr{C} \times 2$	61.41	34.66
$\mathscr{A}$, $\mathscr{B}$, $\mathscr{C} \times$ ½	53.23	34.66

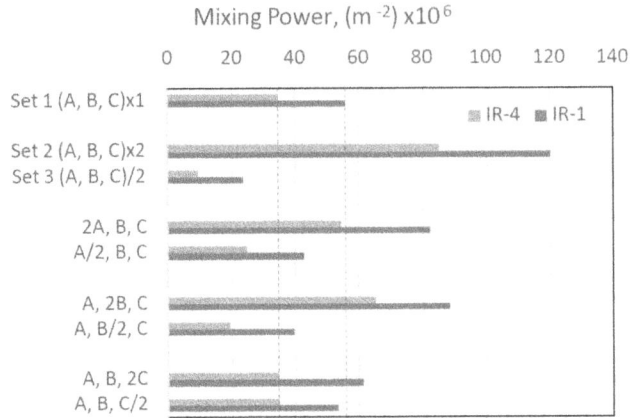

Figure 8.6. Effect of TOECs on mixing power for internal resonance (IR) points 1 and 4.

a factor of two and then give the resultant mixing powers in table 8.4 and figure 8.6. At IR point 1, $\mathscr{B}$ has the largest effect and $\mathscr{C}$ has the smallest. At IR point 4, $\mathscr{B}$ also has the largest effect and $\mathscr{C}$ has no effect. In fact, the TOEC $\mathscr{C}$ has no effect on nonlinearity at any of IR points 4–8 because the primary waves are SH waves and $\mathscr{C}$ has no effect on nonlinearity of shear waves as shown in example 2.4.

8.1.2 Third harmonic generation

Considering third harmonic generation enables use of self-interacting SH waves to generate third harmonic SH waves amongst other things. Chapter 7 showed the holo-internal resonance of SH waves, i.e. any SH mode at any frequency will generate an internally resonant third harmonic. Lamb waves will also generate third harmonics, but as just shown, they generate second harmonics, whereas SH waves can only generate second order Lamb waves. Therefore, we will focus on third harmonics generated by SH waves.

We need FOECs to determine the mixing power associated with third harmonic generation. A search revealed that Erofeyev [11] provides FOECs for aluminum alloy D16T, with the methods used to determine them described in the Russian literature [12]. Erofeyev's table 1.3 provides

$$D = (23 \pm 9) \times 10^{12} \text{ dyne cm}^{-2}$$

$$G = (5.5 \pm 1.7) \times 10^{12} \text{ dyne cm}^{-2}$$

$$H = (3.9 \pm 2.4) \times 10^{12} \text{ cm}^{-2}$$

$$J = (7.2 \pm 3.9) \times 10^{12} \text{ cm}^{-2}.$$

These FOEC symbols correspond to ours as $D \to \mathscr{H}$, $G \to \mathscr{E}$, $H \to \mathscr{F}$, $J \to \mathscr{G}$. Converting to SI units we have

$$\mathscr{E} = (550 \pm 170) \times 10^9 \text{ m}^{-2}$$

$$\mathscr{F} = (390 \pm 240) \times 10^9 \, \text{m}^{-2}$$

$$\mathscr{G} = (720 \pm 390) \times 10^9 \, \text{m}^{-12}$$

$$\mathscr{H} = (2300 \pm 900) \times 10^9 \, \text{N m}^{-2}.$$

There is clearly large uncertainty in these values, thus we should examine how these parameters fit with our other properties for aluminum before we adopt them. Consider first the strain energy density function itself, plotted in figure 8.7 for a uniaxial stress state (e.g. tension or compression test). Figure 8.7 emphasizes the individual terms in the strain energy expansion by plotting as a reference the quadratic strain energy, which is symmetric and yields the linear elastic stress–strain relation. Next, the cubic expansion with the TOECs is shown. Finally, the quartic expansion including the FOECs is also shown. The difference between the cubic and quartic expansions can only be seen in the zoomed-in views.

Plotting the stress–strain curves for the stress states of interest is informative, thus we plot stress–strain curves for the stress states in longitudinal and shear bulk waves in figure 8.8. Responses are plotted for quadratic, cubic, and quartic expansions of the strain energy density giving the linear, quadratic, and cubic stress–strain responses shown. Under normal loading there is a tension/compression asymmetry, with higher stress in compression than in tension for the same strain. Under shear loading there is no nonlinearity for the quadratic stress–strain response because the power flows only to the nonlinear normal stress as already discussed. The difference between the cubic and quadratic stress–strain responses can only be seen by zooming in. The positive values of the FOECs result in slightly higher stress than for the quadratic stress–strain response. Negative FOEC values would reverse this. Clearly, the cubic and quadratic stress–strain curves for both normal and shear loading are only slightly different. The cubic nonlinear stress terms include geometric non-linearity from the finite strain and effects from both TOECs and FOECs. However, the finite strain and TOEC effects are negligible compared to the FOEC effects. As shown in example 3.5, only the FOEC $\mathscr{G}$ plays a role in the shear response. Finally,

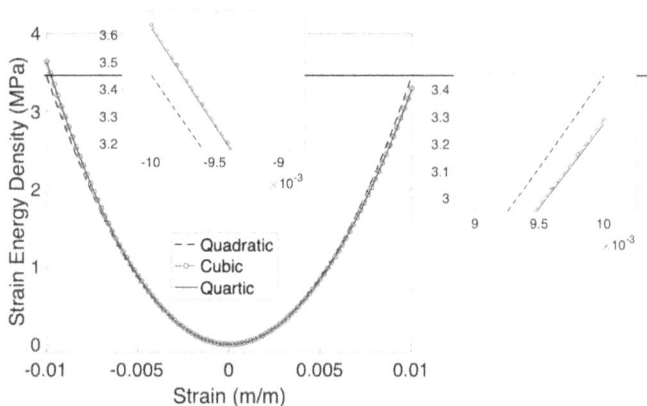

Figure 8.7. Quadratic, cubic, and quartic strain energy density functions.

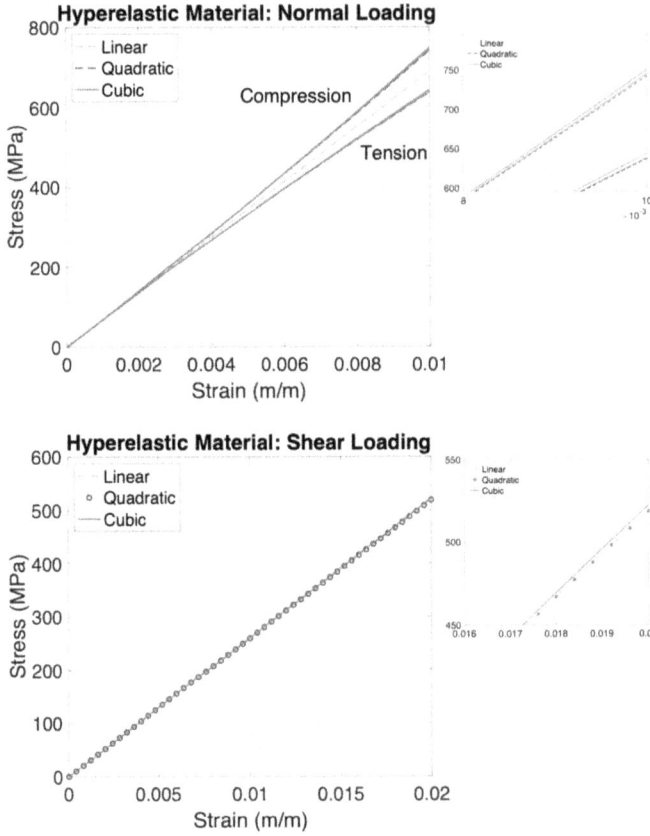

Figure 8.8. Linear, quadratic, and cubic stress–strain curves for quadratic, cubic, and quartic strain energy density functions respectively.

since there is reasonably large uncertainty in the FOEC values, the upper and lower extreme values are used to generate the upper and lower limits for normal loading. The zoomed-in view for normal loading shows that these limits are quite close together, suggesting that more accurate FOEC values may not have much effect, so let's analyse the power flow to third harmonics, which might have a different outcome.

We think of γ defined in example 3.5 as the acoustic nonlinearity coefficient for third harmonic generation. The associated relative nonlinearity coefficient is then

$$\gamma' = \frac{A_3}{A_1^3} = \frac{\max(\mathbf{u}_{aaa})}{(\max(\mathbf{u}_a))^3} = F_n \frac{\max(\mathbf{U}^n(X_3))}{(\max(\mathbf{U}_a(X_3)))^3} X_1 \tag{8.20}$$

leading to a refined mixing power definition for third harmonic generation

$$M_p \equiv F_n \frac{\max(\mathbf{U}^n(X_3))}{(\max(\mathbf{U}_a(X_3)))^3}, \tag{8.21}$$

Table 8.5. Mixing power for third harmonic generation from self-interaction of SH waves.

Waves a	$\mathscr{G}$ (GPa)	F_n^{vol} (N s^{-1}) $\times 10^{-6}$	F_n^{surf} (N s^{-1})	F_n (m^{-1}) $\times 10^{-6}$	M_p (m^{-2}) $\times 10^6$
SH0 @ 1 MHz	330	0.298i	0	0.298	16 400
SH0 @ 1 MHz	720	2.57i	0	**2.57**	**142 000**
SH0 @ 1 MHz	1110	4.85i	0	4.85	267 000
SH0 @ 3 MHz	720	7.72i	0	**7.72**	**3 830 000**
SH0 @ 5 MHz	720	12.9i	0	12.9	17 700 000
SH1 @ 5 MHz	720	−7.11i	0	7.11	4 660 000
SH2 @ 5 MHz	720	10.4i	0	10.4	5 640 000

Note: wavenumbers at 5 MHz for SH1 and SH2 are 9636 and 7953 rad m^{-1}, respectively.

which could be further generalized for wave mixing at third order by accounting for how the waves interact. We demonstrate third harmonic generation through self-interaction of SH waves. First, let's analyse the effect of the FOEC $\mathscr{G}$. Although it has quite a small effect on the stress–strain response, its significant effect on the mixing power is shown in table 8.5. Since SH0 waves are nondispersive and have a uniform wavestructure, it is not difficult to write out the expression for the mixing power,

$$M_p = \mathrm{i}\frac{k_a^3}{\mu}\left[\frac{\lambda}{2} + \mu + \frac{\mathscr{A}}{2} + \mathscr{B} + \mathscr{G}\right]\frac{\max\left(\mathbf{U}^n(X_3)\right)}{\left(\max\left(\mathbf{U}_a(X_3)\right)\right)^3}. \tag{8.22}$$

The mixing power is imaginary because the nonlinear body force is out of phase. Therefore, table 8.5 shows the magnitudes of F_n and M_p. The mixing power values are much larger than for second order interactions because the maximum value of the wavestructure in the denominator is cubed, rather than squared, and these are small values (on the order of 1 nm). We interpret this as an artifact of the wavestructure normalization that makes the different definitions of mixing power for second order and third order interactions incompatible. Thus, mixing power for third order interactions should only be compared with other third order interactions. The effect of frequency on F_n and M_p is clear from the results at 1 and 3 MHz (highlighted in table 8.5). The change in F_n is proportional to the frequency (or wavenumber) ratio, and the change in M_p is proportional to the frequency ratio cubed.

8.1.3 Method of multiple scales

The regular perturbation solution to the nonlinear elastic wave equation was formally applied to the longitudinal wave problem in example 3.2. The nonuniform accuracy of the solution limits it to weak nonlinearity and motivates the use of a more advanced perturbation methodology. The method of multiple scales was demonstrated in example 3.3. The method of multiple scales [13] has been applied to nonlinear Lamb waves by Kanda and Sugiura [14–16] and to linear Lamb waves by Kanda and Maruyama [17, 18]. Likewise, Osika *et al* [19] applied the method of multiple scales to nonlinear SH wave problems. We will provide a brief summary of these modeling efforts.

In their first work [14], Kanda and Sugiura limit the analysis to nonlinearity applied at the boundary through nonlinear springs. They assess the synchronism (phase matching) condition and detuning, i.e. when phase matching is not satisfied. In their implementation of the method of multiple scales (MMS) they eliminate secular terms by writing amplitude equations that yield solvability conditions. In their solution, over long propagation distances, all of the energy in the primary waves eventually transfers to the secondary waves when they are synchronized. This result demonstrated the need to include internal damping (attenuation) in the analysis, which they did in [15, 16] by making the Lamé parameters complex-valued. The former article is devoted solely to the effect of damping on the dispersion curves and attenuation coefficients for the modes, while the latter article tackles the nonlinear Lamb wave problem. In [16] the authors show how both the primary and secondary waves decay with propagation distance when both material and geometric nonlinearities exist.

In [17] Kanda and Maruyama employ several length and time scales to model the evolution of a waveform with propagation distance due to dispersion. Third and fourth order solutions are employed to model symmetric and asymmetric evolution of the waveform, respectively. The results of this analysis of linear guided waves agree well with a numerical Fourier integral method. Likewise, in [18] they extend their analysis to forced Lamb waves by implementing a Green's function approach. Osika *et al*'s [19] work is more closely related to nonlinear ultrasonic guided waves. They use the MMS to formulate the SH wave problem such that it can be solved using a modal decomposition consisting of odd powers. Their results show that only the TOECs $\mathscr{A}$ and $\mathscr{B}$ and FOEC $\mathscr{G}$ play a role in nonlinear SH wave propagation as was shown earlier in this section.

8.2 Mutual interaction in plates

We follow a systematic approach for finding internal resonance points for wave mixing by managing the database obtained from the analysis of the dispersion relations for Lamb (RL) and shear-horizontal (SH) type waves and classify them by their symmetric or antisymmetric nature. Thus, we have wavenumbers for propagating modes as a function of frequency for symmetric and antisymmetric Lamb waves (SRL and ARL) as well as for symmetric and antisymmetric SH waves (SSH and ASH). The database is for aluminum and contains multiple wavenumbers for fd products up to 15 MHz-mm in increments of 0.01 MHz-mm. The discrete nature of the database will cause us to miss some points that satisfy the synchronism criterion, but it is successful in finding a large number of points that at least approximately satisfy the internal resonance condition. We can extrapolate the synchronism results to internal resonance by only assessing wave mode and wave nature combinations that have nonzero power flow from the parity analysis in chapter 7, i.e.:

- Same type same nature $\to$ cumulative SRL second harmonics.
- Same type different nature $\to$ cumulative ARL second harmonics.

- Different type same nature → cumulative SSH second harmonics.
- Different type different nature → cumulative ASH second harmonics.

IR points are identified by imposing a tolerance for the synchronism criterion:

$$\varepsilon = (k_a \pm k_b) - k_n.$$

We imposed a tolerance $\varepsilon = 0.5$ rad m^{-1} to locate a large number of points on the dispersion curves where the criterion was, at worst, approximately satisfied. Further analysis would be necessary to assess whether there is an exact IR point in the vicinity of the found point. We then tightened the tolerance to $\varepsilon = 0.001$ rad m^{-1} to limit the search, recognizing that we may very well be discarding good options. Based on typical wave actuation methods using tonebursts having a finite frequency bandwidth and actuators that excite a wavenumber spectrum, these small tolerances may be irrelevant. The tolerance, sometimes known as detuning, has been analysed in much different ways by Matsuda and Biwa [20] and Kanda and Sugiura [16] for self-interactions.

8.2.1 Co-directional, $\theta = 0°$

The sum and difference frequencies are overlaid on the dispersion curves for the wave type and nature of those second order combinational harmonics satisfying the synchronism criterion in figures 8.9–8.11. The sum and difference frequencies are plotted by points colored for $0 < |\varepsilon| < 0.5$ rad m^{-1} as indicated on the color bar. Sum frequencies are on the left and difference frequencies are on the right of each figure. Each plot in these figures indicates the interacting wave types and natures. The parity analysis results in table 7.4 ensure that the synchronism results are also internal resonance results, hence we reiterate that power only flows as:

Case 1. Same type same nature: SRL–SRL, ARL–ARL, SSH–SSH, ASH–ASH → SRL.
Case 2. Same type different nature: SRL + ARL, SSH + ASH → ARL.
Case 3. Different type same nature: SRL + SSH, ARL + ASH → SSH.
Case 4. Different type different nature: SRL + ASH, ARL + SSH → ASH.

Case 1 results are shown in figure 8.9. The analysis identifies self-interaction points as well as mutual interaction points for co-directional waves. The tolerance of $0 < |\varepsilon| < 0.5$ is satisfied by far more SRL interactions than the other three cases combined (ARL, SSH, and ASH). Altogether there are 19 925 combinations, although this number includes duplicates for sum frequencies because $(f_a + f_b = f_b + f_a)$. The importance of this number is that it is far larger than the eight IR points identified for self-interaction in table 8.1. Some combinations occur at mode cutoffs and are not shown in figure 8.9 because the c_p axis is truncated for clarity. There are many combinations that result in quasi-Rayleigh waves at either the sum or the difference frequency. The mutual interactions of ARL–ARL, SSH–SSH, and ASH–ASH never generate internally resonant S0 waves at the sum frequency and mutual interactions of SRL–SRL only generate internally resonant

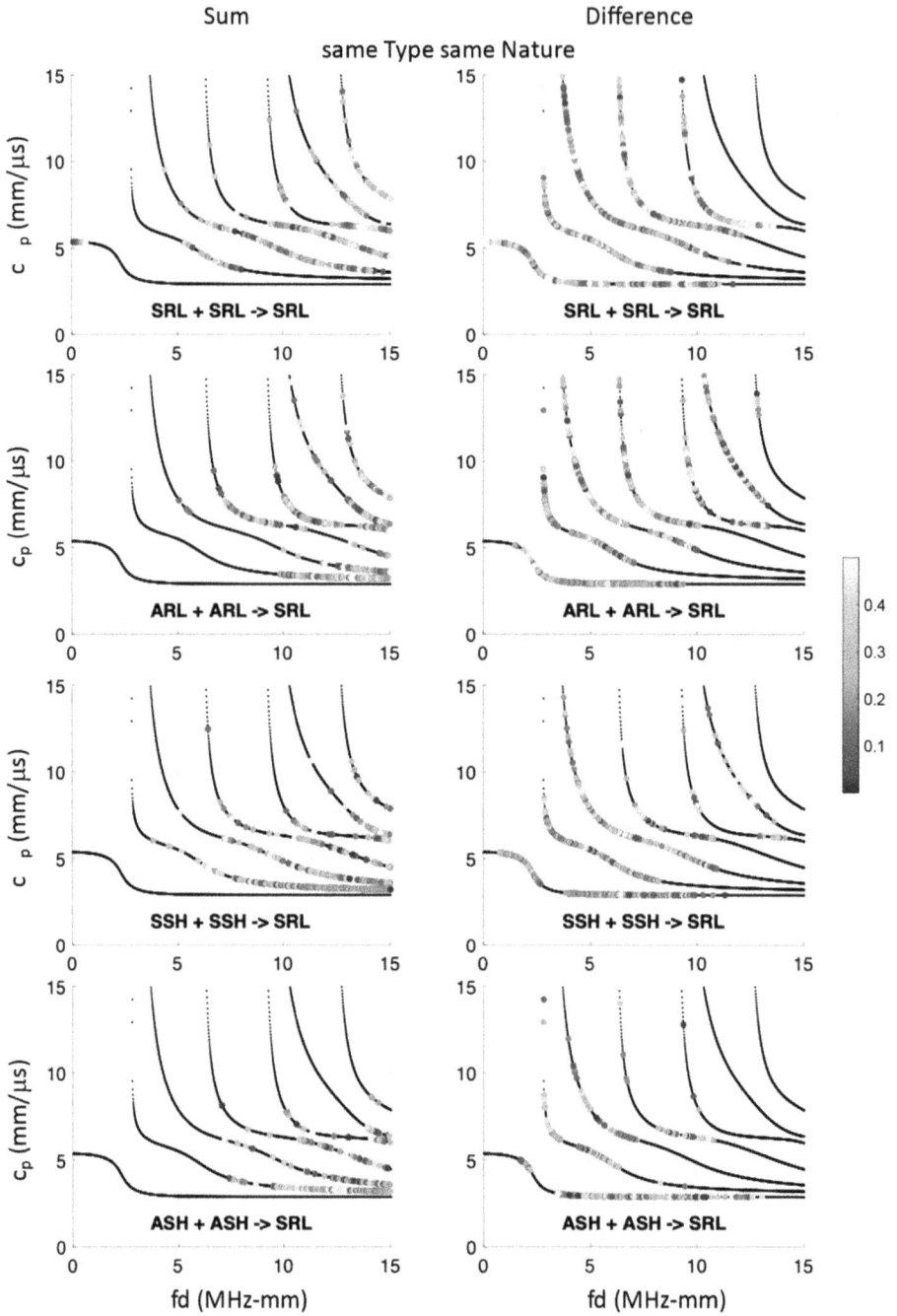

Figure 8.9. Sum and difference frequencies satisfying the synchronism criterion within 0.5 rad m^{-1} for co-directional mixing of wave modes having the same type and same nature.

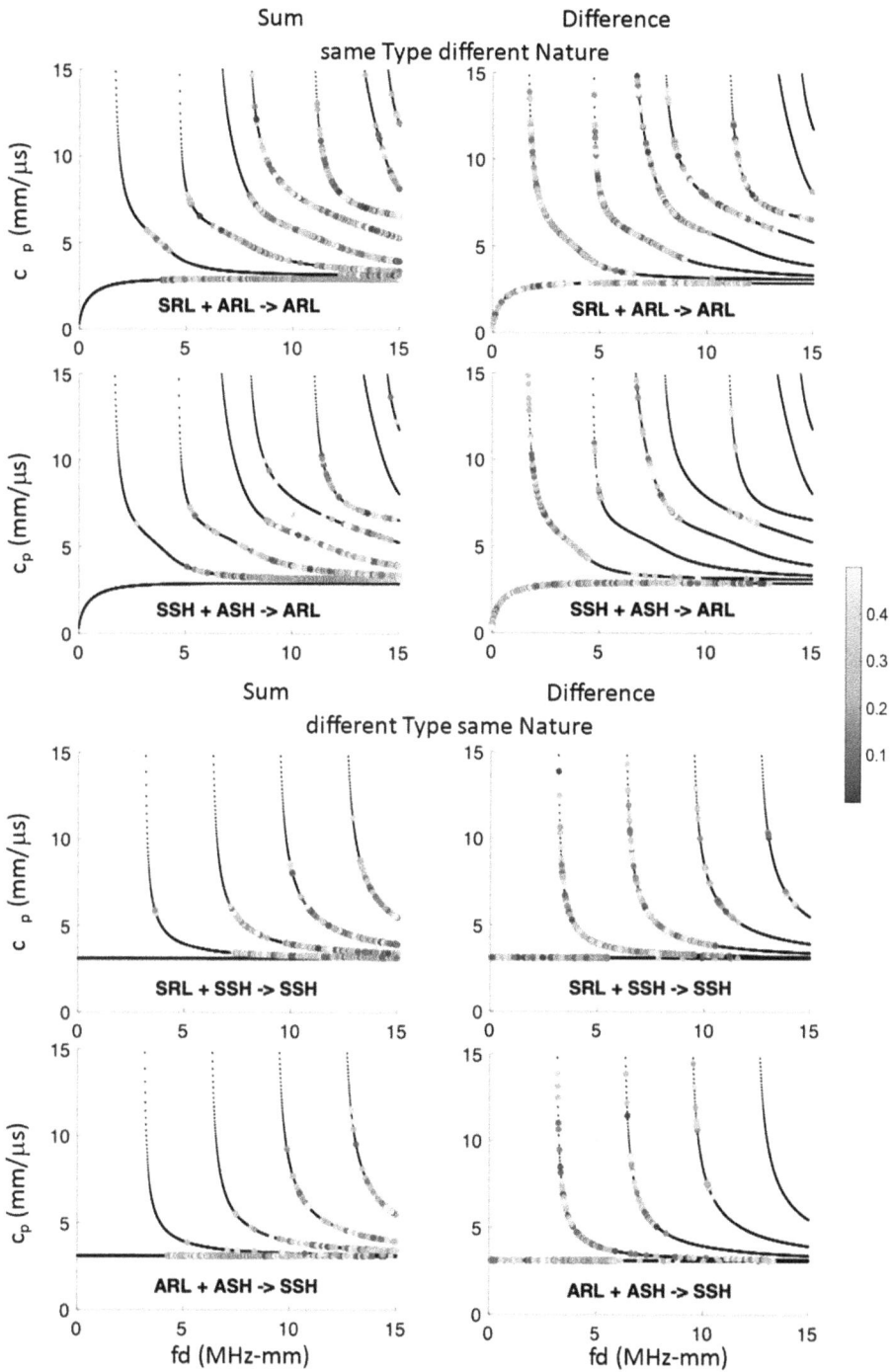

Figure 8.10. Sum and difference frequencies satisfying the synchronism criterion within 0.5 rad m^{-1} for co-directional mixing of wave modes having the same type and different nature (top) and different type and same nature (bottom).

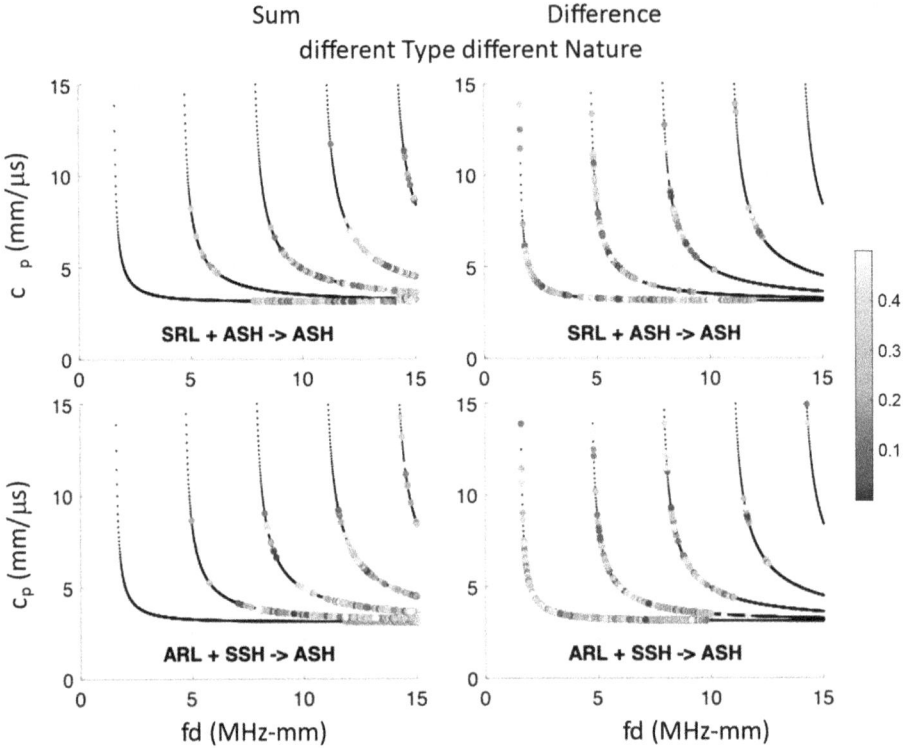

Figure 8.11. Sum and difference frequencies satisfying the synchronism criterion within 0.5 rad m^{-1} for co-directional mixing of wave modes having different type and different nature.

S0 waves at low frequencies. Note that IR point 4 for self-interaction of SH0 waves is missed because there is not dispersion data precisely at the point where the SH0 and S0 waves intersect (it contains $fd = 3.39$, but not $fd = 3.3913$ MHz-mm).

Results for cases 2 and 3 are shown in figure 8.10 (there are 10 824 combinations shown although many of these are in highly dispersive regions). We observe that many options exist to generate internally resonant A0 waves and SH0 waves at the difference frequency in nondispersive regions. However, internally resonant A0 waves at sum frequencies are not generated by SSH and ASH mutual interaction.

Case 4 results are plotted in figure 8.11 (for comparison, there are 3789 combinations). There are many combinations that yield internally resonant SH1 waves having nearly the bulk shear wave speed, c_T.

Having used the dispersion curves to identify wave triplets that are internally resonant (or at least very close to it), we can assess the power flow through the mixing power quantity in equation (6.58). It might be helpful to point out that we are using the eigenvalues for assessing synchronism and then using the eigenvectors for assessing the mixing power. Results for selected wave triplets having $| \varepsilon | < 0.001$ rad m^{-1} are detailed in tables 8.6 and 8.7. There are a few intricacies to point out in these tables. As discussed for table 8.3, negative values for F_n^{vol} and F_n^{surf} are possible, but can always be converted to positive values by multiplying the wavestructure by -1.

Table 8.6. Mixing power for co-directional mutual interactions, $d = 1$ mm, same type, same nature.

IR point	Triplet		f (MHz)	c_p (mm μs^{-1})	c_g (mm μs^{-1})	F_n^{vol} (N s^{-1})	F_n^{surf} (N s^{-1})	F_n (m^{-1})	M_p (m^{-2}) $\times 10^6$
9	S1–S1–S2	a	3.17	6.6234	3.4656	0.0345	0.0163	0.0508	17.7
	sum	b	6.03	4.7912	2.3877				
		n	9.20	5.2960	2.7029				
10	S1–S3–S4	a	10.15	6.2758	4.8645	0.0043	0.0067	0.011	14.3
	sum	b	3.74	6.0584	4.4934				
		n	13.89	6.2158	5.0229				
11	S2–S1–S1	a	9.20	5.2960	2.7029	0.0762	0.0286	0.1048	120
	difference	b	3.17	6.6234	3.4656				
		n	6.03	4.7912	2.3877				
12	S4–S0–S4	a	13.25	6.2737	5.3303	0.156	0.003	0.159	130
	difference	b	1.32	5.1853	4.7744				
		n	11.93	6.4229	4.7996				
13	S4–S1–S3	a	13.89	6.2158	5.0229	0.087	0.0053	0.0923	151
	difference	b	3.74	6.0584	4.4934				
		n	10.15	6.2758	4.8645				
14	A0–A1–S1	a	3.12	5.7620	3.5050	0.0222	0.0111	0.0333	25.2
	sum	b	7.42	2.8847	2.9083				
		n	10.54	3.3851	2.7317				

(Continued)

Table 8.6. (*Continued*)

IR point	Triplet		f (MHz)	c_p (mm μs^{-1})	c_g (mm μs^{-1})	F_n^{vol} (N s^{-1})	F_n^{surf} (N s^{-1})	F_n (m^{-1})	M_p (m^{-2}) $\times 10^6$
15	A3–A2–S1	a	8.84	6.5253	3.0615	0.0211	0.0242	0.0454	166
	difference	b	5.78	6.3495	3.2975				
		n	3.06	6.8853	2.9551				
16	SH0–SH2–S1	a	5.40	3.0996	3.0996	0.0404	0.0172	0.0576	82.9
	sum	b	9.04	3.2996	2.9117				
		n	14.44	3.2218	2.9349				
17	SH2–SH0–S0	a	7.66	3.3895	2.8345	0.0136	0.0166	0.0301	202
	difference	b	5.92	3.0996	3.0996				
		n	1.74	4.9713	3.9371				
18	SH1–SH1–S1	a	2.77	3.7397	2.569	0.0064	0.0236	0.0299	18.7
	sum	b	11.08	3.1303	3.0691				
		n	13.85	3.2358	2.9162				

Table 8.7. Mixing power for co-directional mutual interactions, $d = 1$ mm.

IR point	Triplet		f (MHz)	c_p (mm μs^{-1})	c_g (mm μs^{-1})	F_n^{vol} (N s^{-1})	F_n^{surf} (N s^{-1})	F_n (m^{-1})	M_p (m^{-2}) $\times 10^6$
					Same type, different nature				
19	S1–A2–A3	a	3.06	6.8853	2.9551	0.1191	0.0016	0.1207	69.4
	sum	b	5.78	6.3495	3.2975				
		n	8.84	6.5253	3.0615				
20	S0–A0–A0	a	6.12	2.9026	2.8319	0.4185i	0.3222i	0.7407	932
	sum	b	5.78	2.8741	2.9373				
		n	11.84	2.8887	2.8900				
21	S2–A2–A4	a	10.14	7.7558	2.0638	0.0454	0	0.0455	91.4
	difference	b	5.07	7.7558	2.0638				
		n	5.07	7.7558	2.0638				
22	S2–A1–A0	a	10.54	3.3851	2.7317	0.0149	0.0029	0.0179	32.0
	difference	b	3.12	5.7620	3.5050				
		n	7.42	2.8847	2.9083				
23	SH0–SH1–A1	a	5.04	3.0996	3.0996	0.0226	0.0178	0.0404	40.6
	sum	b	3.95	3.3698	2.8510				
		n	8.99	3.2128	2.9085				
24	SH4–SH1–A1	a	14.00	3.4569	2.7791	0.0213	0.0223	0.0436	624
	difference	b	11.55	3.1279	3.0715				
		n	2.45	6.8589	3.6531				

(*Continued*)

Table 8.7. (*Continued*)

IR point	Triplet		f (MHz)	c_p (mm μs^{-1})	c_g (mm μs^{-1})	F_n^{vol} (N s^{-1})	F_n^{surf} (N s^{-1})	F_n (m^{-1})	M_p (m^{-2}) $\times 10^6$
				Different type, same nature					
25	S0–SH0–SH2	a	1.74	4.9713	3.9371	0.0133	0	0.0133	11.5
	sum	b	5.92	3.0996	3.0996				
		n	7.66	3.3895	2.8345				
26	S1–SH2–SH0	a	14.44	3.2218	2.9349	0.0146	0	0.0146	48.8
	difference	b	5.40	3.0996	3.0996				
		n	9.04	3.2996	2.9117				
27	A1–SH1–SH4	a	2.45	6.8589	3.6531	0.0057	0	0.0057	4.33
	sum	b	11.55	3.1279	3.0715				
		n	14.00	3.4569	2.7791				
28	A1–SH1–SH0	a	8.99	3.2128	2.9084	0.041	0	0.041	48.3
	difference	b	3.95	3.3698	2.8510				
		n	5.04	3.0996	3.0996				

Table 8.7. (*Continued*)

IR point	Triplet		f (MHz)	c_p (mm μs^{-1})	c_g (mm μs^{-1})	F_n^{vol} (N s^{-1})	F_n^{surf} (N s^{-1})	F_n (m^{-1})	M_p (m^{-2}) $\times 10^6$
					Different type, different nature				
29	S0–SH1–SH1	a	3.50	3.0725	2.4380	0.0573i	0	0.0573	42.3
	sum	b	10.13	3.1365	3.0631				
		n	13.63	3.1198	3.0795				
30	S1–SH1–SH1	a	13.85	3.2358	2.9163	0.079	0	0.079	66.6
	difference	b	2.77	3.7397	2.5690				
		n	11.08	3.1303	3.0691				
31	A2–SH0–SH5	a	7.50	5.1650	2.6365	0.0568	0	0.0568	71.2
	sum	b	4.01	3.0996	3.0996				
		n	11.51	4.1918	2.2919				
32	A1–SH0–SH1	a	8.99	3.2128	2.9084	0.0021	0	0.0021	8.87
	difference	b	5.04	3.0996	3.0996				
		n	3.95	3.3698	2.8510				

Therefore, only positive values are reported. There are a few instances when F_n^{vol} and F_n^{surf} are imaginary. Exploration of why this happens points to when the wave-structures are out of phase. Some of the triplets can be explained based on our prior analysis of synchronism. Examples include,

- IR point 19 waves n are generated at the A3 mode crossing point.
- At IR point 20 all the waves are quasi-Rayleigh waves.
- At IR point 21 the three phase velocities and three group velocities are the same, but waves b and n are at the same frequency.
- At IR point 22 the waves n are quasi-Rayleigh waves.

8.2.2 Counter-propagating, $\theta = 180°$

Counter-propagating waves can be analysed the same way as the co-directional waves, the only difference is that $k_b < 0$ if we take $k_a > 0$. The same database used for co-directional waves can be used for counter-propagating waves by assigning a negative sign to k_b. The entirety of the analyses in section 8.2.1 apply to counter-propagating waves, but for brevity the presentation of results is condensed. In figure 8.12 the results for mutually interacting ARL waves are shown for both the sum and difference frequencies.

In co-directional wave mixing, the size of the mixing zone can be maximized by group velocity matching and the use of long toneburst excitations. By maximizing the propagation distance over which the waves interact the cumulative nature of the secondary waves will improve sensitivity to uniform material degradation. However, the drawback is that if the material degradation is localized it may be difficult to detect with a large wave mixing zone because the effect of the localized degradation could be averaged out over the entire mixing zone. On the other hand, counter-propagating waves are ideally suited for detecting nonuniform material degradation because the mixing zone size can be matched to the size of the localized degradation. Smaller mixing zones may be desirable. Group velocity matching becomes

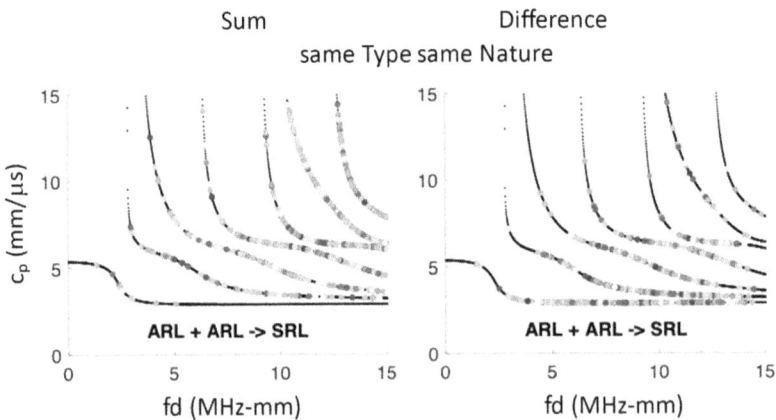

Figure 8.12. Sum and difference frequencies satisfying the synchronism criterion within 0.5 rad m^{-1} for counter-propagating wave modes having same type and same nature (ARL). The secondary waves are SRL.

irrelevant, but the group velocities (along with the toneburst durations) still affect the mixing zone size. The small mixing zone can be scanned over the distance between the two transducer sources by applying appropriate time delays. Various combinations of primary waves exist that generate sum or difference frequencies at mode cutoffs (e.g. IR point 33) or at zero group velocity points (e.g. IR point 41—in fact the group velocity is negative) for SRL waves.

Unlike the results for co-directional wave mixing in table 8.7, for counter-propagating wave mixing we focus on the lower frequency-thickness products by determining the phase velocity dispersion curves for *fd* products in increments of 1 kHz-mm up to 2 MHz-mm. The internal resonance points in this database were identified based on synchronism for wave mode types to which nonzero power flows. We set the detuning threshold to $\varepsilon = 0.5$ rad m^{-1} initially and then tightened it to $\varepsilon = 0.01$ rad m^{-1}. Some of what appear to be the more interesting wave triplets are shown in figure 8.13 and detailed in table 8.8. For all cases in table 8.8 the combinational harmonic propagates in the direction of waves *a*. Unlike the results for co-directional wave mixing not all types of mixing are reported. Figure 8.13(a) shows mixing of antisymmetric Lamb waves (ARL) to generate symmetric Lamb

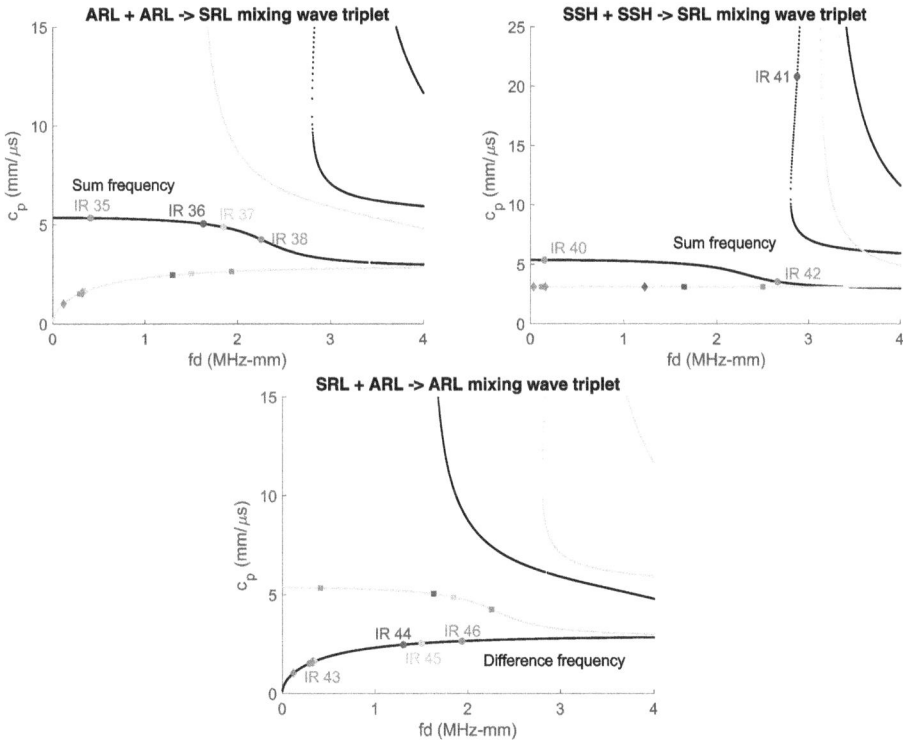

Figure 8.13. Sum and difference frequencies satisfying the synchronism criterion within 0.01 rad m^{-1} for counter-propagating wave modes: (a) mixing two ARL waves to get SRL waves at the sum frequency, (b) mixing two SSH waves to get SRL at the sum frequency, and (c) mixing SRL and ARL waves to get ARL waves at the difference frequency. Symbols: ■ waves *a*, ◆ waves *b*, ● waves *n*.

Table 8.8. Mixing power for counter-propagating mutual interactions, $d = 1$ mm.

IR point	Triplet		f (MHz)	c_p (mm μs^{-1})	c_g (mm μs^{-1})	F_n^{vol} (N s^{-1})	F_n^{surf} (N s^{-1})	F_n (m^{-1})	M_p (m^{-2}) $\times 10^6$
33	S1–S0–S0	a	2.916	28.4694	−1.3578	−0.0145	−0.0025	0.017	8.06
	difference	b	1.226	5.2146	4.8833				
		n	1.690	5.0069	4.0774				
34	S1–S0–S0	a	3.880	5.9851	4.5444	−0.0054i	−0.0086i	0.014	11.5
	difference	b	1.040	5.2609	5.0485				
		n	2.840	3.3571	2.0288				
35	A0–A0–S0	a	0.292	1.5229	2.5650	−0.0504i	0.0523i	0.103	5.87
	sum	b	0.118	1.0261	1.8960				
		n	0.410	5.3424	5.3163				
36	A0–A0–S0	a	1.302	2.4629	3.1339	−0.0381i	0.0444i	0.083	14.4
	sum	b	0.328	1.5956	2.6437				
		n	1.630	5.0451	4.2285				
37	A0–A0–S0	a	1.498	2.5338	3.1308	−0.0424i	0.0464i	0.089	18.0
	sum	b	0.348	1.6332	2.6823				
		n	1.846	4.8818	3.5929				
38	A0–A0–S0	a	1.936	2.6463	3.1077	−0.0563i	0.0463i	0.103	26.8
	sum	b	0.318	1.5761	2.6231				
		n	2.254	4.2543	1.9971				
39	A0–A0–S1	a	2.784	2.7649	3.0529	−1.2E−4i	−0.0468i	0.047	20.8
	sum	b	0.904	2.2531	3.0991				
		n	3.518	6.0887	4.4615				

Table 8.8. (*Continued*)

IR point	Triplet		f (MHz)	c_p (mm μs^{-1})	c_g (mm μs^{-1})	F_n^{vol} (N s^{-1})	F_n^{surf} (N s^{-1})	F_n (m^{-1})	M_p (m^{-2}) $\times 10^6$
40	SH0–SH0–S0 sum	a	0.120	3.0996	3.0996	0.003i	−6.7E−4i	0.0036	0.028
		b	0.032	3.0996	3.0996				
		n	0.152	5.3535	5.3501				
41	SH0–SH0–S1 sum	a	1.650	3.0996	3.0996	0.0034	0.0328	0.0362	19.3
		b	1.222	3.0996	3.0996–1.1055				
		n	2.872	20.7991					
42	SH0–SH0–S0 sum	a	2.498	3.0996	3.0996	0.0466i	−0.0292i	0.0758	7.00
		b	0.162	3.0996	3.0996				
		n	2.660	3.5295	1.8730				
43	**S0–A0–A0 difference**	***a***	**0.410**	**5.3424**	**5.3136**	**0.0099i**	**−0.0166i**	**0.0265**	**10.2**
		b	**0.118**	**1.0261**	**1.8960**				
		n	**0.292**	**1.5229**	**2.5650**				
44	S0–A0–A0 difference	a	1.630	5.0451	4.2285	−0.0173i	−0.0047i	0.0220	7.25
		b	0.328	1.5956	2.6437				
		n	1.302	2.4629	3.1339				
45	S0–A0–A0 difference	a	1.846	4.8818	3.5929	−0.0167i	−0.0037i	0.0204	6.29
		b	0.348	1.6332	2.6823				
		n	1.498	2.5338	3.1308				
46	S0–A0–A0 difference	a	2.254	4.2543	1.9971	−0.0032	0.0030i	0.0064	1.30
		b	0.318	1.5761	2.6231				
		n	1.936	2.6463	3.1077				

(SRL) waves at the sum frequency. Similarly, figure 8.13(b) shows SSH mixing to generate SRL waves at the sum frequency and figure 8.13(c) shows mixing of SRL and ARL waves to generate ARL waves at the difference frequency. The internal resonance points that exist below the cutoff frequencies have the benefit of less, or no, unwanted propagating modes that complicate the signal processing. Internal resonance points 33 and 41 each have the interesting feature that one of the waves, waves a for IR point 33 and waves n for IR point 41, has a negative group velocity. There are more instances of imaginary driving forces in table 8.8 than there are in table 8.7, and while these occur when the wavestructures are out of phase, it is not clear why there are more of these cases for counter-propagating wave mixing than for co-directional wave mixing.

We observe the IR point 43 has an unusually high mixing power (10.2×10^6 m^{-2}) for the fundamental Lamb modes at such low frequencies, making it a good candidate to investigate further for potential simulations and experiments. It is interesting that IR point 35 involves the same three wave modes and frequencies as IR point 43, but in different combination, leading to a much lower mixing power.

8.2.3 Non-collinear, $\theta \neq 0°$ and $\theta \neq 180°$

To analyse the non-collinear wave mixing shown in figure 7.1, we start with equation (7.15), which gives the coefficients of the normal mode expansion for cumulative secondary waves. Equation (7.15) applies when

$$\mathbf{K}_n^* = \mathbf{K}_a \pm \mathbf{K}_b. \tag{8.23}$$

Our first aim is to determine the secondary wavenumber k_n and the direction of secondary wave propagation γ, given that waves a propagate in the X_1 direction and waves b propagate in the direction of angle θ. Substituting into equation (8.23) we get the vector equation

$$k_n^*(\cos \gamma \mathbf{e}_1 + \sin \gamma \mathbf{e}_2) = k_a \mathbf{e}_1 \pm k_b(\cos \theta \mathbf{e}_1 + \sin \theta \mathbf{e}_2),$$

which gives the two scalar equations,

$$k_n^* \cos \gamma = k_a \pm k_b \cos \theta$$

$$k_n^* \sin \gamma = \pm k_b \sin \theta,$$

which in turn give

$$k_n^* = \sqrt{k_a^2 \pm 2k_a k_b \cos \theta + k_b^2} \tag{8.24}$$

$$\tan \gamma = \frac{\pm k_b \sin \theta}{k_a \pm k_b \cos \theta}. \tag{8.25}$$

It is now possible to return to the dispersion curve database and assess which waves a and b, and direction θ for waves b, result in the wavenumber k_n^* corresponding to a synchronized propagating wave mode n in direction γ, that receives nonzero power flow according to table 7.6. When $\theta = 0°$ and $\theta = 180°$ equation (8.24) reduces to $k_n^* = k_a \pm k_b$ and equation (8.25) gives $\gamma = 0$ as they should for co-directional and counter-propagating wave mixing. Let's consider the special case of what we will call cross-propagating waves, where $\theta = 90°$ and equations (8.24) and (8.25) reduce to

$$k_n^* = \sqrt{k_a^2 + k_b^2}$$

$$\tan \gamma = \pm k_b / k_a.$$

The $\pm$ sign is dictated by whether the sum or difference frequency is being analysed. The wavenumber is the same for both sum and difference frequencies.

Of all the possible non-collinear wave mixing possibilities that have nonzero power flow listed in table 7.6, only a subset has been analysed here for brevity, and only for the special case of cross-propagating waves. Both sum frequencies and difference frequencies are considered in each case. The subset contains:

Waves a at $0°$	Waves b at $90°$	Waves n at angle γ
SSH	SSH	SRL
SSH	SSH	SSH
SRL	SSH	SRL
SRL	SRL	SSH
ARL	ARL	SSH

Cases were selected based on the conjecture that they might provide either high or low values of mixing power. Unlike for co-directional and counter-propagating wave mixing it is possible for cross-propagating Lamb waves to generate SSH secondary waves. Driving frequencies were analysed up to 4 MHz for a 1 mm thick plate to focus on the lower frequencies and maintain a reasonably small frequency increment in the dispersion curve database. The results showing where the synchronism criterion is satisfied within a tolerance of 0.5 rad m^{-1} are overlaid on the phase velocity dispersions for the secondary wave modes in figure 8.14. Many synchronism points exist, except for cross-propagating SSH waves to generate secondary SSH waves, where there are no points at the sum frequency and only a limited number at the difference frequency.

A two-step process was followed to construct table 8.9 showing a subset of the internal resonance points for mutual interaction of cross-propagating waves. First the dispersion curve database was analysed for points of synchronism and then the nonlinear forces and mixing power were computed. A tolerance of $\varepsilon = 0.01$ rad m^{-1} was used for synchronization in table 8.9, this is defined as the detuning criterion. This tolerance leads to the nonlinear forces being inexact because the wavenumber from equation (8.24) is used with the wavestructure for k_n^*. All wavestructures are normalized with respect to the average power in the direction of wave propagation. The nonlinear forces are computed from the nonlinear stress components, which are

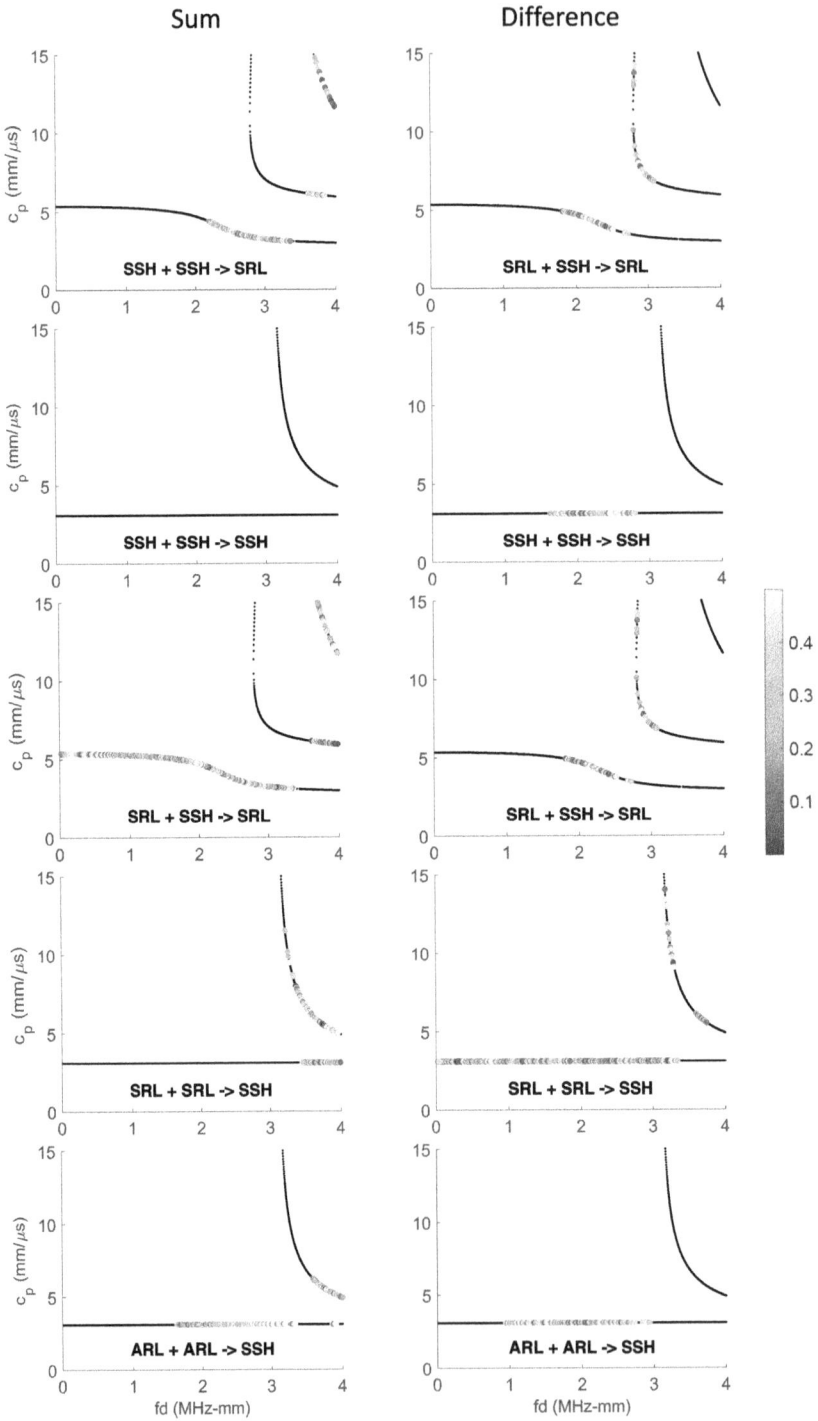

Figure 8.14. Sum and difference frequencies satisfying the synchronism criterion within 0.5 rad m^{-1} for cross-propagating waves, $\theta = 90°$.

Table 8.9. Mixing power for cross-propagating mutual interactions, $d = 1$ mm. Detuning criterion $\varepsilon = 0.01$ rad m^{-1}, unless noted otherwise.

IR point	Triplet		f (MHz)	c_p (mm μs^{-1})	γ (°)	F_n^{vol} (N s^{-1})	F_n^{surf} (N s^{-1})	F_n (m^{-1})	M_p (m^{-2}) $\times 10^6$
47	SH0–SH0–S0	a	0.380	3.0996	80.4	-0.0633	-0.0434	0.107	21.1
	sum	b	2.244	3.0996					
		n	2.624	3.5736					
48	SH0–SH0–S0	a	0.864	3.0996	58.0	-0.0709	0.0380	0.109	30.0
	sum	b	1.382	3.0996					
		n	2.246	4.2713					
49	SH2–SH0–S1	a	3.306	8.9113	-18.3	-0.0144	-0.0189	0.0333	4.47
	difference	b	0.380	3.0996					
		n	2.926	7.4887					
50	SH0–SH0–S0	a	3.912	3.0996	-1.5	$-0.0507i$	$0.0398i$	0.0905	5.62
	difference	b	0.104	3.0996					
		n	3.808	3.0161					
51*	SH2–SH0–SH0	a	3.718	5.6124	-32.2	1.30×10^{-7}	4.99×10^{-11}	1.30×10^{-7}	54.9×10^{-6}
	difference	b	1.292	3.0996					
		n	2.426	3.0996					
52*	SH2–SH0–SH0	a	3.970	4.9605	-26.0	-1.14×10^{-7}	5.06×10^{-11}	1.15×10^{-7}	45.3×10^{-6}
	difference	b	1.210	3.0996					
		n	2.760	3.0996					

(Continued)

Table 8.9. (*Continued*)

IR point	Triplet		f (MHz)	c_p (mm μs^{-1})	γ (°)	F_n^{vol} (N s^{-1})	F_n^{surf} (N s^{-1})	F_n (m^{-1})	M_p (m^{-2}) $\times 10^6$
53	S0–SH0–S0	a	0.156	5.3534	60.2	0.002 00	6.52×10^{-5}	0.002 07	0.0662
	sum	b	0.158	3.0996					
		n	0.314	5.3478					
54	S0–SH0–S0	a	0.346	5.3462	85.0	0.003 09	-1.42×10^{-4}	0.003 23	0.772
	sum	b	2.294	3.0996					
		n	2.640	3.5535					
55	S0–SH0–S0	a	0.482	5.3374	61.5	0.006 32	1.98×10^{-4}	0.006 52	0.683
	sum	b	0.516	3.0996					
		n	0.998	5.2695					
56	S0–SH0–S0	a	0.718	5.3140	65.0	0.009 91	2.55×10^{-4}	0.0102	1.90
	sum	b	0.900	3.0996					
		n	1.618	5.0522					
57	S0–SH0–S0	a	0.740	5.3112	72.3	0.0108	0.856×10^{-4}	0.0108	3.05
	sum	b	1.350	3.0996					
		n	2.090	4.5704					
58	S1–SH0–S0	a	2.938	33.842	−76.5	−0.00564i	$3.04\mathrm{i} \times 10^{-4}$	0.005 95	2.77
	difference	b	1.118	3.0996					
		n	1.820	4.9057					

Table 8.9. (*Continued*)

IR point	Triplet		f (MHz)	c_p (mm μs^{-1})	γ (°)	F_n^{vol} (N s^{-1})	F_n^{surf} (N s^{-1})	F_n (m^{-1})	M_p (m^{-2}) $\times 10^6$
59	S1–SH0–S0	a	3.042	129.53	−86.5	−0.00120i	0.763i $\times 10^{-4}$	0.001 28	0.680
	difference	b	1.182	3.0996					
		n	1.860	4.8683					
60	S1–SH0–S0	a	3.542	6.1867	−30.5	0.0132	0.174 $\times 10^{-4}$	0.0133	9.09
	difference	b	1.044	3.0996					
		n	2.498	3.7606					
61	S0–S0–SH0	a	1.300	5.1919	70.6	−0.004 17	0.001 20	0.005 37	2.77
	sum	b	2.584	3.6272					
		n	3.884	5.1346					
62	S0–S0–SH0	a	0.480	5.3375	−20.7	3.8 $\times 10^{-5}$	−4.7 $\times 10^{-5}$	8.6 $\times 10^{-5}$	0.0175
	difference	b	0.182	5.3528					
		n	0.298	3.0996					
63	A1–A0–SH0	a	1.864	2.6312	17.9	−0.009 58	0.001 15	0.0107	6.14
	sum	b	2.000	8.7542					
		n	3.864	5.1910					
64	A1–A0–SH0	a	3.662	5.1651	−32.4	−0.009 42	2.42 $\times 10^{-4}$	0.009 66	5.16
	difference	b	1.058	2.3475					
		n	2.604	3.0996					

*Detuning criterion = 0.02.

determined from the displacement gradients $\mathbf{H}_a$ and $\mathbf{H}_b$. These displacement gradients have to be transformed to a global coordinate system to properly account for their interactions.

For IR points 47, 48, 61, and 63, waves a and b can be interchanged to get exactly the same nonlinear forces and mixing power at the complementary angle to γ. Cross-propagating waves of the same type that generate SSH secondary waves are unique in that collinear waves of the same type cannot generate secondary SH waves. Additionally, collinear SH waves cannot generate secondary SH waves. Thus, it is interesting to assess mixing of SSH–SSH, SRL–SRL, and ARL–ARL waves to generate SSH waves. Likewise, cross-propagating SSH and SRL waves have particle motion in the same direction, which might promote interactions that increase power flow to secondary waves. Thus, mixing of SSH–SRL waves to generate SRL waves is analysed.

Cross-propagating SSH–SSH waves are shown in figure 8.14 to not generate any synchronized SSH waves at the sum frequency for a tolerance of $\varepsilon = 0.5$ rad m^{-1}. IR points 51 and 52 have a tolerance of $\varepsilon = 0.02$ rad m^{-1}, but the nonlinear forces and mixing power are extremely low, strongly suggesting that no power flows from SSH–SSH wave interaction to SSH waves. A parity analysis could be conducted to assess this if there was sufficient interest in addressing whether the power flow is zero or just extremely small.

IR points 61, 63, and 64 demonstrate that power flows to SH0 secondary waves from both SRL–SRL and ARL–ARL mixing and the mixing power parameter is reasonably large. The mixing power for IR point 60 (mixing S1 and SH0 to generate S0 at the sum frequency) is larger. However, the largest mixing power parameters occur for IR points 47 and 48, where SH0–SH0 mixing generates S0 waves at the sum frequency. It remains for future work to discover why so much power flows from cross-propagating SH0 waves to S0 waves and to analyse mixing angles between 0 and 90 degrees.

8.3 Hollow cylinders

The analysis of self-interactions and mutual interactions for plates in the previous two sections can be extended to hollow cylinders for both circumferential and longitudinal wave propagation. However, doing so requires changing from a Cartesian coordinate system to a more general curvilinear coordinate system, which is beyond the scope of this book. Instead, we will rely upon Liu *et al*'s [21] analysis of self-interaction and mutual interaction of axisymmetric waves propagating in the longitudinal direction of hollow (circular) cylinders. Liu *et al* have also analysed the interaction of flexural waves in hollow cylinders [22, 23]. Here we will highlight the approach and findings for axisymmetric waves propagating in the longitudinal direction.

The nonlinear material behavior is described by the Landau–Lifshitz hyperelastic model where the strain energy function is written to third order in strain, which is paired to the second Piola–Kirchhoff stress. The second Piola–Kirchhoff stress is shifted to the first Piola–Kirchhoff stress and decomposed such that the nonlinear part can be used as the driving force within the framework of the perturbation solution methodology. The orthogonality relation makes the normal mode expansion approach tractable and leads to the definition of nonlinear volumetric and surface driving forces that are normalized with respect to the complex power flow. Recall that for plates a parity analysis was used to identify to which types of waves there is no power flow. Instead, here an analysis based on the structure of the matrices and vectors used to represent the stress, divergence of the stress, and velocity indicates to which wave types no power flows. For self-interaction of either $L(0,n)$ or $T(0,n)$ type waves, nonzero power flows to $L(0,n)$ secondary waves, but no power flows to $T(0,n)$ secondary waves. For mutual interactions between $L(0,n)$–$L(0,n)$ or $T(0,n)$–$T(0,n)$ waves, no power flows to $T(0,n)$ secondary waves. But for mutual interactions between $L(0,n)$-$T(0,n)$ waves, nonzero power flows to $L(0,n)$ and $T(0,n)$ secondary waves. We should mention that third order interactions were not addressed herein.

The synchronism condition in plates could be analysed from the dispersion relations, but for hollow cylinders the dispersion relations in equation (4.111) appear to be much too complicated. Thus, we resort to graphical analysis of the dispersion curves. For self-interaction, we seek synchronism with $L(0,n)$ secondary wave modes for there to be nonzero power flow. The internal resonance points are shown on the phase velocity dispersion curves for a pipe in figure 8.15 and tabulated in table 8.10. The first eight IR points are numbered to match the IR points in a plate (see figure 8.1 and table 8.1). IR points 10–12 occur where the phase velocity is the Lamé wave speed $\sqrt{2}\,c_T$. These IR points can also exist for Lamb waves when $\sqrt{2}\,c_T \geqslant c_L$, which is not the case for aluminum. Note that waves in plates having $c_p = \sqrt{2}\,c_T$ are known as Lamé modes.

Mutual interactions in pipes can be analysed in an analogous way to mutual interactions in a plate, but in a cylindrical coordinate system [21]. The limitation of waves propagating in the longitudinal direction of a hollow cylinder limits mutual interactions to be either co-directional or counter-propagating, unless longitudinal

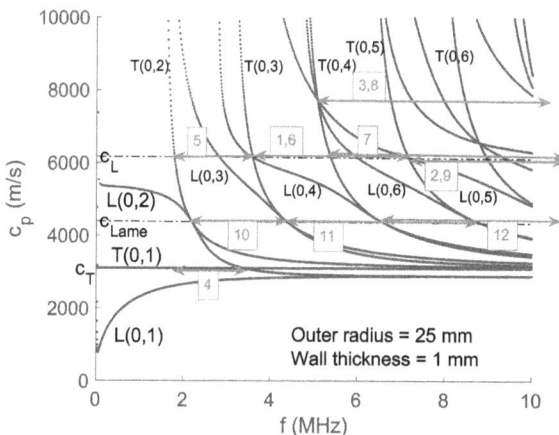

Figure 8.15. Internal resonance points for an aluminum pipe.

Table 8.10. Internal resonance points for axisymmetric waves in hollow cylinders with 1 mm wall thickness.

IR point	Description	Driving frequency f (MHz)	Phase velocity c_p
1	L(0,4)–L(0,5)	3.58	c_L
2	L(0,5)–L(0,9)	7.16	c_L
3	Mode crossing point, L(0,5)–L(0,9) and L(0,6)–L(0,9)	5.05	7790 m s^{-1}
4	T(0,1)–L(0,2)	1.70	c_T
5	T(0,2)–L(0,4)	1.79	c_L
6	T(0,3)–L(0,5)	3.58	c_L
7	T(0,4)–L(0,7)	5.37	c_L
8	Mode crossing point, T(0,4)–L(0,9)	5.05	7790 m s^{-1}
9	T(0,5)–L(0,9)	7.16	c_L
10	T(0,2)–L(0,3)	2.19	c_{Lame}
11	T(0,3)–L(0,6)	4.38	c_{Lame}
12	T(0,4)–L(0,7)	6.58	c_{Lame}

waves and circumferential waves are mixed as cross-propagating or if flexural waves are considered to follow helical wave paths.

8.4 Arbitrary cross-section

The dispersion relations for waveguides having a cross-section of arbitrary shape, as shown in figure 4.4, can be determined using the semi-analytical finite element (SAFE) method as described in section 4.1.4. Srivastava *et al* [24] showed that the SAFE method is amenable to nonlinear analysis. The general approach is the same as for a plate. SAFE is used to determine the wavenumbers of propagating modes and their wavestructures. The normal mode expansion is used to represent the secondary wavefield based on orthogonality of the modes. The nonlinear surface and volumetric forces are computed based on the nonlinear part of the primary stress field and the secondary particle velocity field. Nucera and Lanza di Scalea [25, 26] implemented this analysis in the COMSOL Multiphysics software and demonstrated its application to railroad track for both synchronous and asynchronous wave modes and also to a viscoelastic plate (i.e. with internal damping). The concise nature of this section contrasts with its power and potential impact to apply to virtually any homogeneous waveguide, including the special cases of plates and pipes already discussed.

8.5 Half-space

We turn attention now to Rayleigh waves propagating along the surface of a half-space, which are nondispersive and have elliptical particle motion. How can measurements of wave motion on the surface be related to the material nonlinearity? This nonlinear problem has been solved by Shui and Solodov [27], but here we simply aim to relate the surface displacements to the acoustic nonlinearity coefficient β as in Herrmann *et al* [28]. In chapter 4 the displacements were related to the Helmholtz potential functions in equations (4.14) and (4.15). Thus, the Rayleigh wavefield can be written as

$$u_1 = ikA_1e^{-rX_3}e^{i(kX_1-\omega t)} + sB_1e^{-sX_3}e^{i(kX_1-\omega t)} \tag{8.26}$$

$$u_3 = -rA_1e^{-rX_3}e^{i(kX_1-\omega t)} + ikB_1e^{-sX_3}e^{i(kX_1-\omega t)}, \tag{8.27}$$

where k is the Rayleigh wavenumber and

$$p \equiv ir, \quad q \equiv is \tag{4.11}$$

$$p^2 \equiv \left(\frac{\omega}{c_L}\right)^2 - k^2 = k_L^2 - k^2, \quad q^2 \equiv \left(\frac{\omega}{c_T}\right)^2 - k^2 = k_T^2 - k^2. \tag{4.9}$$

However, the first term for u_1 should be a longitudinal wave, suggesting that we should divide everything by (ik), which then agrees with Viktorov's [29] results:

$$u_1 = A_1e^{-rX_3}e^{i(kX_1-\omega t)} + \frac{s}{ik}B_1e^{-sX_3}e^{i(kX_1-\omega t)} \tag{8.28}$$

$$u_3 = -\frac{r}{ik}A_1e^{-rX_3}e^{i(kX_1-\omega t)} + B_1e^{-sX_3}e^{i(kX_1-\omega t)}. \tag{8.29}$$

The relationship between constants A_1 and B_1 given in equation (4.27) can also be written as

$$A_1 = i\frac{k^2 + s^2}{2kr}B_1 \tag{8.30}$$

enabling us to write the primary wavefield as

$$u_1 = A_1\left[e^{-rX_3} - \frac{2rs}{k^2 + s^2}e^{-sX_3}\right]e^{i(kX_1-\omega t)} \tag{8.31}$$

$$u_3 = \frac{ir}{k}A_1\left[e^{-rX_3} - \frac{2k^2}{k^2 + s^2}e^{-sX_3}\right]e^{i(kX_1-\omega t)}. \tag{8.32}$$

The second harmonic wavefield can be written similarly as

$$u_1(2\omega) = A_2\left[e^{-2rX_3} - \frac{2rs}{k^2 + s^2}e^{-2sX_3}\right]e^{i2(kX_1-\omega t)} \tag{8.33}$$

$$u_3(2\omega) = \frac{ir}{k}A_2\left[e^{-2rX_3} - \frac{2k^2}{k^2 + s^2}e^{-2sX_3}\right]e^{i2(kX_1-\omega t)}. \tag{8.34}$$

In Rayleigh wave testing it is common to measure either the displacement u_3 or particle velocity $\dot{u}_3$ normal to the surface, thus we consider the relative nonlinearity coefficient β' from such measurements,

$$\beta' = \left. \frac{u_3(2\omega)}{[u_3(\omega)]^2} \right|_{X_3=0} = \frac{\frac{ir}{k} A_2 \left[1 - \frac{2k^2}{k^2+s^2}\right]}{\left(\frac{ir}{k} A_1\right)^2 \left[1 - \frac{2k^2}{k^2+s^2}\right]^2} = \frac{A_2}{\frac{ir}{k} A_1^2 \left[1 - \frac{2k^2}{k^2+s^2}\right]}. \tag{8.35}$$

We can now substitute

$$A_2 = \frac{1}{8} \beta k_L^2 A_1^2 X_1, \tag{3.12}$$

where the subscript L has been added to the wavenumber to denote longitudinal bulk waves, to write

$$\beta' = \frac{\frac{1}{8} \beta k_L^2 X_1}{\frac{ir}{k} \left[1 - \frac{2k^2}{k^2+s^2}\right]}. \tag{8.36}$$

Rearranging for the acoustic nonlinearity coefficient for longitudinal bulk waves,

$$\beta = 8\beta' \frac{ir}{k} \frac{1}{k_L^2 X_1} \left[1 - \frac{2k^2}{k^2+s^2}\right]. \tag{8.37}$$

Thus, we can relate the 'absolute' β with the 'relative' β', when the relative one is obtained from Rayleigh wave measurements. It is important to realize however, that this does not address the calibration issue that wave amplitudes are typically measured in volts, which need to be converted into distance units (e.g. nm). The estimation of absolute β from relative β' for bulk longitudinal waves is discussed by Jhang *et al* [30] (their section 2.3.3).

8.6 Closure

This chapter has covered a lot of territory, making it prudent to include more detailed closing remarks. Selection of primary wave modes and frequencies that satisfy the internal resonance conditions and transfer maximal power to secondary waves has been strongly emphasized because the cumulative nature of the generated secondary waves can be attributed to nonlinearity in the waveguide as opposed to measurement system nonlinearities. Moreover, we seek to detect subtle, early, indications of material degradation and therefore aim to maximize detection of the weak material nonlinearity. However, in so doing we do not want geometric nonlinearity from finite amplitude primary waves to mask the material nonlinearity. If due care is taken to avoid making measurements at null points of the bounded oscillation, asynchronous wave self-interactions or mutual interactions can be useful. On the other hand, it is difficult to justify selecting wave modes and frequencies shown by parity analysis to transfer no power to the secondary waves.

It is important to understand that the cumulative nature of the secondary waves occurs only when the waves are actively interacting in the mixing zone of a lossless material. The mixing zone for self-interactions of continuous waves is theoretically infinite. However, if finite duration wave packets are used then the mixing zone is limited by group velocity mismatch. In this respect, the combination of synchronism and group velocity matching of S1–S2 and S2–S4 Lamb waves when $c_p = c_L$ (i.e. IR points 1 and 2) is ideal; it is just difficult to preferentially activate these modes with conventional transducers. The limited number of IR points for self-interaction push us to explore mixing of two waves, where again the size of the mixing zone plays a major role in the rate of accumulation. This chapter shows that many wave mixing combinations (i.e. triplets) exist depending upon the propagation directions. Due to the discrete nature of the dispersion curve database, the synchronism criterion is unavoidably satisfied only approximately. The IR points shown on the dispersion curve figures have a larger detuning tolerance than do the tabulated IR points. Although the cumulative nature of secondary waves is limited when the mixing zone is small, the mixing power parameter seems like a reasonable way to assess energy transfer from primary to secondary waves. The nonlinear forces and associated mixing power parameter are dependent upon the perturbation solution methodology, which while not exact, is a good approximation for weakly nonlinear problems. The method of multiple scales can be applied when that is not good enough, which leads us to the topic of the next chapter, finite amplitude pulse loading. For sufficiently high amplitudes the waveform progressively evolves due to nonlinearities.

One last point should be made before moving on. Much of the analysis and interpretation related to the IR points identified in this chapter is more detailed than in previous publications. Moreover, there is emphasis on SH wave modes and their coupling to secondary Lamb wave modes because they provide very useful opportunities for measurements, as will be discussed subsequently.

References

[1] Deng M 1996 Second-harmonic properties of horizontally polarized shear modes in an isotropic plate *Jpn. J. Appl. Phys.* **35** 4004
[2] Deng M 1998 Cumulative second-harmonic generation accompanying nonlinear shear horizontal mode propagation in a solid plate *J. Appl. Phys.* **84** 6
[3] Deng M 1999 Cumulative second-harmonic generation of Lamb-mode propagation in a solid plate *J. Appl. Phys.* **85** 3051–8
[4] de Lima W J N and Hamilton M F 2003 Finite-amplitude waves in isotropic elastic plates *J. Sound Vib.* **265** 819–39
[5] Matlack K H, Kim J-Y, Jacobs L J and Qu J 2011 Experimental characterization of efficient second harmonic generation of Lamb wave modes in a nonlinear elastic isotropic plate *J. Appl. Phys.* **109** 014905
[6] Rose J L 2014 *Ultrasonic Guided Waves in Solid Media* (Cambridge: Cambridge University Press)

[7] Müller M F, Kim J-Y, Qu J and Jacobs L J 2010 Characteristics of second harmonic generation of Lamb waves in nonlinear elastic plates *J. Acoust. Soc. Am.* **127** 2141–52

[8] Matsuda N and Biwa S 2011 Phase and group velocity matching for cumulative harmonic generation in Lamb waves *J. Appl. Phys.* **109** 094903

[9] 2020 *Measurement of Nonlinear Ultrasonic Characteristics* (Springer Series in Measurement Science and Technology) ed K-Y Jhang, C Lissenden, I Solodov, Y Ohara and V Gusev (Singapore: Springer)

[10] Lissenden C J 2021 Nonlinear ultrasonic guided waves—principles for nondestructive evaluation *J. Appl. Phys.* **129** 021101

[11] Erofeyev V I 2003 *Wave Processes in Solids with Microstructure* (Singapore: World Scientific)

[12] Erofeyev V I and Raskin I G 1991 Propagation of shear waves in nonlinearly-elastic solid *Prikl. Mekh.* **27** 127–9

[13] Nayfeh A H and Mook D T 1979 *Nonlinear Oscillations* (New York: Wiley)

[14] Kanda K and Sugiura T 2018 Analysis of guided waves with a nonlinear boundary condition caused by internal resonance using the method of multiple scales *Wave Motion* **77** 28–39

[15] Kanda K and Sugiura T 2018 Analysis of damped guided waves using the method of multiple scales *Wave Motion* **82** 86–95

[16] Kanda K and Sugiura T 2021 Internally resonant guided waves arising from quadratic classical nonlinearities with damping *Int. J. Solids Struct.* **216** 250–7

[17] Kanda K and Maruyama T 2022 Theoretical analysis of the dispersion of Lamb waves forming a wave packet of finite-bandwidth using the method of multiple scales *Int. J. Solids Struct.* **234–5** 111268

[18] Kanda K and Maruyama T 2023 Theoretical analysis of forced Lamb waves using the method of multiple scales and Green's function method *Acta Mech.* **234** 3533–46

[19] Osika M, Ziaja-Sujdak A, Radecki R, Cheng L and Staszewski W J 2022 Nonlinear modes in shear horizontal wave propagation–analytical and numerical analysis *J. Sound Vib.* **540** 117247

[20] Matsuda N and Biwa S 2014 Frequency dependence of second-harmonic generation in Lamb waves *J. Nondestruct. Eval.* **33** 169–77

[21] Liu Y, Khajeh E, Lissenden C J and Rose J L 2013 Interaction of torsional and longitudinal guided waves in weakly nonlinear circular cylinders *J. Acoust. Soc. Am.* **133** 2541–53

[22] Liu Y, Lissenden C J and Rose J L 2014 Higher order interaction of elastic waves in weakly nonlinear hollow circular cylinders. I. Analytical foundation *J. Appl. Phys.* **115** 214901

[23] Liu Y, Khajeh E, Lissenden C J and Rose J L 2014 Higher order interaction of elastic waves in weakly nonlinear hollow circular cylinders. II. Physical interpretation and numerical results *J. Appl. Phys.* **115** 214902

[24] Srivastava A, Bartoli I, Salamone S and Lanza di Scalea F 2010 Higher harmonic generation in nonlinear waveguides of arbitrary cross-section *J. Acoust. Soc. Am.* **127** 2790–6

[25] Nucera C and Lanza di Scalea F 2012 Higher-harmonic generation analysis in complex waveguides via a nonlinear semianalytical finite element algorithm *Math. Probl. Eng.* **2012** 1–16

[26] Nucera C and Lanza di Scalea F 2014 Nonlinear semianalytical finite-element algorithm for the analysis of internal resonance conditions in complex waveguides *J. Eng. Mech.* **140** 502–22

[27] Shui Y and Solodov I Y 1988 Nonlinear properties of Rayleigh and Stoneley waves in solids *J. Appl. Phys.* **64** 6155–65

[28] Herrmann J, Kim J-Y, Jacobs L J, Qu J, Littles J W and Savage M F 2006 Assessment of material damage in a nickel-base superalloy using nonlinear Rayleigh surface waves *J. Appl. Phys.* **99** 124913

[29] Viktorov I A 1967 *Rayleigh and Lamb Waves—Physical Theory and Applications* (New York: Plenum)

[30] Jhang K-Y, Choi S and Kim J 2020 Measurement of nonlinear ultrasonic parameters from higher harmonics *Measurement of Nonlinear Ultrasonic Characteristics* (Berlin: Springer Nature) pp 9–60

IOP Publishing

Nonlinear Ultrasonic Guided Waves

Cliff J Lissenden

Chapter 9

Finite amplitude pulse loading

The general field of nonlinear acoustics is dominated by applications in fluids such as parametric arrays, acoustic streaming, radiation force, sonic booms, and biomedical applications such as high-intensity focused ultrasound, lithotripsy, and separating soft tissue. In these problems the nonlinearity can be quite strong. This book deals with solid media, wherein both dilatational and shear waves are supported. In all cases the nonlinearity is weak, whether due to the material, the finite amplitude, or surfaces in contact. That is, except for this chapter, where finite amplitude pulses leading to high stresses are considered for waves propagating along the surface of a half-space. Pulse actuation results in a broad frequency bandwidth that makes it difficult, if not impossible, to isolate secondary waves at higher harmonic or combinational harmonic frequencies. That is why this book—again, with the exception of this chapter—focuses on sinusoidal tonebursts or continuous sinusoidal waves and not pulses. However, the current chapter is concerned with progressive waveforms that evolve as they propagate. In the classical sense the progression is from sinusoid to sawtooth. Shock formation is possible. The reader is reminded that Rayleigh waves are nondispersive and are confined roughly within one wavelength of the free surface of a homogeneous isotropic half-space. Relaxing these restrictions dictates that the name be changed to surface acoustic waves (SAWs), for example when the material is anisotropic (e.g. a single crystal), layered, or not actually a half-space.

Consider a finite amplitude pulse loading acting upon the surface of a half-space (or a material thick enough that it can be considered to act like a half-space). While the pulse amplitude is large, we will limit it such that it generates elastic waves useful for nondestructive evaluation rather than elastoplastic waves that damage the material. For simplicity, the half-space is homogeneous and isotropic, although there is a significant body of knowledge on anisotropic and layered half-spaces. Nonlinear Rayleigh waves have important applications ranging from seismic waves on one end of the wavelength spectrum to SAWs for semiconductor devices on the other. Our NDE application is between these two extremes. Short duration laser pulses are capable of providing the desired actuation. The absorption of the electromagnetic energy by a thin skin layer of material creates a large thermal gradient. But the skin layer is constrained by the surrounding material, leading to the generation of elastic waves. As described by Scruby and Drain [1], multiple types of elastic waves are generated in both the thermoelastic and ablative regimes, although we will only analyse the Rayleigh waves. Rayleigh waves propagating along the free surface can be received by many types of sensors and are therefore very useful. Typically, it is the out-of-plane displacement component or particle velocity component that is detected.

9.1 Descriptors of nonlinearity

Since our interest now goes beyond weak nonlinearity, we will introduce additional descriptors. Waveform distortion occurs progressively because the phase speed of the traveling wave depends on the particle velocity as depicted in figure 9.1, which is based on Hamilton and Blackstock's [2] figure 1 in their chapter 4. The peaks (in tension) travel faster than the valleys (in compression) causing the waveform to steepen. When the slope of the waveform becomes infinite a shock instability occurs. The shock formation distance is denoted $\bar{x}$, and as noted in chapter 3 can be computed from $\bar{x} = 1/(\beta \varepsilon k)$ for bulk waves where $\varepsilon = v_o/c_o$ is the acoustic Mach number (v_o is the initial particle velocity and c_o is the small amplitude wave speed). The acoustic Mach number can be as high as 0.01 [3]. Also defined in figure 9.1 is the retarded time $\tau = t - x/c_0$, which enables comparisons of one period at different propagation distances. The shock discontinuity prevents the unstable waveform shown for $x > \bar{x}$ in figure 9.1 from occurring.

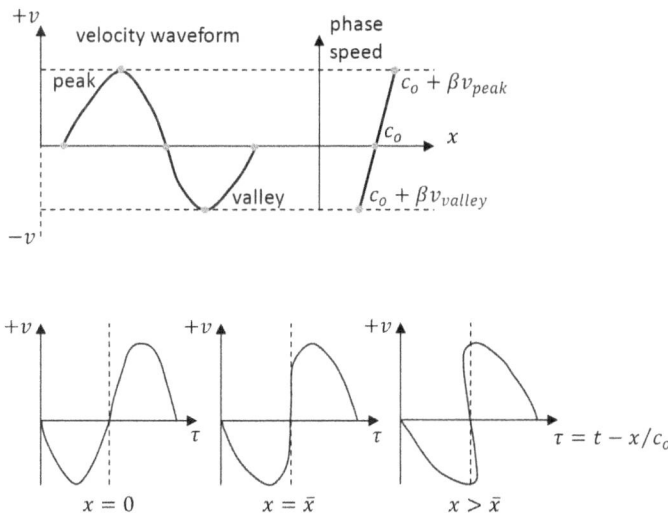

Figure 9.1. Particle velocity waveforms: in the spatial domain showing the variation in phase speed (top) and in the temporal domain at different propagation distances (bottom). After [2].

9.2 Experimental results from laser generation

As already mentioned, nonlinear wave propagation studies are usually performed with sinusoidal continuous waves or long tonebursts, which typically simplifies data analysis and modeling. However, impulse loading, as can be approximated by a short laser pulse, can result in high stress and significantly higher, albeit very localized, nonlinearity.

It appears that the first application of laser generated surface acoustic waves (SAWs) was cleaning dust particles off the surface of silicon wafers by Kolomenskii *et al* [4]. Therein, the authors cite their earlier work published in the Russian literature in the early 1990s. The methodology is typical of a number of similar studies using laser ultrasound to send and receive nonlinear surface waves. A Q-switched Nd:YAG laser emitting nanosecond pulses creates a laser beam that is formed into a point or line on the surface. A fluid layer is often placed on the surface to absorb more energy. Kolomenskii *et al* [5] describe the first experiments, which made use of a 1064 nm Nd:YAG laser with pulse durations of 7 ns and pulse energies up to 130 mJ. The laser beam was focused into a 10 μm wide line that was 8 mm long. Surface waves were detected by the continuous wave laser dual-probe-beam deflection method, which detects the surface slope and therefore gives the out-of-plane component of the particle velocity v_3. The detected v_3 waveform spreads out with propagation distance as diagrammatically depicted in figure 9.2. The in-plane component of the particle velocity v_1 at the surface can be computed using the Hilbert transform due to the elliptical out-of-phase wave motion of a Rayleigh wave as shown in example 9.1. The computed v_1 waveform can exhibit shock formation at the leading and trailing edges of the waveform as shown in figure 9.2. The v_1 waveform has the classical inverted N-shape. The (negative) v_1 values at the bottom of the inverted N are larger in magnitude than the (positive) v_1 values at the top of the inverted N, which causes pulse broadening resulting in a frequency

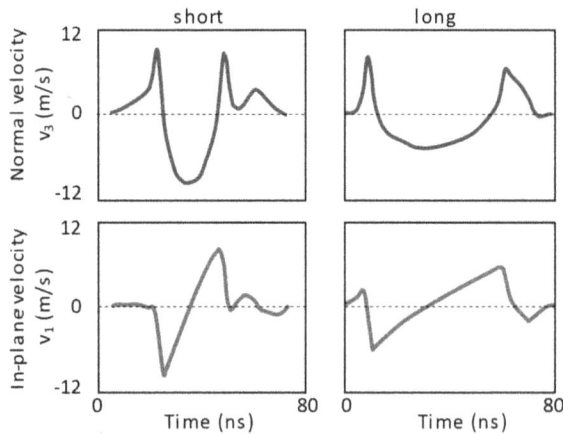

Figure 9.2. Waveforms at short and long propagation distances based on Kolomenskii *et al* [5] results on fused silica.

down-conversion. This waveform type and evolution occurs for fused silica, which has a negative value for the nonlinear parameter ε_1 as per table 9.1.

On the other hand, Kolomenskii and Schuessler [6] found that nonlinear Rayleigh waveforms in aluminum are much different, as depicted in figure 9.3. The out-of-plane component of particle velocity v_3 waveform is predominately a negative spike-like pulse that becomes sharper and has larger amplitude as it propagates. The in-plane component of particle velocity v_1 is N-shaped and exhibits a shock corresponding to the spike in the v_3 waveform. The velocity of the negative peak has a larger magnitude than the positive peaks causing pulse compression resulting in a frequency up-conversion. Unlike fused silica, aluminum has a positive value for the nonlinear parameter ε_1 (table 9.1). Schuessler *et al* [7] report results for stainless steel similar to those for aluminum.

Table 9.1. Nonlinear parameters from Kolomenskii and Schuessler [6].

Material	ε_1	ε_2	ε_3
Fused silica	-0.8 ± 0.1	-0.25 ± 0.005	-4.0 ± 0.5
Polycrystalline aluminum	0.7 ± 0.1	-1.02 ± 0.01	2.5 ± 0.5

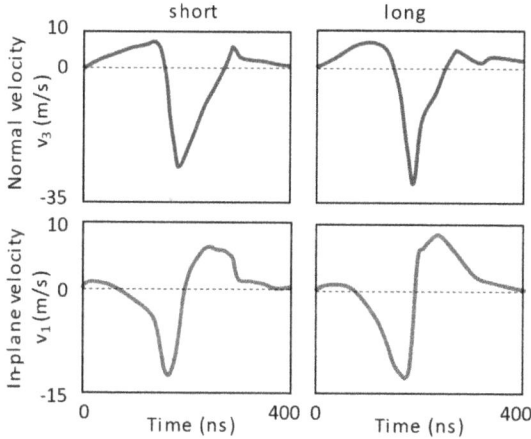

Figure 9.3. Waveforms at short and long propagation distances based on Kolomenskii and Schuessler [6] results on aluminum.

Example 9.1 Relationship between Rayleigh wave components.

We will show the integral transform relationship between surface wave particle velocity components at the free surface of the half-space.

Start by writing the surface waves at the free surface as an infinite series

$$v_i(x_1, \tau) = \sum_{n=-\infty}^{\infty} v_n(x_1) u_{ni} \mathrm{e}^{\mathrm{i}n\omega\tau},$$

where x_1 is the propagation direction, τ is the retarded time, and the coefficients are in terms of their three inhomogeneous partial wave constituents, i.e.

$$u_{ni} = \begin{cases} B_i = \sum_{m=1}^{3} C_m \alpha_i^{(m)} \text{ for } n > 0 \\ B_i^* = \sum_{m=1}^{3} (C_m \alpha_i^{(m)})^* \text{ for } n < 0. \end{cases}$$

Here, $\alpha_i^{(m)}$ is the eigenvector for partial wave m and the C_m are the coefficients required to satisfy the traction free boundary conditions. The Fourier transform is

$$v_j(x_1, \tau) = \frac{1}{2\pi} \int_{-\infty}^{\infty} v_j(x_1, \omega) e^{i\omega t} d\omega$$

and letting

$$v_j(x_1, \omega) = v(x_1, \omega) u_j(\omega),$$

where

$$u_j(\omega) = \begin{cases} B_j \text{ for } \omega > 0 \\ B_j^* \text{ for } \omega < 0 \end{cases}$$

we obtain

$$v_j(x_1, \tau) = \frac{1}{2\pi} \int_{-\infty}^{\infty} \frac{u_j(\omega)}{u_i(\omega)} v(x_1, \omega) u_i(\omega) e^{i\omega t} d\omega.$$

Now letting

$$F(x_1, \omega) = v(x_1, \omega) u_i(\omega) \text{ and } G(\omega) = \frac{u_j(\omega)}{u_i(\omega)}$$

we see that

$$v_j(x_1, \tau) = \frac{1}{2\pi} \int_{-\infty}^{\infty} G(\omega) F(x_1, \omega) e^{i\omega t} d\omega = g * f,$$

where the convolution theorem has been used. Taking the inverse Fourier transform of $F(x_1, \omega)$ gives $f(x_1, \tau) = v_i(x_1, \tau)$ and Kumon [8] has shown the details resulting in

$$g(\tau) = \text{Re}\left(\frac{B_j}{B_i}\right) v_i(x_1, \tau) - \text{Im}\left(\frac{B_j}{B_i}\right) \frac{1}{\pi} \int_{-\infty}^{\infty} \frac{v_i(x_1, \tau')}{\tau' - \tau} d\tau'.$$

For the last step, we use the Hilbert transform

$$\mathcal{H}[h(\tau)] = \frac{1}{\pi} \int_{-\infty}^{\infty} \frac{h(\tau')}{\tau' - \tau} d\tau'$$

to obtain the final result

$$v_j(x_1, \tau) = \text{Re}\left(\frac{B_j}{B_i}\right) v_i(x_1, \tau) - \text{Im}\left(\frac{B_j}{B_i}\right) \mathcal{H}[v_i(x_1, \tau)]$$

relating the particle velocity components at the free surface. The same relation applies for the displacement components [9]. We now specialize this relation for Rayleigh waves presuming that we measure the out-of-plane component and wish to know the in-plane component as follows:

$$v_j(\tau) = -\frac{1}{\gamma}\mathcal{H}[v_i(\tau)]$$

$$\gamma = \left(\frac{1 - \delta/\delta_1}{1 - \delta}\right)^{1/4}$$

$$\delta = \left(\frac{c_R}{c_T}\right)^2, \delta_1 = \left(\frac{c_L}{c_T}\right)^2.$$

The literature investigating the physics of broadband progressive surface wave pulses [3, 5, 10, 11], and their application to (i) surface cleaning [4], (ii) characterization [6, 7, 12], and (iii) fracture [13–15] all employ the dual-probe-beam deflection method to determine surface slope at two points approximately 16 mm apart. A smooth surface is required to measure the slope, which is proportional to v_3. As noted above, v_1 can be computed from v_3 using the Hilbert transform. The displacement components u_1 and u_3 can then be computed by integrating v_1 and v_3. This approach is used to generate figure 9.4 [16] from the data of Kolomenskii and Schuessler [6]. Experimental studies assess fused quartz [10, 11, 14], fused silica [5, 6], aluminum [6], stainless steel [3, 7, 12], and single-crystal silicon [4, 10, 13–15, 17].

Kolomenskii et al [3] noted that high-amplitude pulses in stainless steel propagated as relatively stable solitons as the anomalous dispersion was balanced by the nonlinearity-induced increase in amplitude. Jerebtsov et al [12] later characterized the linear and nonlinear material parameters for stainless steel.

Lomonosov et al's [10, 11] experimental results on single-crystal silicon and amorphous fused quartz clearly show the formation of the leading edge and trailing edge shocks in the v_1 waveform and that they separate as they propagate. A light-absorbing liquid was again used at the source location to intensify the elastic wave generation. Surface cracking was observed near the source, but not in the measurement zone. Particle velocities up to 40 m s^{-1} were measured, which correspond to an acoustic Mach number of 0.0085. The v_1 component displays a characteristic inverted N-shape, as shown in figure 9.5 [11]. The cusps in the vicinity of the shocks are a nonlocal nonlinearity phenomenon observed in solids but not fluids, which were analysed by Hamilton et al [18]. The cusps also distinguish surface waves from bulk longitudinal and shear waves. The pulse expansion leads to a decrease in the primary frequency and secondary peaks at the harmonics. The predictions from the evolution equation are in very good agreement with the experimental results (see Lomonosov et al [11] for example), which will be discussed in more detail in the following section. Lomonosov's group then applied nonlinear surface acoustic waves as a driving force for fracture mechanics [13–15, 17].

Crack nucleation is a critical component of fracture mechanics but is so difficult to deal with in the laboratory that the use of artificial precracks or flaws is

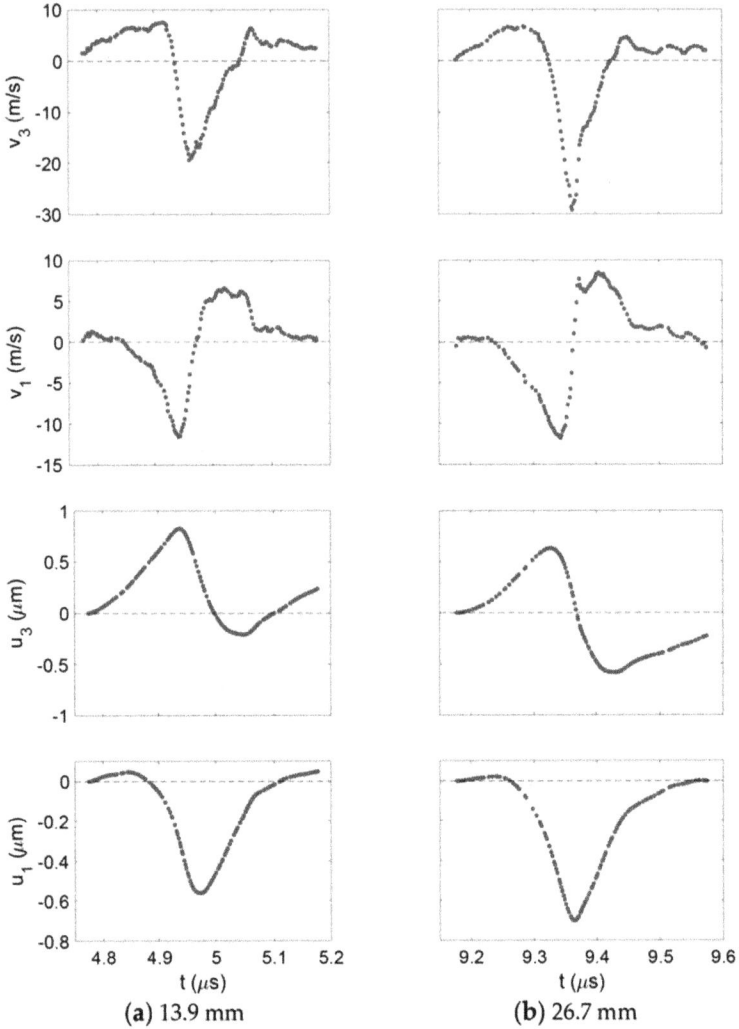

Figure 9.4. Particle velocity and displacement waveforms at (a) short and (b) long propagation distances from Kolomenskii and Schuessler [6] results on aluminum. (Reproduced from [16]. CC BY 4.0.)

commonplace. Finite amplitude surface waves can have sufficiently high localized stresses to nucleate cracks along preferred cleavage planes without the presence of a pre-existing crack. It turns out that crack initiation is different for wave vectors in opposite directions due to the different roles played by tension and compression. That is to say that tensile stress creates a crack, while compressive stress does not. In fused quartz Lehmann *et al* [14] observed a regular pattern of surface cracks that extended away from the surface at an angle of 30°–35°.

Nonlinear Rayleigh waves were used as a noncontact way to initiate cracking in a brittle material (single-crystal silicon) without the benefit of a precrack or notch.

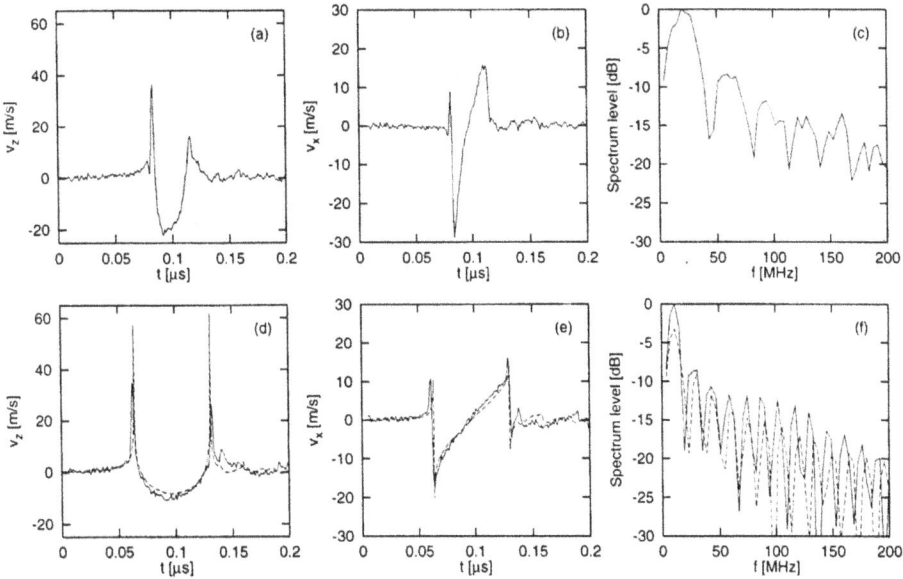

Figure 9.5. Rayleigh surface waveforms and corresponding frequency spectra at distances of 2.3 mm (top) and 18.3 mm (bottom). (Reproduced with permission from [11]. Copyright 1999 AIP Publishing.)

Normal separation (mode I) and mixed mode (tension and shear) were investigated by Kozhushko and Hess [17] and Kozhushko *et al* [15], respectively. For normal separation (mode I) cracking, the critical stress was computed to be in the 5–7 GPa range.

In a more recent article, Li *et al* [19] used laser pulses to generate finite amplitude longitudinal waves in a thermally aged aluminum alloy. They calibrated the interferometer-received u_3 (nm) signal and differentiated to obtain v_3 (m s^{-1}). We now move to modeling the evolution of the surface waveform as it transitions from a sinusoidal shape to a sawtooth shape.

9.3 Modeling waveform evolution

Mayer [20] provides a thorough discourse on mathematical modeling surface waves in nonlinear elastic media as of 1995, which was near the beginning of when most of this research was performed. An early approach used by Kalyanasundaram [21–23] was the method of multiple scales to derive coupled amplitude equations, which was demonstrated in example 3.3 for bulk longitudinal waves. Parker avoided the use of multiple scales by determining an evolution equation for the Fourier transform of the surface displacements [24]. There are two general approaches used to model the progressive waveform evolution that occurs as the Rayleigh wave propagates along the surface. The first approach, originally proposed by Zabolotskaya [25], uses a summation of harmonics and applies the Hamiltonian formalism to avoid the nonlinear boundary conditions. Meegan *et al* [26] indicate that the model results are identical to those of Parker [24]. The second approach, by Gusev and colleagues [9, 27], was developed because the first approach inherently assumes the secondary waveforms take on the shape of the primary waveforms and that evidence in the Russian literature invalidates this assumption. However, in their appendix C, Meegan *et al* [26] take issue with Gusev's model at various depths. Let's look at one of these modeling approaches as an example.

Zabolotskaya [25] describes the approach as follows:

'First, the Hamiltonian function that is of cubic order in the SAW variables is calculated. Second, generalized coordinates that completely determine the state of the system are chosen. Third, the dynamic equations for the SAW field are obtained by differentiation with respect to the generalized coordinates and generalized momenta. The result is a set of coupled equations for the harmonic amplitudes.'

The equations are integrated numerically and the results indicate that shocks appear in the v_3 profile and spikes appear in the v_1 profile at the surface ($x_3 = 0$) as observed in the experimental results.

To derive the equations for slowly varying harmonics of plane waves Zabolotskaya wrote the displacement components for a surface wave in the x_1 direction as expansions of forward and backward propagating modes,

$$u_1(x_1, x_3, t) = \sum_n a_n(t) u_{1n}(x_3) \exp(ink_0 x_1) \tag{9.1}$$

$$u_3(x_1, x_3, t) = \sum_n a_n(t) u_{3n}(x_3) \exp(ink_0 x_1), \tag{9.2}$$

where k_0 is the wavenumber of the primary wave and the wave vectors for the harmonics are $n\mathbf{k_0}$ with $n = \pm 1, \pm 2, \ldots$. The half-space is $x_3 \geqslant 0$ and $u_2 = 0$. The functions $u_{1n}(x_3)$ and $u_{3n}(x_3)$ can be written explicitly [25]. The amplitudes $a_n(t)$ are sought to fully describe the wave field.

The kinetic energy $T = \frac{1}{2} \int_{\mathcal{V}} \rho \mathbf{v} \cdot \mathbf{v} d\mathcal{V}$ is specialized for the displacement field above and the elastic energy density is decomposed into quadratic (for the linear elastic part V) and cubic (for the nonlinear elastic part W) parts. The cubic part gives the interacting terms and is based on the Landau–Lifshitz constants $\mathscr{A}$, $\mathscr{B}$, and $\mathscr{C}$. The Hamiltonian $H = T + V + W$ is formed and the generalized coordinates are taken to be $\mathrm{Re}(a_n)$ and $\mathrm{Im}(a_n)$. The corresponding generalized momenta are

$$p_n = \frac{\partial T}{\partial \dot{a}_n}, \tag{9.3}$$

where again n can be either negative or positive. The amplitudes are written in terms of slowly varying amplitudes b_n

$$a_n = b_n \exp(-i \mid n \mid \omega_0 t) + b_{-n}{}^* \exp(i \mid n \mid \omega_0 t), \tag{9.4}$$

for waves propagating in both directions, which leads to

$$\dot{a}_n = \frac{\partial H}{\partial p_n}, \quad \dot{p}_{-n} = \frac{\partial H}{\partial a_{-n}}. \tag{9.5}$$

To separate waves by their direction, let

$$d_n = \begin{cases} b^*_{-n} \text{ for } n < 0 \\ b_n \text{ for } n > 0, \end{cases} \tag{9.6}$$

in order to find the dynamic equation for the slowly varying amplitudes,

$$\dot{d}_n = -\frac{in}{M\omega_0 \mid n \mid} \frac{\partial W}{\partial d_{-n}}, \tag{9.7}$$

where M can be written explicitly in terms of conventional Rayleigh wave parameters (see equation (8) of [25]).

To compare with nonlinear wave propagation in liquids the particle velocity components are needed. Likewise, we know from previous work that shocks and spikes occur in the velocity and not the displacement. Therefore, the velocities are written

$$v_1 = \dot{u}_1 = \sum_n v_n u_{1n}(x_3) \exp\left(in(k_0 x_1 - \omega_0 t)\right) \tag{9.8}$$

$$v_3 = \dot{u}_3 = \sum_n v_n u_{3n}(x_3) \exp\left(in(k_0 x_1 - \omega_0 t)\right). \tag{9.9}$$

After some more manipulations, Zaboloskaya shows that the dynamic equation (9.7) becomes

$$\left(\frac{\partial}{\partial x} + \frac{1}{c}\frac{\partial}{\partial t}\right)v_n = -\frac{\mu n^2}{Mc^3} \sum_{n=m+l} S_{ml} v_n v_l, \tag{9.10}$$

where M and S_{ml} are defined in [25]. Equation (9.10) describes the evolution of v_n in equations (9.8) and (9.9) and was integrated numerically using a fourth order Runge–Kutta scheme for periodic steady-state waves (i.e. $\frac{\partial v_n}{\partial t} = 0$) with a damping term on the right-hand side of equation (9.10).

The results for steel in figure 9.6 [25] show the predicted shock formation in v_1 and spike-like pulse formation in v_3. The shock formation distance is between $x/x_0 = 1$ and $x/x_0 = 2$. The spike-like pulse is negative in steel based on the negative signs of the TOECs. It is interesting to observe that the v_1 amplitude peak near the shock exceeds the initial amplitude, which is explained by the frequency up-conversion to higher harmonics having smaller wavelengths, confining them closer to the surface and increasing the energy density.

Zabolotskaya's model results in the temporal domain enable comparison with nonlinear acoustics models for fluids (such as the Burgers equation for plane waves and the KZK equation for diffracting beams [2]). It is worth noting that Zabolotskaya [25] reviews earlier analyses, some based on the method of multiple scales, that are not part of the current discussion. Shull et al [28] analyse harmonic generation and show that for planar waves the amplitude of the second harmonic increases linearly with propagation distance. Additionally, they modeled the energy transfer from the primary frequency to the higher harmonics with the result for steel

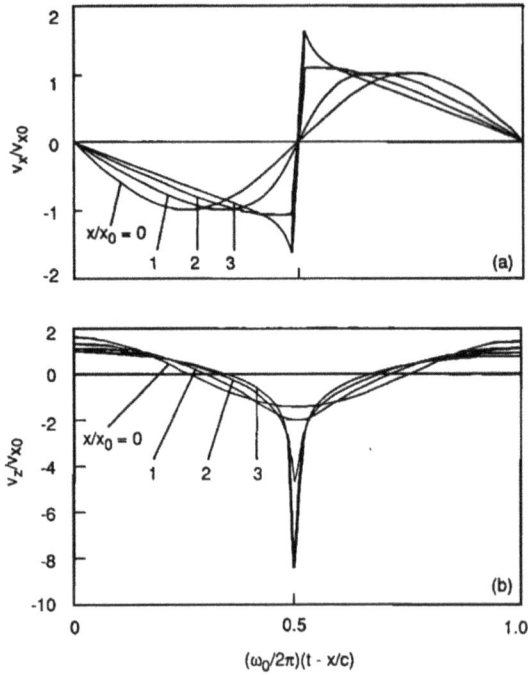

Figure 9.6. Predicted normalized particle velocity waveforms for propagation distances $x/x_0 = 0, 1, 2,$ and 3 in a steel half-space. $v_x = v_1, v_z = v_3$. (Reproduced with permission from [25]. Copyright 1992 AIP Publishing.)

depicted in figure 9.7 [28] where $X = x/x_0$ and $x_0 = Mc^3/\mu v_{10}$. Attenuation is included in these predictions from the full numerical model and an approximate model for the fundamental ($n = 1$). The absorption length is defined by the peak in the second harmonic ($n = 2$) amplitude, which occurs at $X \approx 3$ and the shock region ends at $X \approx 1000$. Shull *et al* [28] also depict the $v_1, v_3,$ and u_3 waveforms for source levels ranging from 10 to 60 dB to assess saturation. The shock formation distance for weakly attenuated plane waves in steel was determined, but not as a function of the TOECs, implying that it could be different for different materials. The complexity of the evolution equations has thwarted derivation of a generic shock formation distance equation that accounts for the nonlinearity of the specific material (i.e. its TOECs), but Shull *et al* [28] leveraged the similarity between nonlinear Rayleigh waves and nonlinear acoustics in fluids to propose what they believe to be a reasonable generic representation of the shock formation distance.

Hamilton *et al* [29] simplified Zabolotskaya's evolution equations for the in-plane displacement component u_1 without compromising the accuracy for most situations. In so doing they were able to distinguish between local and nonlocal nonlinearity effects as described in [18]. Nonlocal nonlinearity, first noted by [30], means that the distortion in one portion of the waveform depends on the evolutionary changes in all parts of the waveform. Nonlocal nonlinearity is a distinguishing feature of nonlinear Rayleigh waves that does not occur in fluids. Shull *et al* [31] added to the model the effects of diffraction of focused and unfocused Gaussian beams. The model was

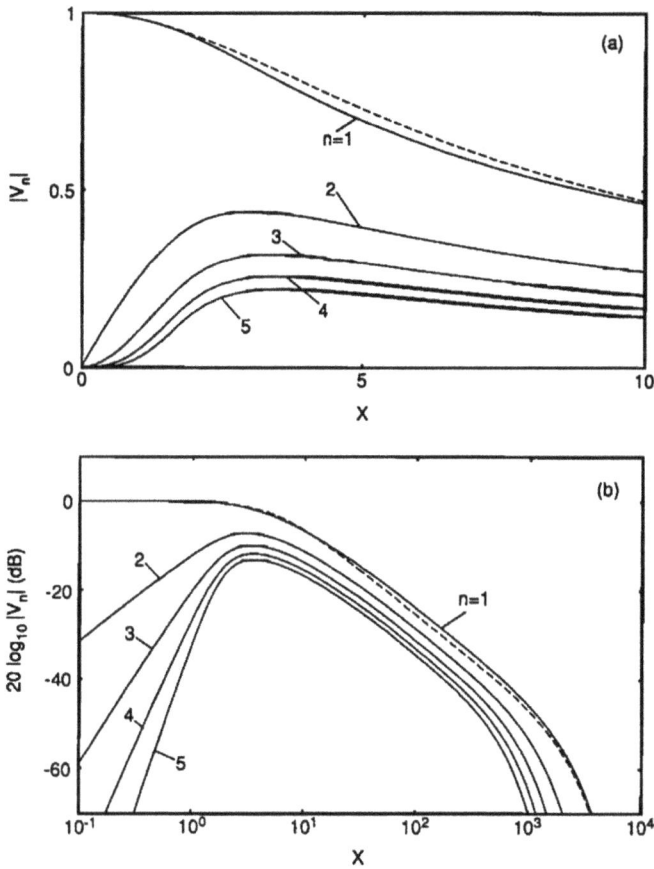

Figure 9.7. Predicted change in first five harmonic amplitudes as a function of dimensionless propagation distance X. Solid lines = numerical; dashed line = approximate model for $n = 1$. (a) Linear scale and (b) same data on a log scale. (Reproduced with permission from [28]. Copyright 1993 AIP Publishing.)

further generalized to include nonperiodic sources and nonplanar spatial distributions by Knight *et al* [32], who also show the equivalence of the theory to that of Parker [24], which is derived using a much different approach. Eckl *et al* [33] also show equivalences between these theories and implement the approach to develop an evolution equation for solitary surface waves.

The research in this area seems to have cooled, but Mora and Spies [34] adjusted Zabolotskaya's model [25] with the aim of applying it to nondestructively evaluate a shot-peened surface having heterogeneous TOECs and uniform SOECs. In so doing they consider first a single driving frequency and second co-directional mutual interactions at two frequencies, hence this research is a misfit with this chapter on broadband pulse loading. Nevertheless, it is an attempt to apply nonlinear Rayleigh waves for nondestructive characterization of plastically deformed surfaces. To succeed it would be necessary to solve the inverse problem for the TOEC distribution in terms of surface-measured wave characteristics. A recent article by

Bakre *et al* [16] proposes that the evolution of a broadband Rayleigh waveform could be used for monitoring the directed energy deposition additive manufacturing process using laser ultrasound. The layer-wise deposition process would be monitored for microstructural anomalies in the top few layers that could affect strength and fracture properties. Detection of the Rayleigh waves requires a laser interferometer to work on a rough surface.

In closing, the recent modeling of Shugaev *et al* [35, 36] is worth mentioning. In [35] they create a molecular dynamics model of progressive nonlinear surface wave propagation including shock formation and the subsequent dissipation. Their movies in the supplemental materials are stunning. In [36] they present a computational model that simulates laser generation of surface waves to assess the limits of the thermoelastic regime, which is of critical importance when the goal is to activate finite amplitude surface waves without ablating (and damaging) the surface.

9.4 Closure

Experimental results from progressive surface waves generated by a pulse laser illustrate that waveform evolution and shock formation depends on the sign of the material properties. Some key aspects of models were summarized. It is suggested that waveform evolution can be used for nondestructive evaluation.

References

[1] Scruby C B and Drain L E 1990 *Laser Ultrasonics Techniques and Applications* (Boca Raton, FL: CRC Press)
[2] Hamilton M F and Blackstock D T 1997 *Nonlinear Acoustics* 1st edn (San Diego, CA: Academic)
[3] Kolomenskii A A, Lioubimov V A, Jerebtsov S N and Schuessler H A 2003 Nonlinear surface acoustic wave pulses in solids: laser excitation, propagation, interactions (invited) *Rev. Sci. Instrum.* **74** 448–52
[4] Kolomenskii A A, Schuessler H A, Mikhalevich V G and Maznev A A 1998 Interaction of laser-generated surface acoustic pulses with fine particles: surface cleaning and adhesion studies *J. Appl. Phys.* **84** 2404–10
[5] Kolomenskii A A, Lomonosov A M, Kuschnereit R, Hess P and Gusev V E 1997 Laser generation and detection of strongly nonlinear elastic surface pulses *Phys. Rev. Lett.* **79** 1325–8
[6] Kolomenskii A A and Schuessler H A 2001 Characterization of isotropic solids with nonlinear surface acoustic wave pulses *Phys. Rev.* B **63** 085413
[7] Schuessler H A, Jerebtsov S N and Kolomenskii A A 2003 Determination of linear and nonlinear elastic parameters from laser experiments with surface acoustic wave pulses *Spectrochim. Acta* B **58** 1171–5
[8] Kumon R E 1999 Nonlinear surface acoustic waves in cubic crystals *PhD thesis* University of Texas at Austin
[9] Gusev V E, Lauriks W and Thoen J 1998 New evolution equations for the nonlinear surface acoustic waves on an elastic solid of general anisotropy *J. Acoust. Soc. Am.* **103** 3203–15

[10] Lomonosov A and Hess P 1996 Laser excitation and propagation of nonlinear surface acoustic wave pulses *Nonlinear Acoustics in Perspective* ed R J Wei (Nanjing: Nanjing University Press)

[11] Lomonosov A, Mikhalevich V G, Hess P, Knight E Y, Hamilton M F and Zabolotskaya E A 1999 Laser-generated nonlinear Rayleigh waves with shocks *J. Acoust. Soc. Am.* **105** 2093–6

[12] Jerebtsov S N, Kolomenskii A A and Schuessler H A 2004 Characterization of a polycrystalline material with laser-excited nonlinear surface acoustic wave pulses *Int. J. Thermophys.* **25** 485–90

[13] Lomonosov A M and Hess P 2002 Impulsive fracture of silicon by elastic surface pulses with shocks *Phys. Rev. Lett.* **89** 095501

[14] Lehmann G, Lomonosov A M, Hess P and Gumbsch P 2003 Impulsive fracture of fused quartz and silicon crystals by nonlinear surface acoustic waves *J. Appl. Phys.* **94** 2907–14

[15] Kozhushko V V, Lomonosov A M and Hess P 2007 Intrinsic strength of silicon crystals in pure- and combined-mode fracture without precrack *Phys. Rev. Lett.* **98** 195505

[16] Bakre C, Afzalimir S H, Jamieson C, Nassar A, Reutzel E W and Lissenden C J 2022 Laser generated broadband Rayleigh waveform evolution for metal additive manufacturing process monitoring *Appl. Sci.* **12** 12208

[17] Kozhushko V V and Hess P 2007 Anisotropy of the strength of Si studied by a laser-based contact-free method *Phys. Rev.* B **76** 144105

[18] Hamilton M F, I'insky Y A and Zabolotskaya E A 1995 Local and nonlocal nonlinearity in Rayleigh waves *J. Acoust. Soc. Am.* **97** 882–90

[19] Li M *et al* 2019 Monitoring of thermal aging of aluminum alloy via nonlinear propagation of acoustic pulses generated and detected by lasers *Appl. Sci.* **9** 1191

[20] Mayer A 1995 Surface acoustic waves in nonlinear elastic media *Phys. Rep.* **256** 237–366

[21] Kalyanasundaram N 1981 Nonlinear surface acoustic waves on an isotropic solid *Int. J. Eng. Sci.* **19** 279–86

[22] Kalyanasundaram N 1981 Nonlinear mode coupling of surface acoustic waves on an isotropic solid *Int. J. Eng. Sci.* **19** 435–41

[23] Kalyanasundaram N, Ravindran R and Prasad P 1982 Coupled amplitude theory of nonlinear surface acoustic waves *J. Acoust. Soc. Am.* **72** 488–93

[24] Parker D F 1988 Waveform evolution for nonlinear surface acoustic waves *Int. J. Eng. Sci.* **26** 59–75

[25] Zabolotskaya E A 1992 Nonlinear propagation of plane and circular Rayleigh waves in isotropic solids *J. Acoust. Soc. Am.* **91** 2569–75

[26] Meegan G D, Hamilton M F, I'inskii Y A and Zabolotskaya E A 1999 Nonlinear Stoneley and Scholte waves *J. Acoust. Soc. Am.* **106** 1712–23

[27] Gusev V E, Lauriks W and Thoen J 1997 Theory for the time evolution of nonlinear Rayleigh waves in an isotropic solid *Phys. Rev.* B **55** 9344–7

[28] Shull D J, Hamilton M F, I'insky Y A and Zabolotskaya E A 1993 Harmonic generation in plane and cylindrical nonlinear Rayleigh waves *J. Acoust. Soc. Am.* **94** 418–27

[29] Hamilton M F, I'insky Y A and Zabolotskaya E A 1995 Evolution equations for nonlinear Rayleigh waves *J. Acoust. Soc. Am.* **97** 891–7

[30] Parker D F and Talbot F M 1985 Analysis and computation for nonlinear elastic surface waves of permanent form *J. Elast.* **15** 389–426

[31] Shull D J, Kim E E, Hamilton M F and Zabolotskaya E A 1995 Diffraction effects in nonlinear Rayleigh wave beams *J. Acoust. Soc. Am.* **97** 2126–37

[32] Knight E Y, Hamilton M F, Il'inskii Y A and Zabolotskaya E A 1997 General theory for the spectral evolution of nonlinear Rayleigh waves *J. Acoust. Soc. Am.* **102** 1402–17
[33] Eckl C, Kovalev A S, Mayer A P, Lomonosov A M and Hess P 2004 Solitary surface acoustic waves *Phys. Rev.* E **70** 046604
[34] Mora P and Spies M 2018 Rayleigh wave harmonic generation in materials with depth-dependent non-linear properties *J. Acoust. Soc. Am.* **143** 2678–84
[35] Shugaev M V, Wu C, Zaitsev V Y and Zhigilei L V 2020 Molecular dynamics modeling of nonlinear propagation of surface acoustic waves *J. Appl. Phys.* **128** 045117
[36] Shugaev M V and Zhigilei L V 2021 Thermoelastic modeling of laser-induced generation of strong surface acoustic waves *J. Appl. Phys.* **130** 185108

Part III

Applications

IOP Publishing

Nonlinear Ultrasonic Guided Waves

Cliff J Lissenden

Chapter 10

Numerical simulations

Finite element analysis is a tremendously powerful tool in science and engineering thanks to advances in both hardware and software. Numerical simulations of nonlinear guided wave propagation are incredibly useful to (i) confirm that assumptions made in modeling are reasonable and (ii) plan experiments and measurement techniques. One of the primary benefits of numerical simulations is that the nonlinearity can be limited to contain only material nonlinearity and nonlinearity from finite strain. Measurements on the other hand are typically encumbered with nonlinearities from the measurement system itself. This chapter overviews finite element analyses pertaining to nonlinear ultrasonic waves, describes some key features of commercially available software commonly used for simulating wave propagation, and then presents some sample problems. Some signal processing tools are employed that are described in more detail in chapters 11 and 12.

10.1 Methods

It is true that there are multiple methods available for numerical simulations of ultrasonic guided waves, namely finite elements, boundary elements, and finite differences. But it is equally true that finite element methods and software capabilities have far eclipsed the competing methods. Thus, we will only describe finite element analyses herein. Furthermore, since there are so many great resources, e.g. [1], available for learning about finite element analysis, the intricacies of element formulation, assembly, application of boundary conditions, etc, are also not discussed here. Finite element analysis of guided wave propagation is discussed by Moser *et al* [2] and Rose [3]. Furthermore, simulation of nonlinear ultrasonic guided waves is discussed by Chillara and Lissenden [4, 5]. In transient dynamics problems like wave propagation two key analysis parameters affecting convergent solutions are the element length, l_{el}, and the time step, Δt. Guidelines for these important parameters are [2]

$$\Delta t \leqslant \frac{1}{20 f_{max}}, \tag{10.1}$$

$$l_{el} \leqslant \frac{\lambda_{min}}{20}, \tag{10.2}$$

where f_{max} and λ_{min} are the maximum frequency and minimum wavelength, respectively, in the simulation. The maximum frequency and minimum wavelength are selected from the entire set of primary and secondary waves. In addition, the time step should not exceed the time for the fastest wave to travel between the nodes of an element. These recommendations are intended to provide accurate results without excessive computational cost. We will start by discussing the model of a waveguide.

There are many nuances associated with modeling the waveguide and actuating the guided waves. In many cases finite element analysis (FEA) is selected because the geometry of the object being modeled is too complicated for analytical modeling to be sufficiently accurate. However, this is not necessarily the present case. We are turning to FEA because wave propagation is a transient dynamics problem that requires time marching solutions, which has been coded into existing FEA software. It bears mentioning that this use of FEA is completely different than the semi-analytical finite element (SAFE) discussed in chapter 4 for determining dispersion curves. If we want to model nonlinear Lamb wave propagation, it may suffice to use a 2D plane strain model of the waveguide. However, SH waves require a 3D model. For waves propagating in the X_1 direction with a plate bounded by free surfaces at $X_3 = \pm h$ the Lamb wave model would be of the X_1–X_3 plane. Although analytical models could consider the plate to be infinite in the X_1 direction, that is not possible in FEA. If the FEA is intended to represent an infinite plate, then infinite elements can be used to absorb incoming waves. As already shown in chapter 5 for finite-width plates, the plane strain assumption is not always appropriate. Thus, 3D models may be necessary for finite-width plates, or if it is desired to model non-planar waves, or if the interaction between SH and Lamb waves is important. Figure 10.1 shows an example 3D model to simulate co-directional mixing of planar waves in a wide plate. Discretization of the model with a uniform mesh of

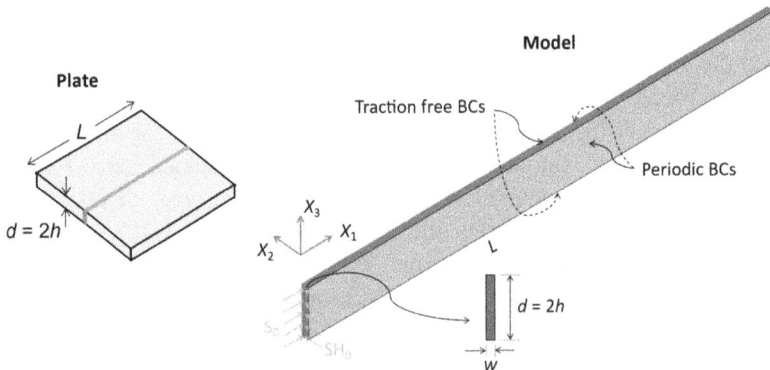

Figure 10.1. 3D finite element model of a plate waveguide for co-directional wave mixing, hexahedral elements are not shown.

hexahedral elements is not shown. The dimension L, in the wave propagation direction, must be sufficiently long that end-wall reflections do not interfere with the signals of interest, or infinite elements need to be used. The choice of whether to model in 2D or 3D is important with respect to the desired results, and also to the run time required to solve the problem. 3D problems will have many many more degrees of freedom, which dictates the size of the global matrix that must be solved for each time increment. Some global matrices may be too large for the software, or the hardware on which it is run, to solve. If not too large, it could still be very time consuming to solve. Fortunately, in many cases the size of a 3D domain to be discretized can be managed by applying periodic boundary conditions.

Unlike SAFE, the waves must be actuated as applied boundary conditions or applied loads. This can be accomplished in various ways depending on the objective of the analysis. If the objective is to understand the nonlinear wave propagation resulting from a specific wave mode or mixing of specific wave modes, then *wavestructure loading* is appropriate. Wavestructure loading is when the dynamic complex-valued displacement components (or the associated stress components) of the wavestructure are applied to a cross-section of the waveguide. In some cases, it is beneficial to transform the displacement wavestructure into body forces that act over a prescribed subvolume of the waveguide. If the objective is to understand the nonlinear wave propagation from a certain transducer, then either the transducer can be explicitly modeled or the surface traction it creates can be applied to the plate. Modeling the transducer could include its own nonlinearities, e.g. wedge material, piezoelectric material, and input voltage signal. Most FEA simulations of nonlinear guided wave propagation up to this point in time take advantage of the material nonlinearity being the only nonlinearity simulated, but future work could address nonlinearities in the measurement system itself. Likewise, current simulations have taken advantage of the received signals being free of incoherent noise, but noise could be injected into the input signals to assess the signal-to-noise ratio.

Another benefit of FEA compared to experiments is that waveform signals can be received anywhere, making it easy to apply a 2D fast Fourier transform (2D-FFT) to assess the wavenumbers and frequencies of the propagating modes. Signals can be received at a specific node or averaged over the contact area of a sensor. Signals can be received in the interior of the waveguide if desired, while sensors are typically relegated to the waveguide surface. Localized material degradation can be imposed by changing the elastic constants. Material nonlinearity can be increased by changing the third order elastic constants (Landau–Lifshitz: $\mathscr{A}$, $\mathscr{B}$, $\mathscr{C}$) without changing the second order elastic constants (Lamé: λ and μ, or E and ν). Clearly, there is plenty of flexibility.

10.2 Software tools

Three commercially available software packages are mentioned herein. Without doubt, there are others.

COMSOL Multiphysics® (https://www.comsol.com/) is designed to simplify the analysis of problems having coupled physics. The user interface is easily navigable. Modeling the waveguide along with the transducers is streamlined. Direct explicit integration is used for elastic wave problems. Comsol's site recommends using elements with quartic shape functions with 1.5 elements/wavelength. The Murnaghan nonlinear elastic model can be used for nonlinear ultrasonic guided wave simulations.

Abaqus Explicit (https://www.3ds.com/products-services/simulia/products/abaqus/) uses a conditionally stable direct integration approach to perform the time marching solution in small steps. Accelerations are determined using a diagonal lumped mass matrix that can quickly be inverted. The stiffness matrix is not inverted. Many material models are available, but neither the Landau–Lifshitz nor the Murnaghan model are included. However, these hyperelastic models can be implemented through a user defined material Fortran subroutine, known as a VUMAT for the Explicit version of Abaqus.

Pogo (http://www.pogo.software/index.html) is an explicit time domain-based finite element solver for problems in linear elastodynamics. Its unique way of parallelizing the problem for solution on graphical processing units makes it faster and able to solve larger problems than other commercial software. However, it is currently limited to linear problems. We will hope this changes in the future.

10.3 Sample problems

10.3.1 Reported in the literature

Some examples of finite element simulations of nonlinear guided wave propagation reported in the literature are summarized below.

Chillara and Lissenden [4] used COMSOL to investigate non-synchronized second order harmonics. The case closest to being synchronized was analysed first —the S0 mode at 0.5 MHz. By comparing (i) linear elastic infinitesimal-strain, (ii) nonlinear, and (iii) purely geometric nonlinearity solutions both finite strain (geometric nonlinearity) and material nonlinearity are observed to have distinct contributions on the amplitude of the second harmonic and the quasistatic pulse. The non-synchronized primary A0 mode continuously generated an S0 second harmonic that propagates on its own and is faster than the A0 mode, although it is not cumulative. Mutual interaction of the S0 mode at 0.4 and 1.1 MHz generated a combinational harmonic at the sum frequency.

One-way wave mixing is the interesting case where mutual interactions between co-directional waves generates secondary waves at the difference frequency that back-propagate towards the source of the primary waves. Hasanian and Lissenden [6, 7] identified six wave triplets (10–14 and 16 in table III [6] and 11–14 in table IV [7]) that are one-way mixing.

Ding *et al* [8] used Abaqus to simulate one-way mixing with S0 waves at 480 kHz and A0 waves at 366 kHz to generate back-propagating A0 waves at 114 kHz. The material nonlinearity is not due to a nonlinear elasticity model, but rather due to a region having a distribution of microcracks defined by a crack density parameter. The crack contact model includes clapping and slipping. See also [9] for distributed microcrack domains within Lamb wave propagation models. Wave mode and frequency selection is based on trying to maximize power flow to the secondary waves. Sun and Qu [10] investigated the resonance conditions and mixing zones for one-way mixing considering toneburst durations and group velocities. They simulated mixing S0 waves at 7.03 MHz and A0 waves at 5.43 MHz to generate back-propagating A0 waves at 1.60 MHz in Abaqus. Li *et al* [11] used COMSOL to simulate one-way mixing of longitudinal waves L(0,2) at 1.4 MHz and L(0,1) at 1.102 MHz in a pipe, which are similar to the S0 and A0 lamb waves, respectively, in a plate. The amplitude of the back-propagating wave decreased to about 10% of its peak value in the mixing zone, but remained nearly constant thereafter. The effect of localized material degradation was studied. Lissenden *et al* [12] used Abaqus to simulate one-way mixing of S0 waves at 679 kHz and SH0 waves at 541 kHz to generate back-propagating SH0 waves at 138 kHz in a 1.8 mm thick plate. Variations in the size of the mixing zone and localization of TOECs were studied. In addition, the mutual interaction of counter-propagating SH0 waves to generate S1 Lamb waves at the mode cutoff was simulated. Since the S1 waves are standing waves, they could be used for volumetric inspection by using time delays to move the mixing zone along the length of the plate. The latest one-way mixing simulations, by Liu *et al* [13], use Abaqus to simulate S0 waves at 431 kHz mixing with SH0 waves at 341 kHz to generate back-propagating SH0 waves at 90 kHz. Localized material degradation was simulated, and for the first time, experiments were conducted. Material degradation was induced artificially as corrosion.

Jiang *et al* [14] simulated the generation of the quasistatic pulse from the fundamental S0 and A0 lamb-like waves in unidirectional composites. Li *et al* [15] used COMSOL to simulate the mutual interactions of counter-propagating waves to generate zero-group velocity Lamb waves. Counter-propagating S0 waves at 1.98 and 0.84 MHz and counter-propagating SH0 waves at 1.81 and 1.01 MHz (identified in [7]) were analysed and localized changes in the TOECs are shown to proportionally increase the amplitude of the relative nonlinearity coefficient.

10.3.2 Lamb wave analyses using commercial software

Finite element simulations of nonlinear guided wave propagation are conducted with the COMSOL Multiphysics software to demonstrate some of the more interesting aspects and important analysis methods. Self-interactions for IR point 1 are analysed first and then mutual interactions from counter-propagating low-frequency fundamental Lamb waves (IR point 43) are analysed. Both cases are internally resonant. Waves propagate in the X_1 direction, the lateral faces normal to the X_2 direction are fully restrained ($u_2 = 0$) to provide plane strain conditions, and the top and bottom surfaces at $X_3 = \pm h$ are traction free. The 2D quadrilateral quadratic elements are used. The same material properties for aluminum are used in these analyses as elsewhere:

$$\rho = 2700 \text{ kg m}^{-3}$$

$$E = 69 \text{ GPa}, \ \nu = 0.33$$

$$\mathscr{A} = -350 \text{ GPa}, \ \mathscr{B} = -155 \text{ GPa}, \ \mathscr{C} = -95 \text{ GPa}$$

$$l = -250 \text{ GPa}, \ m = -330 \text{ GPa}, \ n = -350 \text{ GPa},$$

where Murnaghan's constants have been computed from equation (2.48). The results include secondary waveforms and mixing power computations. I want to thank Hamidreza Afzalimir for conducting these analyses.

10.3.2.1 Self-interactions

Consider self-interactions of the S1 Lamb wave mode having the phase velocity $c_p = c_L$, which is IR point 1 (see table 8.3). The second harmonic S2 Lamb waves are synchronized and have matching group velocities, thus the primary S1 waves and the secondary S2 waves propagate together and interact continuously.

The 6 mm thick plate is 400 mm long. The primary frequency of the S1 mode is $3.588/6 = 0.598$ MHz. The S2 second harmonic is 1.196 MHz and its wavelength is $6153/1.196 \times 10^6 = 5.14$ mm. Thus, the maximum element size is 0.25 mm (from equation (10.2)). The time steps need to be smaller than both $\frac{1}{20 f_{max}} = 42$ ns and $\frac{0.000\,25}{4376} = 57$ ns. Therefore, we use 10 ns time steps to be conservative. The finite element mesh is 1 element wide, 24 elements high, and 4000 elements long, having a total of 96 000 elements. Multiple simulations are conducted to emphasize different effects. We start by applying a 25 cycle toneburst having a maximum amplitude of 100 nm. The applied u_1 and u_3 amplitudes are a function of X_3 based on the S1 wavestructure at 0.598 MHz.

Animations of the wave propagation along the plate are given in figure 10.2, in which the u_1 displacement component along the top surface of the plate ($X_3 = h$) is plotted from the finite element analysis results. (figure 10.2 is derived from a companion analysis in Abaqus/Explicit for visualization purposes.) The u_1 displacement component at $X_3 = h$ (figure 10.2(a)) dominates the S1 mode wavestructure as shown in figure 8.2. The second harmonic waves in figure 10.2(b) are the S2 mode, and are also represented by u_1 along the top surface of the plate. The waves in figure 10.2(b) are obtained using the phase inversion technique, which will be

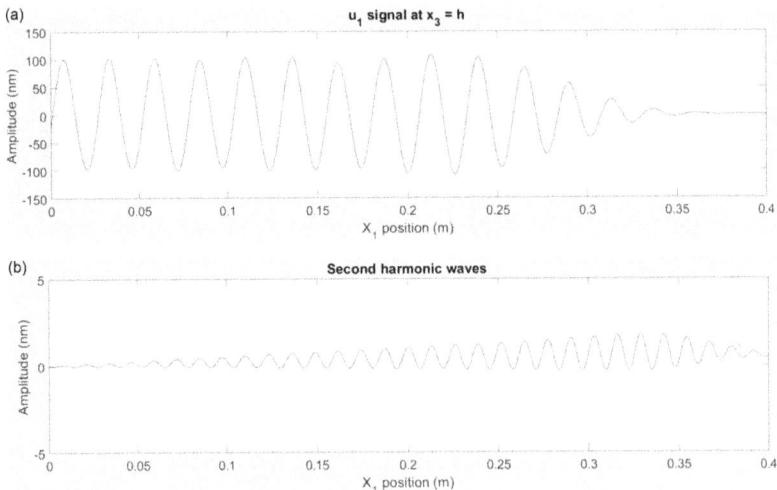

Figure 10.2. Animations of displacement components along the top of the plate for $0 < t < 130$ ms: (a) u_1 and (b) second harmonic waves. Animation available at https://doi.org/10.1088/978-0-7503-4911-6.

described subsequently. The cumulative nature of the second harmonic waves is depicted nicely in the figure 10.2(b) animation (less well in the still image). Also notice that the second harmonic waves have an increasing offset as they propagate, which is due to the zero frequency nonlinearity (i.e. the quasistatic pulse).

Figure 10.3 shows that the wavestructure loading is effective at actuating a dominant S1 mode and that the FFT is sufficient for determining the second harmonic if there is a single dominant mode (and the secondary mode has the same group velocity). The wave energy evident in the u_3 component and the u_1 waveform not having the same trapezoidal shape as was applied (see the inset in figure 10.3) indicate that other modes are present. We assess the presence of multiple modes by using a 2D-FFT along the top of the plate. The 2D-FFT results are overlaid on the dispersion curves in figures 10.4(a) and (b). There are 10 000 points in the spatial domain ($0.1 < X_1 < 0.2$ m) and 13 000 points in the temporal domain ($0 < t < 130$ μs). Zero padding is used in both spatial and temporal domains. The spectral density of the second harmonic is not visible at the scale shown in figure 10.4(a), thus the scale is changed in figure 10.4(b). The primary waves are predominantly the S1 mode and the second harmonic waves are predominantly the S2 mode.

It is not currently possible to physically apply wavestructure loading to a plate. Normally, an angle-beam transducer is used with the angle computed by Snell's law. We simulated a contact transducer coupled to an acrylic wedge with an angle of 25.7 degrees. Attempting to enhance excitation of the u_1 displacement at the surface of the

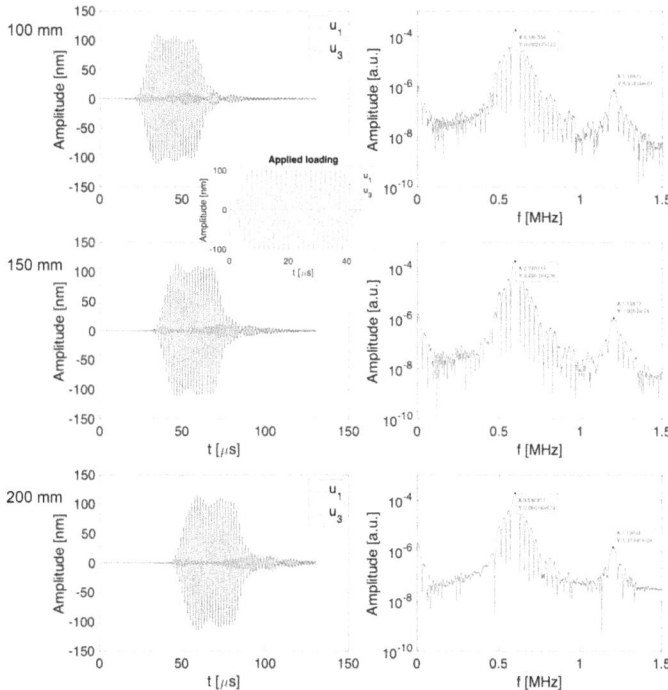

Figure 10.3. A-scans for u_1 and u_3 and FFTs of u_1 data at $X_1 = 100$, 150, and 200 mm. $X_3 = h$. The inset shows the applied u_1 displacement at $X_3 = h$.

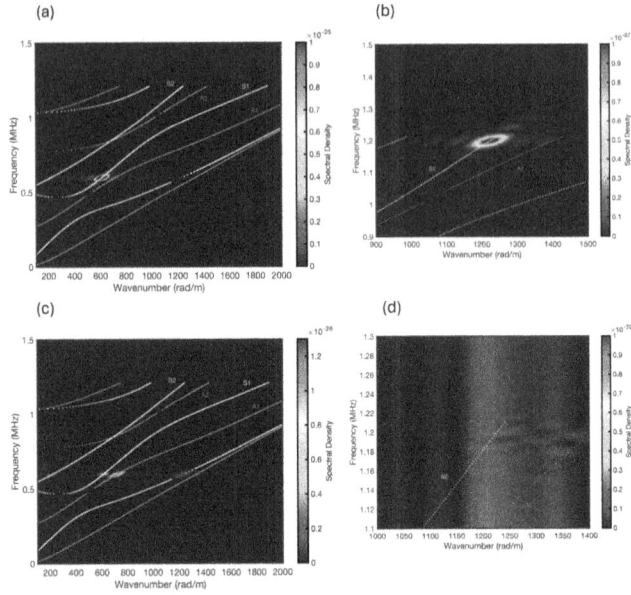

Figure 10.4. Spectral density obtained from 2D-FFT overlaid on dispersion curves for IR point 1: (a) wavestructure loading, (b) wavestructure loading, but zoomed in on second harmonic frequency, (c) angle-beam transducer loading, and (d) angle-beam transducer loading, but zoomed in on second harmonic frequency.

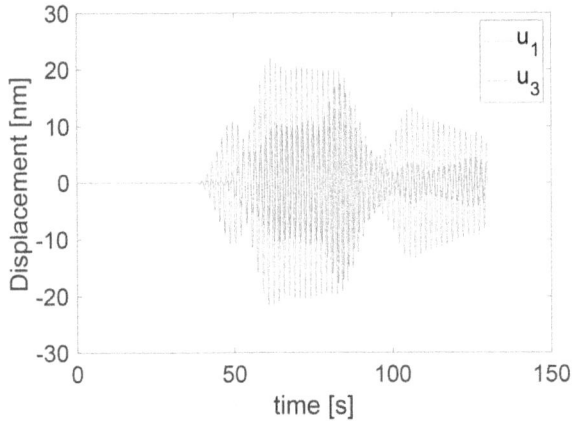

Figure 10.5. A-scans for u_1 and u_3 at $X_1 = 124$ mm. $X_3 = h$.

plate, we perfectly bonded the wedge to the plate. The A-scans received at $X_1 = 124$ mm are plotted in figure 10.5. Although the wedge angle was designed to actuate the S1 mode, the complicated shape of the waveform indicates the presence of multiple modes and the u_3 component is much larger than the u_1 component. The 2D-FFT was similarly applied to the data and the results are plotted in figures 10.4(c) and (d). Figure 10.4(c) indicates that the A1 mode is dominant, with S0, S1, S2, and even A0 modes also present at the driving frequency of 0.598 kHz. Figure 10.4(d) does not show any significant spectral density for the S2 mode, suggesting that an angle-beam transducer is not a good way to excite IR point 1. Nonetheless, we will see in chapter 12 that such experiments were successful.

The 2D-FFT requires time series data for an array of spatial positions, which can be difficult to acquire from some transducers. The short-time Fourier transform (STFT) can be used to separate modal content via a spectrogram when time series data are only available at one point, but the analyst must decide upon the appropriate window size and the overlap. However, we have already shown that the wavestructure loading predominantly excites the S1 mode, which in turn generates the S2 second harmonic waves. Instead of demonstrating the STFT we will use the phase inversion (or phase reversal) technique to determine the second harmonic waves in the temporal domain, without having to transform to the frequency domain. Two separate loadings are conducted sequentially. First, the desired loading is applied and signal $S_{(a)}$ received. Second, a loading obtained by inverting the phase of the desired loading is applied and signal $S_{(-a)}$ received. By adding together the signals received from these two loadings we obtain $S_{(a)+(-a)}$. The odd powers of the signal $S_{(a)}$ are canceled and the even powers are doubled in amplitude; see Zhu *et al* [16] and the references therein for applications. The results for the 25 cycle toneburst are shown in figure 10.6(a). The simulations were repeated with a two cycle toneburst and the results are shown in figure 10.6(b). Since we are

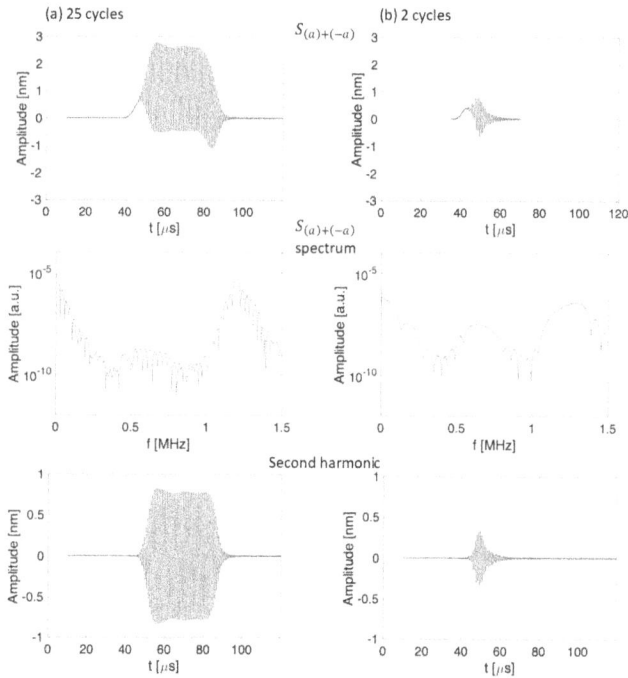

Figure 10.6. Phase inversion signal $S_{(a)+(-a)}$, its FFT, and filtered second harmonic signal from u_1 at $X_1 = 200$ mm and $X_3 = h$: (a) 25 cycles and (b) two cycles.

interested in the second harmonic waves, we need to separate them from the quasistatic pulse (zero frequency). Thus, we apply a high-pass filter to $S_{(a)+(-a)}$ at 250 kHz and then divide by two to obtain the second harmonic waves; the second harmonic waves are also plotted in figure 10.6. We note that figure 10.2(b) was obtained by phase inversion, but was not filtered to remove the quasistatic pulse.

The cumulative nature of the filtered S2 second harmonic is shown in figure 10.7 for both 25 cycle and two cycle tonebursts. The rate of cumulative growth is determined by linear curve fitting the relative nonlinearity coefficient β' to the propagation distance. The slope is the mixing power M_p defined in chapter 6. The simulation results for a 25 cycle toneburst are linear with $M_p = 333\,328$ m^{-2}. The results for a two cycle toneburst are not linear, which is not fully understood,

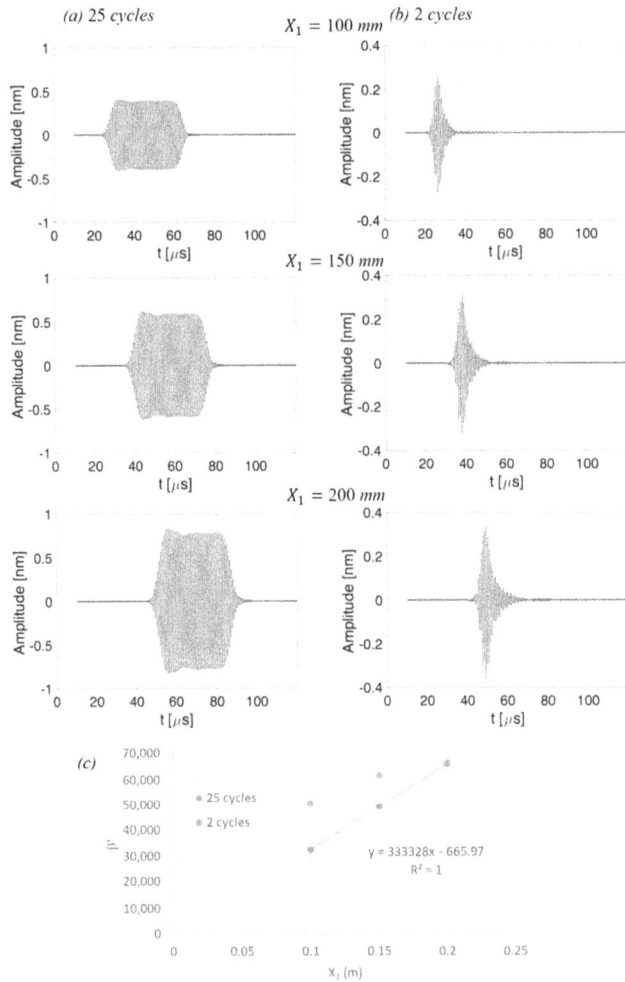

Figure 10.7. Filtered second harmonic signal from u_1 phase inversion for (a) 25 cycles and (b) two cycles. (c) Relative nonlinearity coefficient as a function of propagation distance.

but may be due to difficulties capturing the sharp peaks in the time domain. The mixing power quoted in chapter 8 for the S1–S2 mode pair is $M_p = 55.96 \times 10^6$ m^{-2}. We provide comments on the mixing power comparison after the mutual interactions example problem.

10.3.2.2 Mutual interactions

As another example, consider the mixing of counter-propagating waves at IR point 43 (see table 8.8), where the mutual interaction of S0 waves a and A0 waves b generates the secondary A0 waves n that propagate in the direction of the S0 waves at the difference frequency. The wave mixing parameters are provided in table 10.1 and the dispersion curves and wavestructures are shown in figure 10.8 along with a schematic of the waves propagating in the plate. The 1 mm thick plate is 1000 mm long. The element edges need to be less than $\frac{\lambda_{min}}{20} = 0.26$ mm. Therefore, we use 0.25 mm square elements. The time steps need to be smaller than both $\frac{1}{20 f_{max}} = 122$ ns and $\frac{0.00025}{5313.6} = 47$ ns. Therefore, we use 10 ns time steps. The finite element mesh is

Table 10.1. IR point 43 wave mixing parameters.

	Mode	f (MHz)	λ (mm)	k (rad m^{-1})	c_p (m s^{-1})	c_g (m s^{-1})	θ (°)
Waves a	S0	0.410	13.0	482.2	5342.4	5313.6	0
Waves b	A0	0.118	8.70	722.6	1026.1	1896.0	180
Waves n	A0	0.292	5.22	1204.7	1522.9	2565.0	0

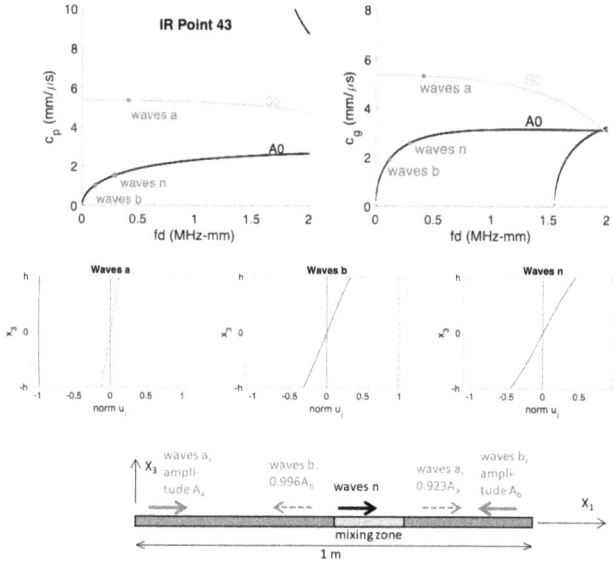

Figure 10.8. Dispersion curves, wavestructures (blue = u_1, red = u_3; solid = real, dashed = imaginary), and wave propagation schematic.

four elements high and 4000 elements long, having a total of 16 000 elements. The sources are modeled by applying uniform u_1 displacements on the left face to actuate the S0 mode and uniform u_3 displacements on the right face to actuate the A0 mode. Specifically, waves a are generated by an applied dynamic displacement loading in the X_1 direction at $X_1 = 0$ and waves b are generated by an applied dynamic displacement loading in the X_3 direction at $X_1 = 1000$ mm. In this way pure S0 and A0 waves are generated at opposite sides of the plate even without exactly applying the actual wavestructures. This only works when the S0 and A0 dispersion curves are sufficiently separated below the mode cutoffs. Infinite elements are used to create non-reflecting boundaries at each end, although these were found to be imperfect.

Multiple simulations are conducted to emphasize different aspects of the mutual interactions. We start by applying Tukey-windowed tonebursts of 25 cycles, with the center 17 cycles having the constant amplitude of 1 nm. We treat this as a 17 cycle toneburst, since the tapered four cycles at each end are used to ease computational challenges associated with the boundary conditions. For counter-propagating waves the size of the mixing zone is dictated by the shorter primary wave packet, which is waves b. The mixing zone is 17×8.70 mm $= 148$ mm. A 200 μs delay is applied to waves a to force the mixing zone to start near the center ($X_1 = 457.5$ mm) and enable tracking the secondary waves n as they propagate to the right. Signal processing is done by computing the difference signal as described in section 12.2, i.e.

$$S_{\text{Diff}} = S(a + b) - [S(a) + S(b)], \tag{12.1}$$

which requires three distinct simulations, waves ($a + b$), waves a alone, and waves b alone. Apologies to the readers for the reference forward to equation (12.1); you may want to jump ahead and read a portion of section 12.2 now. Note that the phase inversion used in the self-interaction example problem can also be applied to mutual interaction problems, but that the self-interactions are not canceled [16] as they are when (12.1) is used. Hence, phase inversion is not used for this mutual interaction problem.

Figure 10.9 shows animations of the S0 waves a and A0 waves b approaching, interacting to generate the secondary A0 waves n, and then separating. These three animations are produced by plotting u_1 and u_3 displacements for waves ($a + b$) and also u_3 for S_{Diff} along the top surface of the plate. Although S0 waves are primarily u_1 and A0 waves are primarily u_3, as shown in figure 10.8, both displacement components exist for both waves a and b and are visible. The S_{Diff} animation shows nicely how the secondary waves n are generated in the mixing zone and grow in number and amplitude with time until the mixing zone diminishes to zero and they simply propagate to the right. At the end of the animation the waves n are overwhelmed by waves encroaching from each end of the plate. These waves are thought to be associated with variable end-wall reflections. Companion analyses were performed with Abaqus/Explicit and the animations in figure 10.9 were created more readily from Abaqus than COMSOL.

The animations in figure 10.9 appear to show that the amplitudes of waves a and b return to their original values after interacting, but the secondary waves are generated from the energy in the primary waves, thus a closer examination is

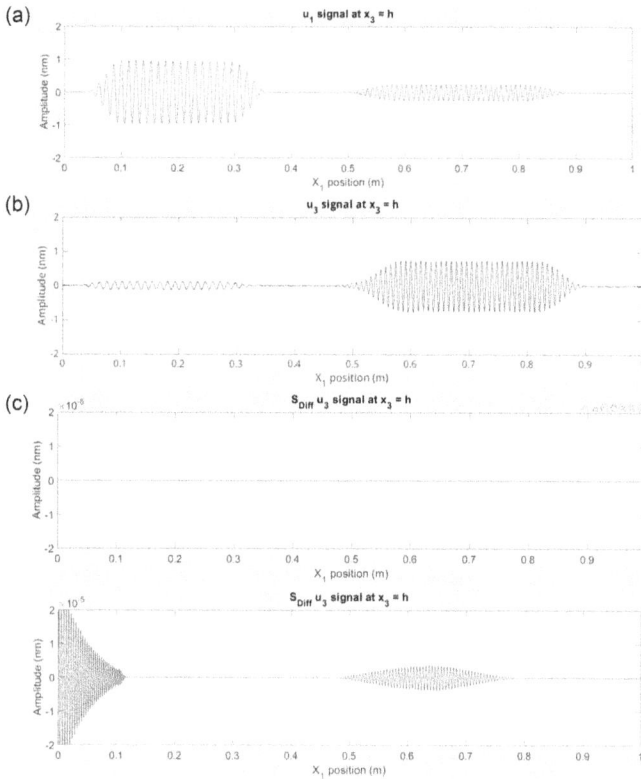

Figure 10.9. Animations of displacement components along the top of the plate for $0 < t < 500$ ms: (a) u_1, (b) u_3, and (c) S_{Diff}. Animation available at https://doi.org/10.1088/978-0-7503-4911-6.

warranted. A-scans at $X_1 = 325$ and 750 mm were windowed and transformed by fast Fourier transform (FFT), see section 11.4, to compare the spectral amplitudes and ensure that only waves a and b were being analysed. The results are shown in the wave propagation schematic in figure 10.8. The amplitude of waves a decreases to 0.923 of the initial value, while the amplitude of waves b decreases to 0.996 of its initial value. It seems reasonable that more energy is converted from the S0 waves a because the A0 waves n are traveling in the direction of waves a, while the A0 waves b are traveling in the opposite direction.

Waveforms and frequency spectra obtained at the point ($X_1 = 600$ mm, $X_3 = h$) are shown in figure 10.10 to decipher the essence of the mutual interactions. We start with the A-scans for waves a only (figure 10.10(a)), waves b only (figure 10.10(b)), and waves ($a + b$) (figure 10.10(c)). Even though the wavelength of waves b is smaller than that of waves a, the period of waves b is longer than that of waves a because their wavenumbers are so different. The modulation of waves ($a + b$) is apparent in figure 10.10(c), which also shows that the frequency peaks are at f_a and f_b. The presence of the difference frequency $f_n = f_a - f_b$ is not visible due to the side lobes of the primary frequencies. However, by computing S_{Diff} from equation (12.1) and taking its FFT, the difference frequency f_n is observed to have the dominant spectral

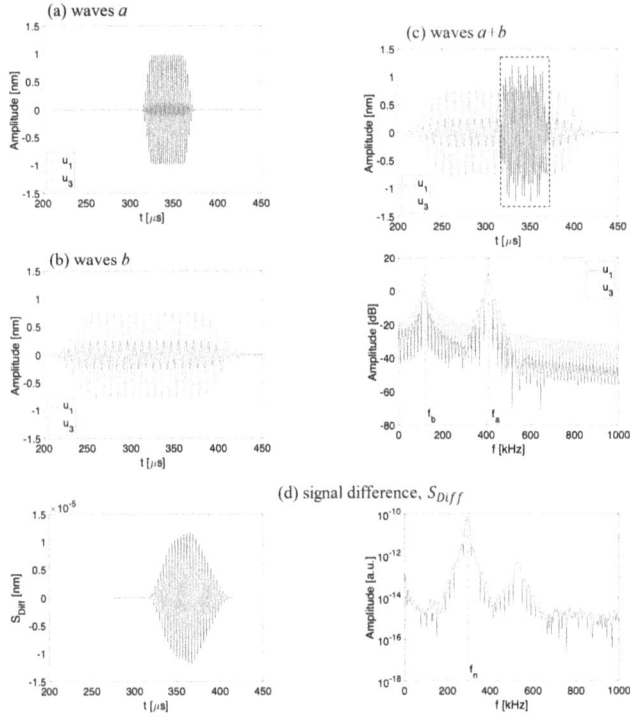

Figure 10.10. A-scans and FFTs of the wave packets at X_1 = 600 mm for 1 nm amplitude: (a) waves a, (b) waves b, (c) waves $a + b$, and (d) the difference signal S_{Diff}.

peak in figure 10.10(d). Note that the sum frequency $f_a + f_b = 0.528$ MHz also exists (having a peak two orders of magnitude less than f_n), even though the wavenumbers at the sum frequency are acutely mismatched. Although S_{Diff} is a very clean noise-free signal (it is finite element simulation after all!), its amplitude is extremely small at $\sim 10^{-5}$ nm, explaining why the f_n peak is not evident in the figure 10.10(c) spectrum.

Let's investigate the effect of the wave amplitude by simply increasing the applied amplitude from 1 to 100 nm and keeping everything else the same. The results analogous to figure 10.10 for a 1 nm amplitude are provided in figure 10.11 for a 100 nm amplitude. The shapes of the S_{Diff} waveforms are virtually identical, but the amplitudes differ by four orders of magnitude. Recall that the change in loading amplitude is just two orders of magnitude. The difference frequency f_n is not visible in the figure 10.11(c) frequency spectrum, but notice that for the 100 nm amplitude, the second harmonic S0 peak is evident at 0.820 MHz.

The amplitude increase of S_{Diff} with propagation distance in the mixing zone is given in table 10.2 for both applied amplitudes of 1 and 100 nm. Here we denote the peak amplitude of S_{Diff} as A_n. The relative nonlinearity coefficient is

$$\beta' = A_n/(A_a \cdot A_b). \tag{10.3}$$

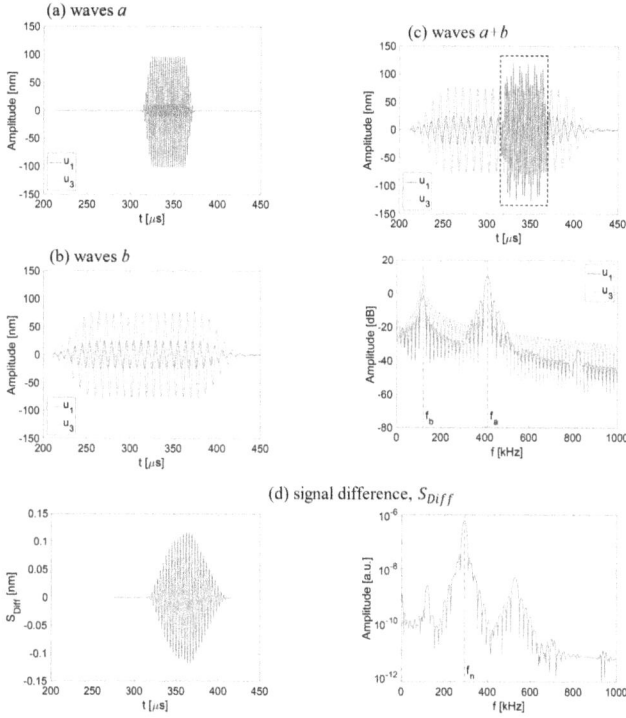

Figure 10.11. A-scans and FFTs of the wave packets at $X_1 = 600$ mm for 100 nm amplitude: (a) waves a, (b) waves b, (c) waves $a + b$, and (d) the difference signal S_{Diff}.

Table 10.2. IR point 43 wave amplitudes in the mixing zone from time domain waveforms.

	1 nm amplitude				100 nm amplitude			
X_1 (m)	A_a (nm)	A_b (nm)	A_n (nm)	$A_n/(A_a \cdot A_b)$ (m^{-1})	A_a (nm)	A_b (nm)	A_n (nm)	$A_n/(A_a \cdot A_b)$ (m^{-1})
0.550	0.988	0.762	8.63×10^{-6}	11 458	96.5	76.2	0.0866	11 772
0.575			1.00×10^{-5}	13 354			0.100	13 638
0.600			1.17×10^{-5}	15 574			0.117	15 955

There are several important assumptions necessary to compute β' from (10.3):

- A_n is associated only with the secondary waves of interest—in this case, at the difference frequency and the combined wavefield n (u_1 and u_3) is described by a single value (here we only use the u_3 component because it dominates);
- A_a is a single value to represent the combined wavefield a (u_1 and u_3)—here we only use the u_1 component because it dominates;
- A_b is a single value to represent the combined wavefield b (u_1 and u_3)—here we only use the u_3 component because it dominates.

As described in relation to (6.58), $\beta' = M_p X_1$, thus we can estimate the mixing power from the slope of the cumulative β' (or in other words 'the rate of accumulation'). Figure 10.12 does just that for both applied loading amplitudes (1 and 100 nm).

Figure 10.12 demonstrates that the rate of accumulation is linear for this range of propagation distances in the mixing zone and that the mixing power is quite insensitive to the incident wave amplitude as the slopes are M_p= 82 320 and 83 667 m^{-2} respectively for 1 nm and 100 nm. The difference between M_p values is just 1.6%. Does the toneburst duration affect the rate of accumulation? We reduced the number of cycles from 17 to 2 to find out. Keeping everything else in the analysis the same, we obtained M_p= 48 862 m^{-2}. Clearly, the size of the mixing zone affects the rate of accumulation. In addition, the β' values were significantly lower (ranging from 821–1991 m^{-1}). If we were to re-do the analysis, we would use tonebursts having the same time duration for waves a and b instead of the same number of cycles.

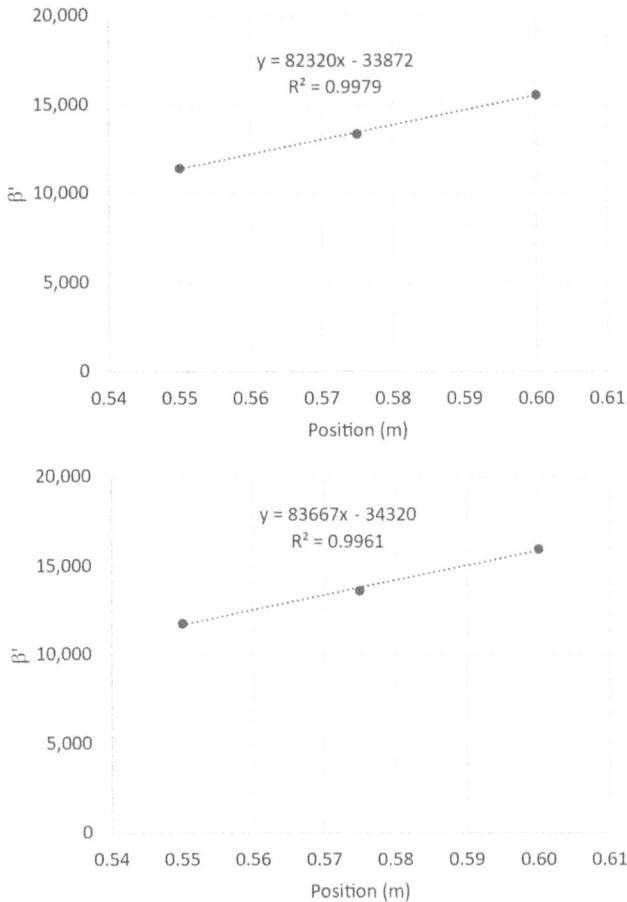

Figure 10.12. Relative nonlinearity plotted as a function of propagation distance in the mixing zone for IR point 43 and applied amplitudes: (a) 1 nm and (b) 100 nm. The slope is the mixing power.

We also want to assess how well the mixing power characterizes different TOEC values. We repeated the simulation of 17 toneburst cycles with a 100 nm amplitude using TOEC values multiplied by a factor of two. We obtained a mixing power, $M_p = 269\,376$ m^{-2}, compared to 83 667 m^{-2}, which is a factor of 3.22 higher. The increase in M_p ($\times$ 3.22) is larger than the increase in TOEC ($\times$ 2). Again, the S_{Diff} is dominated by the spectral peak at the difference frequency, but figure 10.13 illustrates that there are other peaks at f_b, $f_a + f_b$, and even $2f_a - f_b$.

A final comment on the mixing power relates to comparison of the results from the simulations with the analytical results in table 8.8, where M_p is given to be 10.2×10^6 m^{-2} for IR point 43. The mixing power values in table 8.8 are based on continuous waves and wavestructures normalized with respect to the average power flow. Clearly, the mixing power is not a material property since it depends on the loading, but it may be possible to bring the model and simulation results into better agreement by better accounting for the wavestructure in (10.3).

Our final simulation is of localized high material nonlinearity. For $550 < X_1 < 575$ mm, the TOECs are increased by a factor of two, elsewhere they retain their nominal values. The affected region is much smaller than the mixing zone. The amplitude of the u_3 component of S_{Diff} is plotted at various points along the top surface of the plate in figure 10.14 for the case of localization (blue) and the reference state with no localization

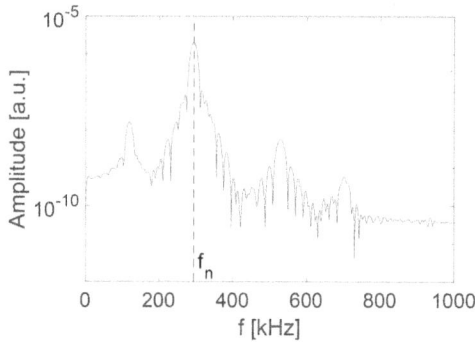

Figure 10.13. S_{Diff} frequency spectrum received at $X_1 = 600$ mm for TOEC $\times$ 2.

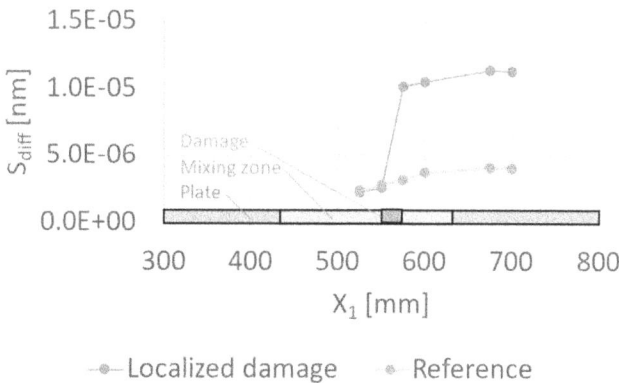

—●— Localized damage   ● Reference

Figure 10.14. Increased S_{Diff} accumulation due to localized material nonlinearity, 25 mm $\times$ d.

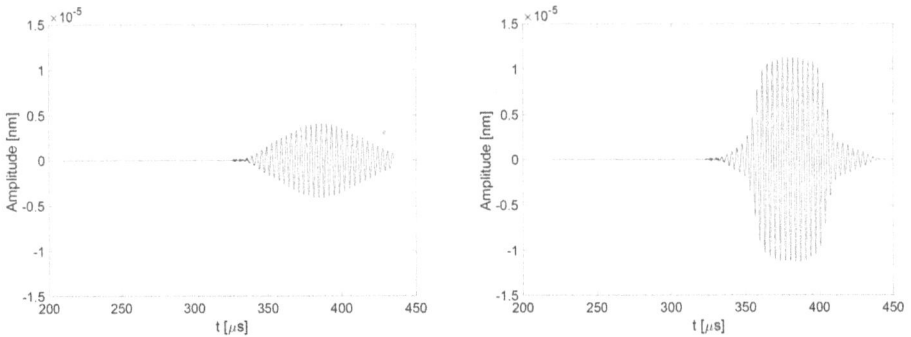

Figure 10.15. Stable S_{Diff} waveforms at $X_1 = 675$ mm, beyond the mixing zone: (a) baseline and (b) after localized material nonlinearity.

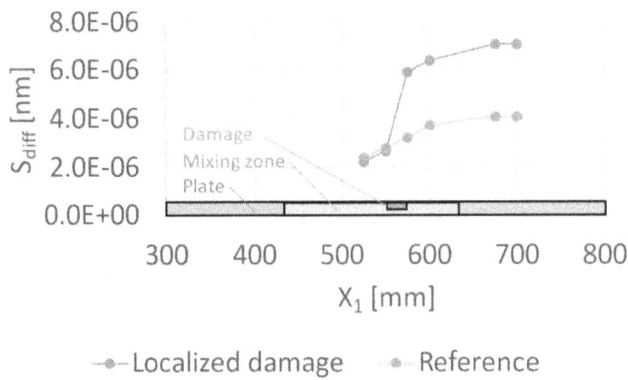

Figure 10.16. Increased S_{Diff} accumulation due to localized material nonlinearity, 25 mm × d/4.

(orange). The increase in S_{Diff} is large in the localized material, and importantly, the amplitude remains high as the secondary waves propagate on their own. The S_{Diff} waveforms received at $X_1 = 675$ mm and shown in figure 10.15 are stable once the secondary waves leave the mixing zone (remember that the model does not include attenuation). Buoyed by this success, we were curious if a localized region reduced from the plate thickness to one-fourth of the plate thickness (d/4) could be detected. Figure 10.16 indicates that the S_{Diff} amplitude is sensitive to this change. These results illustrate the strong potential that guided wave mixing has for detecting localized precursors to material degradation. Note that these localization results are from Abaqus.

10.4 Closure

Finite element simulations provide a valuable tool for confirming expectations for an experiment.

References

[1] Hughes T J R 1987 *The Finite Element Method. Linear Static and Dynamic Finite Element Analysis* (Englewood Cliffs, NJ: Prentice-Hall)

[2] Moser F, Jacobs L J and Qu J 1999 Modeling elastic wave propagation in waveguides with the finite element method *NDT E Int.* **32** 225–34

[3] Rose J L 2014 *Ultrasonic Guided Waves in Solid Media* (Cambridge: Cambridge University Press)

[4] Chillara V K and Lissenden C J 2014 Nonlinear guided waves in plates: a numerical perspective *Ultrasonics* **54** 1553–8

[5] Chillara V K and Lissenden C J 2016 Review of nonlinear ultrasonic guided wave nondestructive evaluation: theory, numerics, and experiments *Opt. Eng.* **55** 011002

[6] Hasanian M and Lissenden C J 2017 Second order harmonic guided wave mutual interactions in plate: vector analysis, numerical simulation, and experimental results *J. Appl. Phys.* **122** 084901

[7] Hasanian M and Lissenden C J 2018 Second order ultrasonic guided wave mutual interactions in plate: arbitrary angles, internal resonance, and finite interaction region *J. Appl. Phys.* **124** 164904

[8] Ding X, Zhao Y, Deng M, Shui G and Hu N 2020 One-way Lamb mixing method in thin plates with randomly distributed micro-cracks *Int. J. Mech. Sci.* **171** 105371

[9] Zhao Y *et al* 2017 Generation mechanism of nonlinear ultrasonic Lamb waves in thin plates with randomly distributed micro-cracks *Ultrasonics* **79** 60–7

[10] Sun M and Qu J 2020 Analytical and numerical investigations of one-way mixing of Lamb waves in a thin plate *Ultrasonics* **108** 106180

[11] Li W, Lan Z, Hu N and Deng M 2021 Modeling and simulation of backward combined harmonic generation induced by one-way mixing of longitudinal ultrasonic guided waves in a circular pipe *Ultrasonics* **113** 106356

[12] Lissenden C J, Guha A and Hasanian M 2022 Mutual interaction of guided waves having mixed polarity for early detection of material degradation *J. Nondestruct. Eval. Diagn. Progn. Eng. Syst.* **5** 041001

[13] Liu Y, Zhao Y, Deng M, Shui G and Hu N 2022 One-way Lamb and SH mixing method in thin plates with quadratic nonlinearity: numerical and experimental studies *Ultrasonics* **124** 106761

[14] Jiang C, Li W, Deng M and Ng C-T 2022 Quasistatic pulse generation of ultrasonic guided waves propagation in composites *J. Sound Vib.* **524** 116764

[15] Li W, Zhang C and Deng M 2023 Modeling and simulation of zero-group velocity combined harmonic generated by guided waves mixing *Ultrasonics* **132** 106996

[16] Zhu H, Ng C T and Kotousov A 2022 Low-frequency Lamb wave mixing for fatigue damage evaluation using phase-reversal approach *Ultrasonics* **124** 106768

IOP Publishing

Nonlinear Ultrasonic Guided Waves

Cliff J Lissenden

Chapter 11

Making measurements

Reliable measurements must be designed utilizing the sharp modeling tools described in the preceding chapters. Using specific instrumentation and well-documented methodology makes these measurements repeatable and reliable. After describing aspects of the instrumentation, we discuss generation of primary waves and reception of the secondary waves. Since improper processing of multimodal signals can be disastrous, we close with a concise description of some useful signal processing tools.

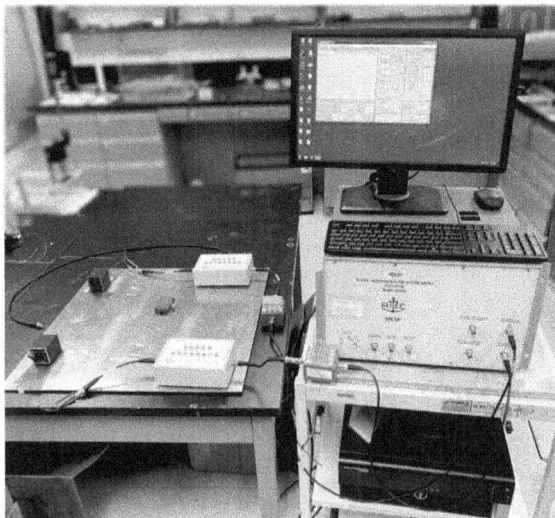

Nonlinear ultrasonic guided waves test setup.

This chapter tries to faithfully represent the numerous applications for nonlinear ultrasonic guided wave measurements and the diverse methods that are available.

The focus of the chapter is clearly on detecting slight changes in received signals from self-interaction and mutual interaction of ultrasonic guided waves with the aim of detecting material degradation at an early stage (please refer back to figure 1.2). The very nature of this objective requires measurements of weak nonlinearity, where the higher harmonics or combinational harmonics generated are necessarily small. A long-term goal is to find the threshold below which the material can be considered to be in pristine condition, and then to shift that threshold earlier in the service life to provide time for decision-making and the logistics of maintenance. A thorough discourse on the measurement of nonlinear ultrasonic parameters from higher harmonics is provided in the book chapter by Jhang *et al* [1]. Measurement of nonlinear guided waves is described in Lissenden and Hasanian [2]. This chapter is intended as a supplement that introduces instrumentation and methods that are useful for nonlinear guided wave measurements. In it we discuss several important aspects involved with making guided wave measurements in preparation for chapter 12, which provides a sampling of laboratory results.

11.1 Instrumentation

The maintenance and life management of fleets of vehicles, power and industrial plants, infrastructure, and other structures and machines is performed in vastly diverse ways. Nonlinear ultrasonic guided wave-based nondestructive evaluation could ultimately be impactfully implemented as online structural health monitoring, offline structural health monitoring, or nondestructive inspection in the field. In the case of quality assurance testing, it could be done in a manufacturing environment or in a controlled laboratory environment. Most of the instrumentation described here targets the controlled laboratory environment, which seems appropriate given the state of the art.

Applications of ultrasound owe their development to the discovery of piezo-electricity (conversion of elastic energy to electrical energy for sensing) by Pierre and Jacques Curie in 1880 and discovery of its inverse effect (conversion of electrical energy to elastic energy for wave generation) by Gabriel Lippmann in 1881. Piezoelectric transducers rely on two amazing characteristics: (i) the electric dipoles associated with the crystal structure can easily change polarization, which is responsible for the energy conversion and (ii) the deformation of elastic materials is instantaneous, enabling piezoelectric materials to respond at extremely high frequencies. To this day, ultrasound applications are still dominated by the transduction of electrical energy into elastic energy and vice versa by piezoelectric elements, although electromagnetic transducers and lasers are also used for specific niche applications.

The fundamental components, labeled A–E, of a generic nonlinear ultrasound system depicted in figure 11.1 are described below.

A. Function generator. Ultrasound systems use signals that could be a continuous wave (CW), a radio frequency (RF) toneburst with a prescribed number of cycles, or a broadband spike pulse. Occasionally, a broadband chirp signal is used. Ultrasonic

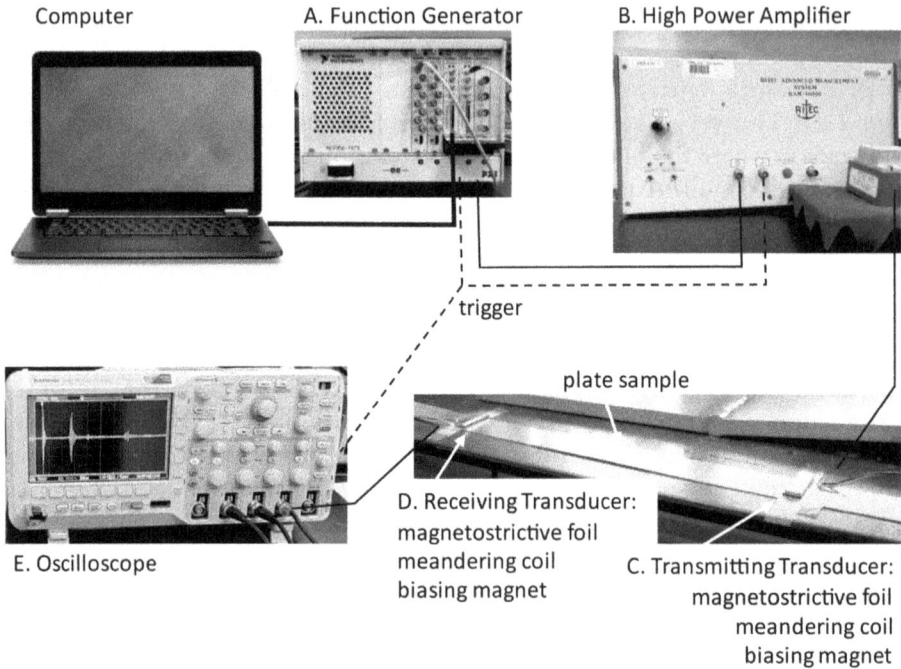

Figure 11.1. Generic components of nonlinear ultrasound measurement instrumentation.

guided wave testing typically relies upon RF toneburst signals and tries to distinguish the incident wave packet from wave scattering and reflections, which can be complicated if multiple modes are generated. It is almost always strongly preferred to excite a single wave mode, but this is often not possible due to the source influence [3]. While it is common to smooth the start and end of the toneburst with a Hanning window for linear guided wave testing, windowing the toneburst is uncommon for nonlinear guided wave testing because it is desirable to have a long stable waveform with a narrow frequency bandwidth. To demonstrate why, the frequency spectra for 2 and 20 cycle numerically generated sinusoidal toneburst signals having a central frequency of 1 MHz are shown in figure 11.2. Too short a toneburst can result in the bandwidth being too wide or sidebands of the primary frequency encroaching upon the second harmonic frequency in the case of self-interaction or combinational harmonics in the case of mutual interactions. We simply say for now that the signal processing methods matter, and that these are described in section 11.4. A single signal generator is sufficient for self-interaction testing, but two signal generators are needed for mutual interactions. Having the goal of early detection of material degradation, we seek to detect very small signals at frequencies other than the driving frequency. Thus, we need to be very cognizant of the noise floor and we want to minimize the noise starting with the generated signal being a sinusoid when on and nil when off.

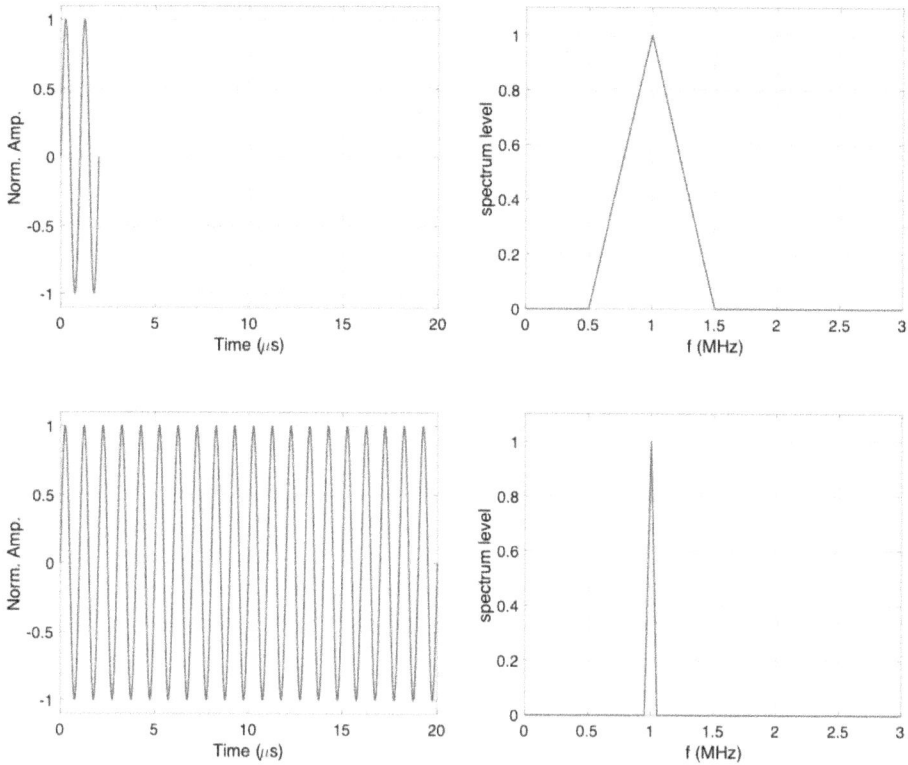

Figure 11.2. Numerically generated toneburst signals and their frequency spectra.

B. High power amplifier. The generated signal is amplified to enable the transducer to actuate finite amplitude waves. Signal distortion is the first concern and should be minimized. Likewise, the voltage sent to the transducer is important because too high a voltage can cause a piezoelectric element to breakdown, in particular for polycrystalline materials such as the ever-popular lead zirconate titanate (PZT), that are not naturally polarized, but must be poled as part of the manufacturing process. Thus, breakdown voltage is an important aspect in transducer selection.

C. Transmitting transducer. A variety of transmitting transducers are described in section 11.2 along with methods that are unique to guided waves and help mitigate unwanted source influence effects such as multiple mode generation (see also, for example, [4]). Techniques used to couple the transducer to the waveguide are a major determining factor of the measurement repeatability. The couplant could be a fluid, such as a gel or oil, or an adhesive. Figure 11.1 shows magnetostrictive transducers because they are well-suited for SH wave actuation and reception. When conducting mutual interaction tests that are co-directional, we may use a separate generator-amplifier-cabling-transducer system for each driving frequency to avoid mutual interactions between the electrical signals.

With this preferred set-up the only way for mutual interactions to occur is between the elastic waves in the waveguide.

D. Receiving transducer. A variety of receiving transducers are described in section 11.3 along with methods intended to promote sensitivity and repeatability. For self-interaction tests the primary and secondary waves are received at the same location and time if they have matching group velocities. When SH waves self-interact to generate symmetric Lamb waves having a completely different polarity, it may be necessary to use different types of receiving transducers. Even if the secondary waves have the same polarity, such as S1–S2 lamb waves or SH0 waves, different receiving transducers may be necessary based on their bandwidth. Selecting a receiving transducer at the secondary frequency and using it to also detect the primary waves may act like a partial filter if the bandwidth is insufficient. Typically, self-interaction tests aim to compute the relative nonlinearity through A_2/A_1^2 or A_3/A_1^3 for second order and third order interactions, respectively. Thus, it is important to measure the amplitudes of both primary and secondary waves, particularly in the presence of attenuation and diffraction.

E. Oscilloscope. Finally, the signal is sent to the oscilloscope for viewing, digitizing, and recording. See Jhang *et al* chapter 2 for a detailed discussion of digitization for nonlinear ultrasonics [1].

Optional components that can positively impact measurements in at least some set-ups are described below.

B.1. Matching network placed between B and C. Signal generators and amplifiers for ultrasound systems are often designed for 50 Ω transducers. If the transducer impedance is different than what the ultrasound system is designed for, then an impedance matching network consisting of capacitors and inductors can increase the voltage to the transducer. A matching network is not always necessary.

B.2. Attenuator placed between B and C. The system nonlinearity depends on the output level of the amplifier. By keeping the output level the same for all tests, the system nonlinearity will be the same. Attenuators can be used to reduce the voltage sent to the transducer without changing the amplifier output level.

B.3. Filter placed between B and C. For self-interaction testing a high power low pass filter can be placed between the amplifier and the transducer to reduce the distortion that occurs at integer multiples of the driving frequency. Alternatively, mutual interaction tests can be designed such that the frequencies of interest (e.g. the sum and difference frequencies) are sufficiently far away from integer multiples of the driving frequencies making filters less important. A low pass filter is not always necessary.

D.1. Receiver placed between D and E. The receiving transducer sends the received signal to a receiver where, optionally, a broad bandpass filter can eliminate the extreme frequency components or a narrow bandpass filter can filter everything except the frequency of interest. The signal may also be preamplified.

A schematic and photo of a typical set-up are shown in figure 11.3.

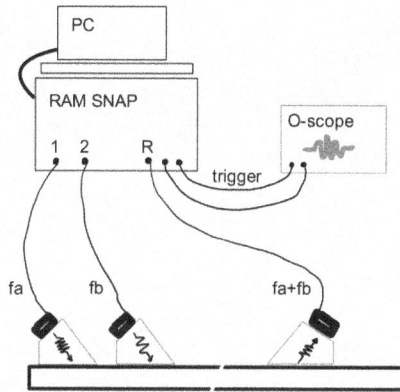

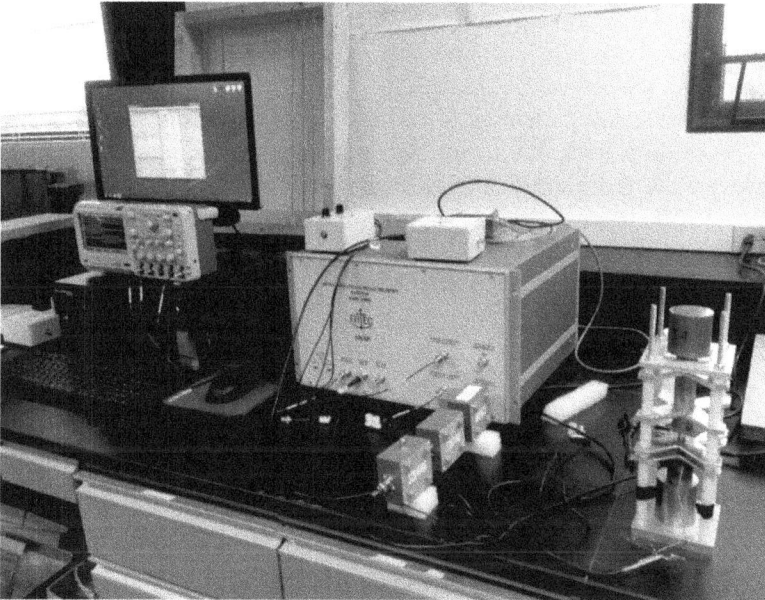

Figure 11.3. (a) Schematic of nonlinear ultrasonic guided wave instrumentation and (b) a typical system set-up for bulk wave testing.

11.2 Generation

11.2.1 Transmitting transducers

As previously discussed, piezoelectric materials are by far the most common active elements in transducers for ultrasound actuation. Contact transducers utilizing the d33 piezoelectric coupling coefficient and the resonant thickness mode of the element are commonly used for ultrasonic NDE. Contact transducers typically have an acoustic impedance matching layer, backing material to provide damping, and a protective housing. The most important design characteristic is the resonant frequency, but the frequency bandwidth can be tailored as well. The size (typically the radius) of the piezoelectric element dictates to a large extent the beam spreading and source influence. Liquid (or gel) couplant between the transducer and plate replaces air and reduces the acoustic impedance match. Liquid couplant, unless highly viscous such as baked honey or molasses, transmits only out-of-plane displacement to the plate. A solid couplant, such as an adhesive, transmits both in-plane, u_1, and out-of-plane, u_3, displacement components to the plate. As will be described in the methods section below, it is rare to couple a contact transducer directly to a waveguide because that provides very little control over which modes are generated.

Piezoelectric wafer active sensors (PWAS), so named by Giurgiutiu [5], are valuable for structural health monitoring (SHM) of hot spots where material degradation is anticipated. The usual wafer geometry is a thin circular disc. The radial resonance of the disc is typically excited by the d31 piezoelectric coupling coefficient, and the wave energy is transmitted to the structure by shear, thus the disc is typically adhesively bonded to the surface. Commercialization of PWAS-based systems for guided wave SHM has been advanced by companies like Acellent Technologies Inc. (Sunnyvale, CA, USA).

The active piezoelectric element for both contact transducers and PWAS is most often polycrystalline PZT due to its high coupling coefficients and relatively low cost. However, there are applications (in harsh environments for example) where other piezoelectric materials outperform PZT, as well as ongoing efforts to eliminate lead from products. Important considerations for selecting the piezoelectric material for nonlinear ultrasonic testing are the piezoelectric coupling coefficients, the Curie temperature, the maximum voltage that can be applied to the element, and the material nonlinearity. Lithium niobate is often used instead of PZT for nonlinear ultrasonic testing [1, 6] because it typically exhibits less nonlinearity and permits higher voltages despite its lower piezoelectric coupling coefficient. Kim *et al* [7] compared second harmonic results for a thermally aged aluminum alloy from PZT and lithium niobate-based contact transducers and report no significant difference.

Electromagnetic acoustic transducers (EMATs) are versatile noncontact transducers that can generate many ultrasonic wave types [8]. EMATs can be subdivided into two types: Lorentz force EMATs and magnetostrictive EMATs. Lorentz force EMATs actuate waves with limited amplitudes, thus we will only describe magnetostrictive transducers (MSTs) [9] in this section. Magnetostrictive materials parallel piezoelectric materials as active materials, but the interacting fields are magnetic and elastic instead of electric and elastic. Most materials that we would want to monitor have insufficient magnetostriction to provide useful ultrasonic actuation, thus a layer or patch of material having high magnetostriction (e.g. iron-cobalt and specifically remendur, iron-gallium and specifically galfenol) is affixed to the surface. The magnetic domains in the patch are aligned by a biased (static) magnetic field and then the patch is actuated by a dynamic magnetic field created by a meandering electric coil. The MST can generate SH waves given one polarization of the bias magnetic field and Lamb waves given an orthogonal polarization. The wavelength generated is dictated by the spacing of the meanders in the electric coil. MSTs are simple devices that can be self-made with a magnetostrictive patch, an electric coil (e.g. printed circuit board), and a rare earth magnet. Innerspec Technologies (Forest, VA USA) and Guidedwave (Bellefonte, PA USA) market inspection products for SH waves in plates and torsional waves in pipes based on magnetostrictive transducers. They can generate both SH and Lamb waves [10] in plates and torsional waves in pipes. In essence, they make the use of SH waves practical. Compared to PZT-based transducers, EMATs have a low electrical impedance, which makes the impedance matching network described above more beneficial or even necessary. EMATs are noncontact devices as there can be a gap, or liftoff, between electric coil and surface. However, the coupling decays exponentially with liftoff distance, making it desirable to simply place the coil on the surface. Nevertheless, the lack of a coupling agent can improve the repeatability of nonlinear measurements.

Aspects of transmitting transducers that must be carefully considered include directionality, bandwidth, and coupling. All the modeling previously described in this book is based on planar (straight-crested) wavefronts. Small PWAS, such as a point-source, typically generate an omnidirectional wavefield with circular-crested wavefronts, where the amplitude of the primary waves is decreasing as the wavefront advances. Thus, the cumulative nature of the secondary waves and the use of the amplitude ratios A_2/A_1^2 or A_3/A_1^3 must be assessed if the wavefront is not planar [11]. By using a narrowband transducer, the harmonics in the electrical signal may be filtered out. If couplant is necessary, then to promote repeatability it should be uniform from measurement to measurement, which can be achieved by applying a uniform pressure.

11.2.2 Transmitting methods

The simplest ways to preferentially activate guided wave modes are with an angle-beam transducer or a comb transducer. The phase velocity dispersion curves provide the theory of operation, as depicted in figure 11.4. An angle-beam transducer is

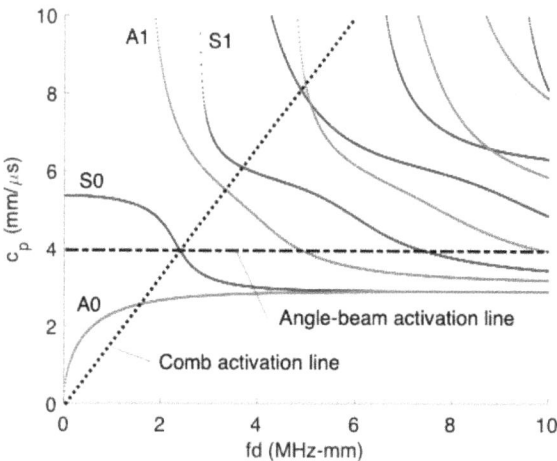

Figure 11.4. Actuating guided waves in a plate.

comprised of a contact transducer mounted on a wedge, often machined from acrylic, to enable the generated elastic wave to impinge on the plate surface at an oblique angle. Snell's law provides the relationship between the bulk waves in the wedge and the phase velocity of the resulting guided wave in the plate:

$$\sin \theta_w = \frac{c_w}{c_p} \sin (90°),$$ (11.1)

where θ_w is the angle of the wedge, c_w is the longitudinal wave speed in the wedge, and c_p is the phase velocity of the Lamb wave mode at the driving frequency. Snell's law results in the angle-beam activation line being a horizontal line in the c_p–fd plane, and where it intersects a dispersion curve is preferentially excited. However, for a toneburst excitation there are a range of frequencies excited, and the finite transducer size results in a range of phase velocities. Therefore, there is a zone of excitation instead of a true excitation point [3]. As equation (11.1) implies, the wedge material must be chosen such that $c_w < c_p$, which can be a limiting factor.

A comb transducer is a linear array of active elements having a fixed spacing, or pitch p. Unlike a phased array, there are no phase or time delays. The pitch sets up the preferred wavelength as $\lambda = p$, and thus the activation line for a comb transducer is diagonal,

$$c_p = \lambda f = p\frac{(fd)}{d}.$$ (11.2)

Like the angle-beam transducer, the comb transducer does not activate a single point on the dispersion curve, in this case because the array spacing permits wavelengths of $\lambda_n = p/n$, where $n = 1,2,\ldots$ creating side lobes for $n > 1$. The larger the number of elements in the array, the smaller are the side lobes [3].

In figure 11.4, at $fd = 2.4$ MHz-mm the angle-beam and comb transducers will both preferentially excite the S0 mode. But for these activation lines the S1 mode would be excited by the angle-beam transducer at 7.5 MHz-mm and by the comb transducer at 3.7 MHz-mm.

11.3 Reception

11.3.1 Receiving transducers

There is a diverse range of transducers capable of receiving ultrasonic guided wave signals, including the angle-beam and comb transducers described above that can help filter out wave modes not on the activation lines. The biggest difference between receiving guided waves and receiving bulk waves is that guided waves are propagating along the surface, while bulk waves are traveling toward that surface. Thus, if it physically contacts the surface, the sensor could affect the signal it is designed to receive (which is known as crosstalk). The length along the surface from which the signal is received should be small compared to the wavelength to maximize sensitivity. The nonlinearity of the active element is less important for reception than transmission because the voltage is much smaller.

Contact transducers can detect the u_3 (out-of-plane) displacement component of Lamb and Rayleigh waves, shear transducers can detect the u_1 (in-plane) displacement component of Lamb and Rayleigh waves or the u_2 component of SH waves presuming they are properly oriented, and PWAS are 'tuned' based on their size to a specific wavelength. Polyvinylidene fluoride (PVDF) is a piezoelectric polymer available in sheet form that makes an excellent receiver because it is very thin, compliant, broadband, and can be stretched to have significant in-plane anisotropy that can be used to receive both Lamb waves and SH waves [12].

The above sensors require coupling to the surface; typically contact transducers are coupled with liquid or gel, shear transducers with shear couplant (e.g. molasses or baked honey), PWAS with adhesive, and PVDF with adhesive. Cyanoacrylate and epoxy are widely used adhesives, and phenyl salicylate (salol) is a low melting point crystalline organic material often used as a convenient temporary bonding agent.

Noncontact transducers eliminate the need for couplant, but in some cases liftoff variability could lead to inconsistent reception if not adequately controlled. As discussed for generation, both types of EMAT can be used for reception although the Lorentz force type may have a lower signal to noise ratio. For guided wave reception, both types of EMATs typically use a meander coil whereby they function like comb transducers based on the meander spacing. Air-coupled transducers have piezoelectric active elements and carefully designed impedance matching layers to functionally grade the acoustic impedance of the piezoelectric element to air. Thus, the Lamb waves or Rayleigh waves must leak out of the waveguide into the air to be detected.

Laser interferometers and laser Doppler vibrometers (LDVs) are optical devices that detect motion of a surface in the direction of the laser beam. The LDV uses the Doppler shift associated with the surface velocity relative to a fixed surface to determine amplitude and frequency of velocity. 3D-LDVs use three laser beams having different orientations to determine velocity components in the three Cartesian coordinate directions. The laser interferometer splits the output beam into a signal beam and a reference beam. After the signal beam reflects from the moving surface it is re-combined with the reference beam creating an interference pattern on a photo detector, enabling the surface displacement waveform to be determined. Unique aspects of laser-based elastic wave detection include:

- The laser head can be meters away from the surface, making this a truly noncontact method. The laser beam can propagate through glass and water as well as air.
- Light can be transported to a remote laser head through an optical fiber enabling the lasing cavity to be in a remote location.
- The frequency bandwidth is essentially flat, making it possible to receive signals at all ultrasonic frequencies without bias.
- The spot size of the laser beam can be as small as 10 μm, leading to very localized measurements, which can be both positive and negative, depending on the application.
- Surface roughness can scatter the laser light creating a noisy signal.

- Retro-reflective tape on the surface enables laser beams to impinge on the surface at oblique angles and light to reflect back to the laser head making possible detection of in-plane displacement components.
- The sample or the laser head can be mounted on a translational stage to scan a subdomain of the surface, and since these are noncontact measurements they can be done very rapidly.

Lastly, the directivity of the receiver needs to be considered when detecting in-plane displacement components or using a comb transducer.

11.3.2 Receiving methods

The wavefields for guided waves are distinctly different than those for bulk waves. The ultrasonic nonlinear coefficient (or parameter, if you prefer) is defined for planar 1D bulk waves in terms of material parameters in equation (3.9) and then related to wave characteristics in equation (3.12). But guided waves have a more complicated 3D displacement field that varies with position, although for the idealized cases of Lamb waves and Rayleigh waves the displacement field is 2D due to the plane strain condition. Even SH waves, which have just one component u_2, have wavefields that vary sinusoidally through the thickness. In the absence of an ultrasonic nonlinearity coefficient for guided waves, researchers typically use simple adaptations such as $(k^2 A_2)/(X_1 A_1^2)$ or more commonly and more simply A_2/A_1^2 or A_3/A_1^3. As defined by equation (3.9), β is a genuine material parameter, because it depends only on elasticity tensor coefficients. On the other hand, β given by equation (3.12) depends on measurement parameters (k and X_1) and therefore is not a material parameter, nor is its relative value β'. In fact, Jhang et al [1] show the calibration procedure (from equation [13]) to relate β' measured from voltages to β in equation (3.12) and then the relationship between equations (3.12) and (3.9).

Demonstrating measurement repeatability provides confidence in the results by quantifying the variability. Thus, it is important to fully understand all the variables for a given measurement set-up and recognize that different set-ups will have different variables. Example variables are pressure acting on couplant between a contact transducer and a wedge, position of the wedge on the sample, extent of damping material used at a boundary (if any), temperature, cables, connectors, and grounding. Standard operating procedures for controlling these variables are defined and faithfully followed to limit uncertainties, but do not drive the uncertainty to zero. Thus, repetitive tests are performed to characterize the uncertainty in the measurements.

11.4 Signal processing

Signal processing techniques specifically for guided waves are reviewed by Diogo et al [14]. The common data obtained from a transducer receiving an ultrasonic waveform are the voltages at a predefined sample rate for a prescribed period of time. Depending on the transduction method, the voltage (V) may or may not be calibrated to the actual wave amplitude (nm). The plot of this time series of received voltages is known as an A-scan. The most basic signal processing method is to simply average together a series of A-scans with the aim of reducing the incoherent noise in the data. Any number of averages from 32 to 1064, or more, may be used depending on the significance of the noise in the data and the time it takes to collect the data, which is dictated by the frequency (of the duty cycle) used to repeatedly send the signal to the transmitter. It is a basic oscilloscope function to average the data before recording it for further analysis. Averaging the waveforms does not ensure that the data are repeatable the next day or in another lab because there may be either random or systematic errors associated with transducer positioning, couplant, etc. Repetitive testing, where the test set-up is disassembled and reassembled, is used to confirm that the test set-up is repeatable within some acceptable tolerance.

The results of nonlinear ultrasonic testing are based on the wave distortion. In second harmonic, third harmonic, and wave mixing experiments the frequency spectrum is typically used as an indicator of the wave distortion and thus the nonlinearity. Given that the data are in a finite time series, the discrete Fourier transform implemented with a fast Fourier transform (FFT) algorithm is an obvious choice. The spacing of the points in the frequency spectrum, Δf, is dictated by the sampling rate, Δt, and the number of points in the selected time series, N, through $\Delta f \Delta t = 1/N$. Commonly, the waveform is windowed and zero padded. As shown in example 3.1, window functions affect the spectral amplitude, and thus consistency of use is critical. Using a consistent window size for reception points over a range of propagation distances requires adequate planning due to dispersion and mismatching group velocities.

The utility of an FFT is limited when there are multiple modes having different group velocities. In these cases, which are a challenging hallmark of guided waves, and often unavoidable, we can turn to the 2D-FFT to transform from the spatial domain to the wavenumber domain and the temporal domain to the frequency domain [15]. Specific details on frequency analysis can be found in sources such as [16]. However, the 2D-FFT requires time series data for spatially arrayed points. Noncontact transducers such as laser Doppler vibrometers and air-coupled transducers can be used to scan over the surface, or an array transducer (e.g. [17, 18]) can be used. In lieu of acquiring data along a spatial array of points, the short-time Fourier transform (STFT) can be employed to separate waves having different group velocities as described by Neithammer *et al* [19].

Sometimes it is necessary to use relatively short toneburst excitations, for example if the goal is to have a small wave mixing zone to detect localized material degradation. Short bursts have broad bandwidths that could overlap the low amplitude secondary waves in the frequency spectrum, or the sidebands could overlap the secondary waves making it impossible to identify the secondary amplitudes. This is an example of when the phase inversion method, e.g. [20, 21], is very useful. It was used in section 10.3.2.1 for the self-interactions example. When using the phase inversion method, the frequency spectrum does not need to be determined. First, the signal is sent and data recorded from the receiver. Then the same signal is sent again, but this time 180 degrees out of phase. The phase inversion is just a superposition; adding the out-of-phase signals together eliminates the primary waves and odd harmonics, while the amplitudes of the even harmonics are doubled. In a similar vein, the nonlinearity associated solely with the mutual interaction part of a wave mixing measurement can be approximated well using a very simple signal difference method as used in section 10.3.2.2 and described in section 12.2. Both the phase inversion and the signal difference methods are based on the breakdown of the principle of superposition due to nonlinearity.

11.5 Closure

Instruments and methods for making nonlinear ultrasonic guided wave measurements were described. Their implementation is exemplified in chapter 12.

References

[1] Jhang K-Y, Choi S and Kim J 2020 Measurement of nonlinear ultrasonic parameters from higher harmonics *Measurement of Nonlinear Ultrasonic Characteristics* (Measurement Science and Technology) ed K-Y Jhang, C J Lissenden, I Solodov, Y Ohara and V Gusev (Singapore: Springer Nature) pp 9–60

[2] Lissenden C J and Hasanian M 2020 Measurement of nonlinear guided waves *Measurement of Nonlinear Ultrasonic Characteristics* (Measurement Science and Technology) ed K-Y Jhang, C J Lissenden, I Solodov, Y Ohara and V Gusev (Singapore: Springer Nature) pp 61–108

[3] Rose J L 2014 *Ultrasonic Guided Waves in Solid Media* (Cambridge: Cambridge University Press)

[4] Khalili P and Cawley P 2016 Excitation of single-mode lamb waves at high-frequency-thickness products *IEEE Trans. Ultrason. Ferroelectr. Freq. Control* **63** 303–12

[5] Giurgiutiu V 2007 *Structural Health Monitoring with Piezoelectric Wafer Active Sensors* (New York: Academic)

[6] Williams C, Borigo C, Rivière J, Lissenden C J and Shokouhi P 2022 Nondestructive evaluation of fracture toughness in 4130 steel using nonlinear ultrasonic testing *J. Nondestruct. Eval.* **41** 13

[7] Kim J, Lee K-J and Jhang K-Y 2016 Comparison of ultrasonic nonlinear parameters measured by PZT and LiNbO₃ transducers *AIP Conf. Proc.* **1706** 060008

[8] Hirao M and Ogi H 2017 Electromagnetic acoustic transducers *Measurement Science and Technology* 2nd edn (Toyonaka: Springer Nature)

[9] Kim Y Y and Kwon Y E 2015 Review of magnetostrictive patch transducers and applications in ultrasonic nondestructive testing of waveguides *Ultrasonics* **62** 3–19

[10] Sha G and Lissenden C J 2021 Modeling magnetostrictive transducers for structural health monitoring: ultrasonic guided wave generation and reception *Sensors* **21** 7971

[11] Hurley D C and Fortunko C M 1997 Determination of the nonlinear ultrasonic parameter using a Michelson interferometer *Meas. Sci. Technol.* **8** 634–42

[12] Ren B, Cho H and Lissenden C J 2017 A guided wave sensor enabling simultaneous wavenumber-frequency analysis for both lamb and shear-horizontal waves *Sensors* **17** 488

[13] Dace G E, Thompson R B and Buck O 1992 Measurement of the acoustic harmonic generation for materials characterization using contact transducers *Review of Progress in Quantitative Nondestructive Evaluation* (New York: Plenum) pp 2069–76

[14] Diogo A R, Moreira B and Gouveia C A J 2022 A review of signal processing techniques for ultrasonic guided wave testing *Metals* **12** 936

[15] Alleyne D and Cawley P 1991 A two-dimensional Fourier transform method for the measurement of propagating multimode signals *J. Acoust. Soc. Am.* **89** 1159–68

[16] Randall R B 1987 *Frequency Analysis* (Naerum: Bruel and Kjaer)

[17] Minonzio J-G, Talmant M and Laugier P 2010 Guided wave phase velocity measurement using multi-emitter and multi-receiver arrays in the axial transmission configuration *J. Acoust. Soc. Am.* **127** 2913–9

[18] Ren B and Lissenden C J 2016 PVDF multielement lamb wave sensor for structural health monitoring *IEEE Trans. Ultrason. Ferroelectr. Freq. Control* **63** 178–85

[19] Niethammer M, Jacobs L J, Qu J and Jarzynski J 2001 Time-frequency representations of Lamb waves *J. Acoust. Soc. Am.* **109** 1841–7

[20] Kim J-Y, Jacobs L J, Qu J and Littles J W 2006 Experimental characterization of fatigue damage in a nickel-base superalloy using nonlinear ultrasonic waves *J. Acoust. Soc. Am.* **120** 1266–73

[21] Zhu H, Ng C T and Kotousov A 2022 Low-frequency Lamb wave mixing for fatigue damage evaluation using phase-reversal approach *Ultrasonics* **124** 106768

Chapter 12

Highlights of experimental testing

All the modeling and numerical simulation in the preceding chapters are no substitute for making actual measurements, whether in the laboratory or ultimately in the field. Rather, the modeling and simulations enable experiments to be conducted with some measure of confidence that they will be successful. This chapter highlights experimental testing, starting with self-interactions, moving on to mutual interactions and wave mixing, and then closing with quasi-Rayleigh waves.

In keeping with our approach to expose the fundamental concepts associated with nonlinear ultrasonic guided waves, this chapter tries to emphasize the key aspects of experimentation. With apologies to those whose hard work is not highlighted herein, the intent was to provide a sampling as opposed to a comprehensive review.

12.1 Self-interaction

Key experiments demonstrating self-interaction of Lamb waves are discussed in this section and selected details are provided in table 12.1. Although the focus is squarely on internal resonance (IR) points 1–3, a few other studies are discussed because they provide results showing other important aspects of self-interaction.

The first experiments showing the cumulative behavior of second harmonic Lamb waves were conducted by Deng *et al* [1, 2]. Glass wedges and contact transducers were used to excite the IR point 3 at the S2–A2 mode crossing point (where the frequency is 2.7 MHz for a 1.85 mm thick aluminum plate). The angle-beam transducer could generate mutual interactions between the S2 and A2 Lamb modes, as well as self-interaction of the S2 mode and self-interaction of the A2 mode. Unfortunately, we do not know what combination of S2 and A2 modes is excited here, partially because the honey couplant will promote better generation of the S2 mode than would gel or less-viscous couplants (which is a good thing for this application) and partially because of the source influence [3]. This ambiguity is inherent to IR point 3 being at a mode crossing point. The S4 mode is the internally

Table 12.1. Summary of experiments on Lamb wave self-interactions.

Mode pair	Plate specifics	Measurement details	Signal processing	Comments
Deng et al [1, 2] show for the first time the cumulative behavior of second harmonic Lamb waves.				
IR point 3 S2/A2–S4, mode crossing point	Aluminum 1.85 mm thick 500 mm by 500 mm $fd = 5.0$ MHz-mm	T/R: glass wedges with 1.7 [1] or 2.0 [2] MHz 20 mm diam. contact transducers coupled by honey toneburst: 2.28–4.0 MHz central frequency, 25 μs duration 70–180 mm propagation lengths	Five averages, used the integrators in the Ritec SNAP system to acquire primary and second harmonic signals	• Both A_1 and A_2 decrease with propagation distance, but A_2/A_1^2 increases • Show that system nonlinearity exists only in the vicinity of the transducer resonant frequency. Thus, their measurements are obtained at frequencies above the resonant frequency. Putting wedges back-to-back shows minimal system nonlinearity at the driving frequencies. • Attenuators are used to show that the nonlinearity is constant over a range of driving voltages given constant amplifier output.
Deng and Pei [5] show that second harmonic Lamb waves generated by exciting the mode crossing point is sensitive to tensile fatigue damage.				
IR point 3 S2/A2–S4 mode crossing point	Aluminum 1.85 mm thick, 60 mm wide, by 500 mm long. Two interrupted tensile fatigue tests (up to 10,000 cycles)	T/R: 2 MHz 20 mm diam. immersion transducers at 11.5 deg. angle in water. Toneburst: 2.5–3.2 MHz central frequency, 12 μs duration. Reception: Ch1 full signal and Ch2 has a HPF@4.7 MHz 130 mm propagation length	Stress wave factor (SWF) computed	• Peak second harmonic received at 2.95 MHz, which is $fd = 5.5$ MHz-mm, which is a little higher than 5 MHz-mm (where the crossing point is located). • Nonlinearity (SWF) decreases monotonically with fatigue damage. • Ch1 received waveform is not shown making it impossible to know which modes are present in the signal.

Table 12.1. (*Continued*)

Mode pair	Plate specifics	Measurement details	Signal processing	Comments
Bernes et al [6] develop a method to measure cumulative second harmonic Lamb waves in a plate using IR point 1, having matching phase and group velocities.				
IR point 1 $S1$–$S2$ @ $c_p = c_L$	Aluminum 6061-T6 and 1100-H14 1.6 mm thick reception between 200–500 mm $fd = 3.44$ MHz-mm	T: Angle-beam actuator (2.25 MHz narrowband, 12.5 mm diam. bonded to wedge); acrylic wedge fluid-coupled to plate; R: same as T or heterodyne laser interferometer. Toneburst: 20 cycle 2.15 MHz toneburst	1000 signals averaged spectrogram from STFT squared ampl. 25 MHz sampling, three repeats	• The angle-beam actuator excited the A1 mode more so than the S1 mode. • The interferometer detects out-of-plane displacement but the S2 mode has only a very small out-of-plane displacement at surface. • STFT enables mode identification and provides the ability to use the spectral amplitudes of specific modes for A_1 and A_2.
Pruell et al [7–9] improve activation of IR point 1 and show that the second harmonic Lamb waves are sensitive to plastic strain and fatigue damage.				
IR point 1 $S1$–$S2$ @ $c_p = c_L$	Aluminum 1100-H14 1.6 mm thick reception between 200–600 mm dog-bone samples 16 mm wide and 430 mm long, $fd = 3.57$ MHz-mm	T: Angle-beam actuator (2.25 MHz 12.5 mm diam. narrowband Panametrics X-1055) acrylic wedge. Toneburst: 25 cycles, 2.225 MHz. R: 5 MHz, 12.5 mm diam. Panametrics A-109S or heterodyne laser interferometer	1000 signal averaged spectrogram from STFT squared ampl. 25 MHz sampling. Three fatigue samples. Five plastic strain samples	• The angle-beam actuator excited the A1 mode more so than the S1 mode. • The interferometer detects only out-of-plane displacement. • The narrow dog-bone samples will have much different dispersion curves than Lamb waves, which are for plane strain. • Acrylic wedge design was improved relative to Bermes et al [6].

(*Continued*)

Table 12.1. (*Continued*)

Srivastava and Lanza di Scalea [14] show that even harmonics are comprised solely of symmetric modes while odd harmonics are comprised of both symmetric and antisymmetric modes.

Matlack et al [15] characterize the efficiency at which two different Lamb wave mode pairs generate second harmonic waves.

Liu et al [18] show that A1 antisymmetric Lamb waves at the Lame wave speed generate cumulative second harmonic antisymmetric waves that are negligibly small, in agreement with parity analysis of power flow.

Mode pair	Plate specifics	Measurement details	Signal processing	Comments
Not internally resonant, A0–A0 and S0/A0–A1/ A0/S0	Aluminum plate 2.54 mm thick. Under load/no load. 250 mm propagation length	T/R: out-of-plane Pico 0.543 MHz, Pinducer; in-plane macrofiber composite. Toneburst: 0.320 MHz	FFT, Morlet wavelets	• A static load increases the nonlinearity. • Both antisymmetric and symmetric modes are generated at odd harmonics (3, 5, 7). • Only symmetric modes are generated at the second harmonic.
IR point 1 S1–S2. IR point 2 S2–S4	Aluminum 6061-T6 1.6 mm thick, 200–500 mm propagation length	T/R: Angle-beam actuator coupled with salol (T) and oil (R). Toneburst: 35 cycles, 2.25 and 4.5 MHz for S1–S2 and S2–S4, narrowband 6.25 mm diam. received by narrowband 5 or 10 MHz	1000 signals averaged, STFT	• A fluid couplant used for R instead of solid because it is produces less variability according to the authors. • Diffraction accounted for using Achenbach and Xu model [19].
IR point 1 S1–S2 plus A1–A2 and A2–A4	Aluminum 6061-T6 and 1100-H14 1.6 and 3.175 mm thick, 225–450 mm propagation lengths	T/R: Variable angle-beam transducers (KB-Aerotech Gamma C07416 2.25 MHz actuator, Panametrics A405S 5 MHz receiver)	1000 signals averaged, 100 MHz sampling. Three repeats, Morlet wavelets	• Antisymmetric mode pairs have matching phase and group speed, but parity analysis indicates no power flows to antisymmetric second harmonics. • Nonlinearity coefficients are correlated for different frequencies, modes, and propagation distances. • Morlet wavelets used in lieu of STFT to identify frequency content of different modes.

resonant second harmonic mode and there are three challenges associated with it: (i) it is highly dispersive at this frequency, (ii) it has the lowest group velocity of any mode at this frequency—making signal processing more difficult, and (iii) because its group velocity does not match that of the primary waves the wave mixing zone is finite and thus the linearly cumulative growth of the second harmonic is limited. Nonetheless, the computed A_2/A_1^2 ratio is shown to increase between 60 and 180 mm in figure 12.1 [1], indicating that the nonlinearity is attributable to the material and not the instrumentation. It should be noted however, that both the amplitudes A_1 and A_2 decrease with propagation distance as shown in figure 5 of Deng *et al* [2]. The perturbation method used in the nonlinear theory is based on A_1 remaining constant and A_2 increasing linearly for planar waves in a lossless media. It seems reasonable to neglect attenuation for these propagation distances (70–180 mm) in an aluminum plate, but diffraction from the finite-sized source will result in beam spreading and decreasing amplitudes. The lack of linearity in the cumulative A_2/A_1^2 ratio could be attributable to the absence of planar waves or the mismatched group velocities. Since we do not know what combination of S2 and A2 modes are being generated by the angle-beam transducer it is interesting to analyse the power flow for the extreme cases of a pure S2 mode and a pure A2 mode. According to table 8.3, a pure S2 mode has a nonlinear driving force less than a pure A2 mode (0.0583 compared to 0.0808 m^{-1}), but a higher mixing power (67.81 × 10^6 compared

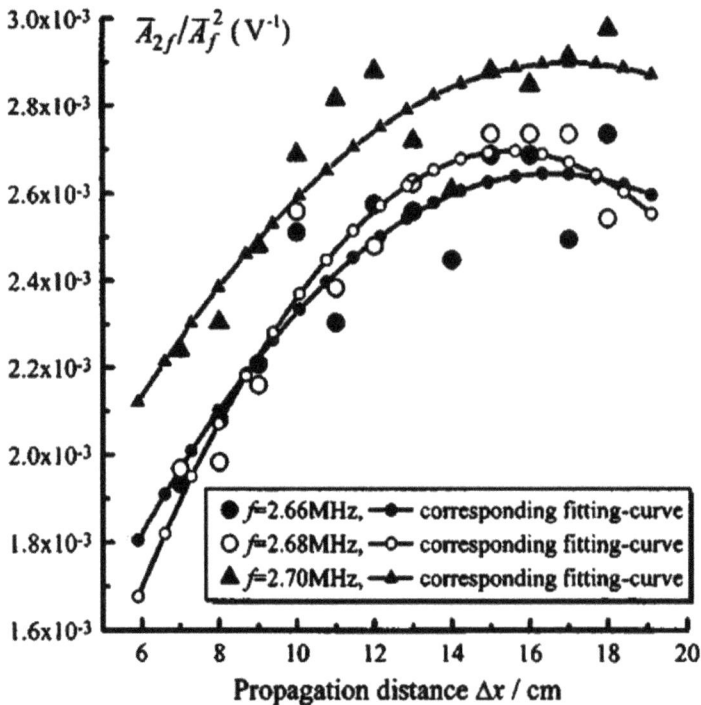

Figure 12.1. Cumulative-like nature of the relative nonlinearity coefficient for IR point 3. (Reproduced with permission from [1]. Copyright 2005 AIP Publishing.)

to 40.69×10^6 m^{-2}). No power flows to the S4 mode from the mutual interaction of S2 and A2 at IR point 3, but rather it flows to the A4 mode as indicated by IR point 21 shown in table 8.7 ($F_n = 0.0455$ m^{-1} and $M_p = 91.4 \times 10^6$ m^{-2}). Possible enhancements to the measurement methods for IR point 3 are: (i) to preferentially excite the S2 mode over the A2 mode, which could be done by bonding the wedge to the plate to enable continuity of in-plane displacements in the plate and the wedge or using a transducer such as a macrofiber composite with dominant in-plane excitability; (ii) limit the beam spreading by changing the aspect ratio of the source; and (iii) account for diffraction in either the data analysis or the model [4].

Deng and Pei [5] went on to show that second harmonic Lamb waves are sensitive to fatigue damage. The same mode crossing IR point 3 is used to assess fatigue damage. Angle-beam transducers are used, but the wedge medium is water rather than acrylic, thus it is most likely that the A2 mode is preferentially activated over the S2 mode based on the A2 and S2 wavestructures. Likewise, the S4 mode does not have good receive-ability through the water since the in-plane displacement at the plate surface is quite small. The instrumentation splits the received signal into two: one was simply received and in the other a high pass filter was used to isolate the second harmonic. A stress wave factor computed around the second harmonic frequency, $\int_{f_1}^{f_2} [A(f)]^2 df$, where $f_1 < 2f_0 < f_2$, borrowed from the acousto-ultrasonics literature, was used to characterize the nonlinearity. However, the primary wave amplitudes were not reported, nor were the modes identified. Surprisingly, the stress wave factors decrease monotonically with fatigue cycling. Early fatigue damage in metals entails dislocations leading to persistent slip bands, which typically increase the material nonlinearity. These results demonstrate that it is possible to make the difficult nonlinear ultrasonic guided wave measurements that show early indications of material degradation. In hindsight, they also show how difficult it is to use IR point 3 for these measurements because of the challenging mode excitability, dispersion, diffraction, and mismatched group velocities. Thus, it was time to try a different IR point.

A method to characterize nonlinearity in plates that addresses the dispersion and multi-modal nature of Lamb waves was proposed by Bermes et al [6]. IR point 1 is used, which is the S1–S2 mode pair having phase velocities equal to the longitudinal wave speed and matching group velocities. Although both the S1 and S2 modes are dispersive, their matching group velocities result in no limitations on the wave mixing zone regardless of the duration of the driving signal. Furthermore, their group velocities are higher than the other propagating modes at these frequencies, making them the first arriving signal and simplifying the signal processing. It is important to realize that at the primary frequency there are many propagating modes (i.e. A0, S0, A1, S1, and S2). A drawback of this mode pair is that the S1 mode at this frequency has no out-of-plane displacement component at the surface, making it generally more difficult to excite. A contact transducer was bonded to an acrylic wedge that was fluid-coupled to the plate. The fluid coupling makes it difficult to excite the dominant in-plane displacement of the S1 mode. A heterodyne laser interferometer was used to detect the out-of-plane displacement of the second harmonic waves. It is the second harmonic S2 mode that is synchronized, but its out-of-plane displacement component is also small at the surface, making reception challenging. The laser interferometer provides noncontact reception and makes it easy to scan along the propagation distance to characterize the cumulative nature of the internally resonant second harmonic. A spectrogram obtained from the STFT with group velocity dispersion curves overlaid upon it is used to identify primary and secondary modes. From the spectrogram, the spectral amplitude of specific modes (e.g. S1 and S2) can be identified based upon arrival time and group velocity. It appears that the A1 mode dominates the S1 mode at the primary frequency and the A2 mode dominates the S2 mode at the second harmonic frequency. A cumulative-like secondary A2 mode could be generated by mutual interaction between the A1 and S1 modes even though not perfectly synchronized, or non-cumulative secondary A2 mode could arise from system nonlinearity. The cumulative nature of the second harmonic is observed and it is interesting to note that the ratio of the nonlinearities determined for two different aluminum alloys is similar to the ratios of their absolute β values obtained from the literature. Apparently, despite the presence of other dominant wave modes at both the primary and secondary frequencies, the large nonlinear force and mixing power for IR point 1 make the cumulative nature measurable. It seems likely that transducers able to preferentially actuate the S1 mode over all the others could improve this measurement dramatically.

Pruell et al [7–9] showed that the S1–S2 mode pair is sensitive to both quasi-static tensile plastic strain in the range 0.005–0.02 mm^{-1} [7, 9] and low cycle fatigue damage (5–50 cycles) [8]. The acrylic wedge used for actuation was re-designed to have a clamp that provides uniform pressure on the couplant and a more narrow section that the bulk waves traverse, with the result that the wave amplitude was increased by a factor of five over Bermes et al [6] according to Pruell et al [9]. Other studies using IR point 1 include Xiang et al [10] for creep damage, Liu et al [11], Metya et al [12] for the effect of tempering temperatures, and Wang et al [13] for contact acoustic nonlinearity.

Srivastava and Lanza di Scalea [14] conducted experiments to demonstrate that their mathematical induction result is correct, i.e. even harmonics are comprised solely of symmetric modes while odd harmonics are comprised of both symmetric and antisymmetric modes. They cleverly applied a quasi-static load to increase the strain energy in the plate in order to make the problem more nonlinear. First, transducers sensitive only to out-of-plane displacement were used in pitch-catch mode to predominantly send and receive antisymmetric Lamb waves. An FFT revealed the presence of the primary frequency plus the third, fifth, and seventh harmonics. Signal processing using Morlet wavelets indicates that the primary waves and the third harmonic waves are the A0 mode. If we look at the phase velocity dispersion curves (see figure 12.2) for this experiment we see that the driving frequency is below all cutoff frequencies and only the A0 and S0 modes exist. The third harmonic A0 is not exactly synchronized, but the phase velocities are not terribly different (2187 and 2727 ms^{-1}). The group velocities match very well. Second, transducers sensitive to in-plane displacement were used in pitch-catch to

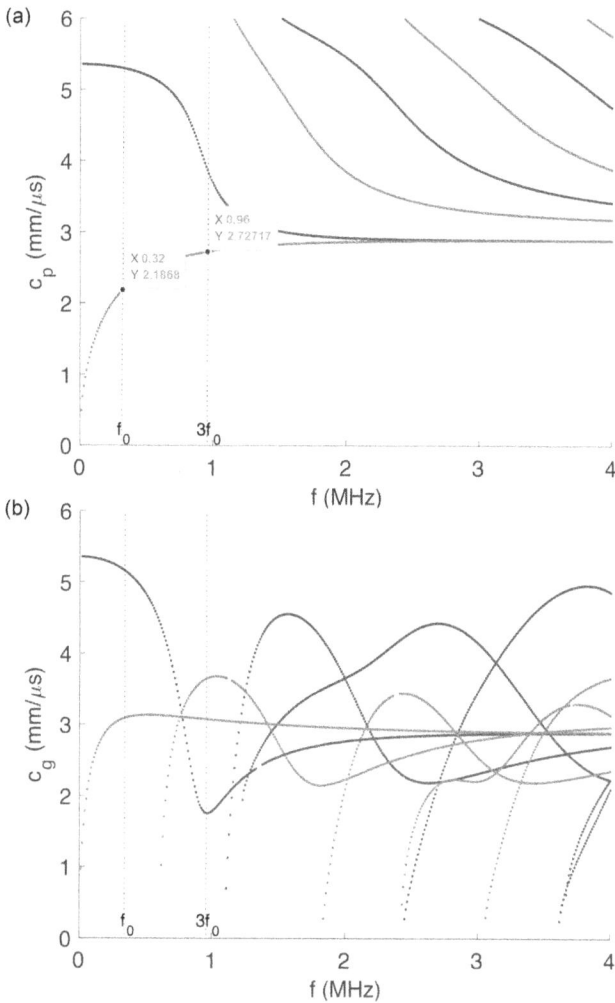

Figure 12.2. (a) Phase velocity and (b) group velocity dispersion curves for 2.54 mm thick aluminum plate.

send and receive predominantly symmetric Lamb waves. The same driving frequency was used in both tests. The transducers are known as macrofiber composites and have aligned piezoelectric fibers that provide good in-plane excitation and reception. An FFT from these test results reveals second, third, and fifth higher harmonics. Signal processing using Morlet wavelets indicates that the transducers excite both S0 and A0 waves, that the second harmonics are S0 waves, and that the modal content of the third harmonics is dominated by A1, followed by A0, as well as some S0 waves. Even though the third harmonic wave modes are far from synchronized these are significantly larger than the second harmonics. Another interesting result is that the higher harmonics are significantly larger under static load than when the load is removed.

Matlack *et al*'s [15] experiments focus on IR points 1 and 2, which both have synchronized phase and group velocities that are the fastest of all the propagating modes. The 'rate of accumulation of the second harmonic' is introduced as the constant of proportionality between the relative nonlinearity $\beta' = \frac{A_2}{A_1^2}$ and the propagation distance, and serves as the basis for the mixing power parameter M_p defined in equation (6.57). We note that Zhu *et al* [16] introduced a parameter described as the 'efficiency of cumulative second harmonic generation' at the same time that Hasanian and Lissenden [17] introduced the mixing power parameter. As was discussed in section 8.1.1, the mixing powers for IR points 1 and 2 differ by a factor of 4, which is comparable to longitudinal waves, where the factor of 4 difference is due to the doubling of the wavenumber. Unlike previous experiments, the wedge used for the transmitter is coupled to the plate with salol (phenyl salicylate), a solid, to more effectively actuate the S1 (IR point 1) and S2 (IR point 2) wave modes. However, oil couplant is used for coupling the wedge receiver because they felt it gives better repeatability. Moreover, liquid couplant is more conducive to scanning the wedge receiver along the propagation path. Unfortunately, a spectrogram showing the modal content (i.e. A0, S0, A1, S1, S2) is not provided to confirm the effectiveness of the transducers to predominately actuate the intended modes. A slice through the spectrogram for IR point 2 at the primary frequency shows the S2 amplitude to be about half that of another (unidentified) peak, suggesting only limited success. The cumulative nature of the second harmonics are shown in figure 12.3 [15], where we can observe the relationship between relative nonlinearity and propagation distance to be roughly linear up to 200 mm. The measured mixing power ratio, IR point 2 to IR point 1, is 4.23, which is slightly higher than the predicted value of 4. Matlack *et al* [15] conclude that IR point 1 is preferred to IR point 2 because there are fewer extraneous modes and it is at a lower frequency that

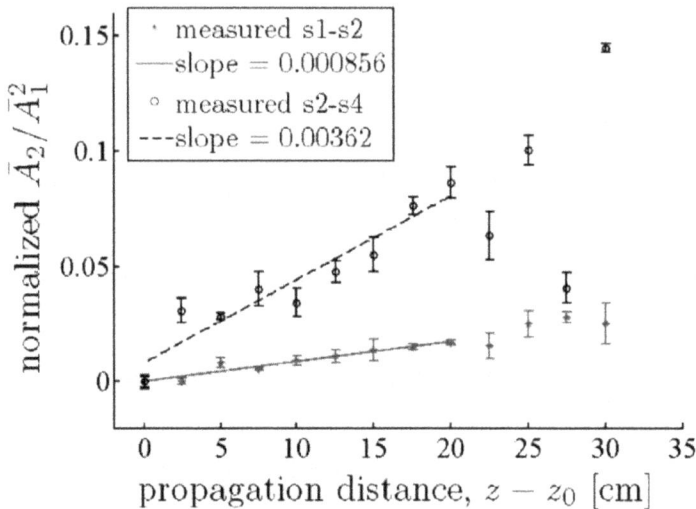

Figure 12.3. Cumulative nature of the relative nonlinearity coefficient for IR points 1 and 2. (Reproduced with permission from [15]. Copyright 2011 AIP Publishing.)

is easier to excite, despite the factor of 4 difference in mixing power. Experimental results for IR point 1 are also provided by Liu *et al* [11].

As Liu *et al* [18] describe in detail, there was at one time conflicting information about the effect of symmetry on second harmonic generation. Therefore, they conducted experiments using synchronized S1–S2 and A1–A2, A2–A4 mode pairs. The conclusive results are that the generated antisymmetric second harmonics were negligibly cumulative compared to the symmetric second harmonics. The parity analysis of power flow in chapter 7 indicates that there is no power flow to antisymmetric second harmonics. The very small, measured nonlinearity was further investigated and found to be the same for different materials, leading to the conclusion that it is not attributable to material nonlinearity. Since the group velocities of the antisymmetric mode pairs are smaller than the other propagating modes at these frequencies (as evident on the group velocity dispersion curves), signal processing is difficult, thus Morlet wavelets were used.

IR points 4–8 are all comprised of SH primary waves that generate second harmonics that are symmetric Lamb waves. The change in polarity from primary waves to secondary waves may be advantageous from a measurement perspective because it is straightforward to measure primary and secondary waves separately. However, IR points 5–7 have large group velocity mismatches, as shown in table 8.2, which severely limits the size of the wave mixing zone and thus the cumulative nature of the second harmonics. IR point 8 has less group velocity mismatch between the SH3 and S4 Lamb waves and although table 8.3 shows the nonlinear force and mixing power to be zero when integrated through the thickness of the plate, experiments conducted by Liu *et al* [11] resulted in a cumulative second harmonic between 400 and 800 mm. In these experiments primary waves were actuated by a magnetostrictive transducer with a biased magnetic field in the X_2 direction and second harmonic waves were received by a magnetostrictive transducer with a biased magnetic field in the X_1 direction and meander coil spacing half as large as for actuation. While we are unaware of experimental results for IR point 4 (SH0 and S0 Lamb waves), closely related wave mixing experiments with co-directional SH0 waves at frequencies that sum to $fd = 3.39$ MHz-mm are presented by Shan *et al* [20] and will be discussed in section 12.2.

Low frequency S0 primary waves do not generate phase-matched S0 second harmonics, and thus the second harmonic amplitude is a bounded oscillation along the propagation distance. Despite this, Zuo *et al* [21] showed that the relative nonlinearity coefficient increases with distance up to 400 mm for the primary frequency of 300 kHz. This occurs because, while not exactly phase-matched, the mismatch is small. In addition, Hu *et al* [22] showed how low-frequency leaky S0 waves can be used in plates with water loading on one side.

The simplicity of the fundamental SH wave mode, SH0, is appealing for nonlinear guided wave testing; it is nondispersive, does not leak into adjacent fluid, and has a single displacement component that is uniform throughout the thickness of the plate. Although its self-interaction does not generate second harmonics, it does generate third harmonics. Furthermore, if the driving frequency is below the first cutoff frequency, then it is assured that the SH0 mode is the only mode actuated.

In general, guided wave NDE applications for SH waves trailed Lamb waves until magnetostrictive transducers (MSTs) were found to function simply, efficiently, and economically. Designing the meander coil spacing to match the desired wavelength improves transduction. Thus, to provide the best reception of the third harmonic, the meander coil spacing of the receiver should be one-third that of the transmitter. The SH0 third harmonic has been shown to be sensitive to plastic deformation [23] and fatigue damage [24] in aluminum plates. Let's examine these tests below.

In [23] dog-bone shaped 2024-T3 aluminum plates 1 mm thick and 610 mm long having gage sections 51, 102, 229, and 457 mm long were subjected to quasi-static tension tests in displacement control to plastically deform the gage section. The different gage lengths localized the plastic strain over different lengths. Post-test plastic strains were computed from a grid of inscribed hash marks. Maximum plastic strain values varied between 0.047 and 0.084 m m^{-1}. After measuring the grid marks, the plates were trimmed down to the gage section widths between 34.8 and 37.2 mm for guided wave testing.

In [24] six dog-bone shaped 2024-T3 aluminum plates 1 mm thick and 605 mm long were fatigued in tensile load control with the maximum stress of 345 MPa and a fatigue ratio of 0.1. One plate was loaded to failure in $N_f = 4539$ cycles, while cycling of the other plates was interrupted after 0%, 20%, 40%, 60%, and 80% of N_f, after which the plates were trimmed down to the gage section width of 64 mm for guided wave testing. Hence no failure statistics are available. No fatigue cracks were visible by unaided eye in any of the plates, except the one that fractured.

Nonlinear guided wave testing was conducted using MSTs to send and receive the SH0 mode in through-transmission. The primary frequency is 0.830 MHz and the third harmonic is 2.49 MHz. The MSTs contain a 25 mm by 35 mm Fe–Co foil bonded to the plate with cyanoacrylate, meander coil, and a biasing magnet. The five-turn meander coil spacings gave preferred wavelengths of 3.6 and 1.2 mm, respectively, for the transmit and receive MSTs, respectively. The transmit and receive MSTs were spaced either 400 mm [23] or 240 mm [24] apart. Recent unpublished research indicates that Fe–Co foils can be sufficiently coupled to the plate with double-sided tape, which gives repeatable results and could be very useful for measuring the cumulative effect of secondary waves. The high-power RAM-5000 SNAP system (Ritec, Warwick, RI, USA) was used to generate 15 cycle tonebursts. The comb-like nature of the MST lengthens the burst and effectively windows it, while also sending waves in both positive and negative directions. The waves propagating in the negative direction reflect off the backwall and are received later in time. The plates are narrow enough that the SH0 waves interact with the edges and these interactions arrive after the incident wave packet. The average of 32 received signals is recorded and ten replicate tests are conducted by removing and re-installing the meander coil and biasing magnet.

Figure 12.4 [23] shows a sample A-scan (resulting from 32 averages) and its frequency spectrum computed using the fast Fourier transform (FFT) with a Tukey window fit to the primary wave packet. The third harmonic amplitude is nominally two orders of magnitude less than the peak at the primary frequency. In lieu of changing the wave propagation distance, the amplifier output level was changed

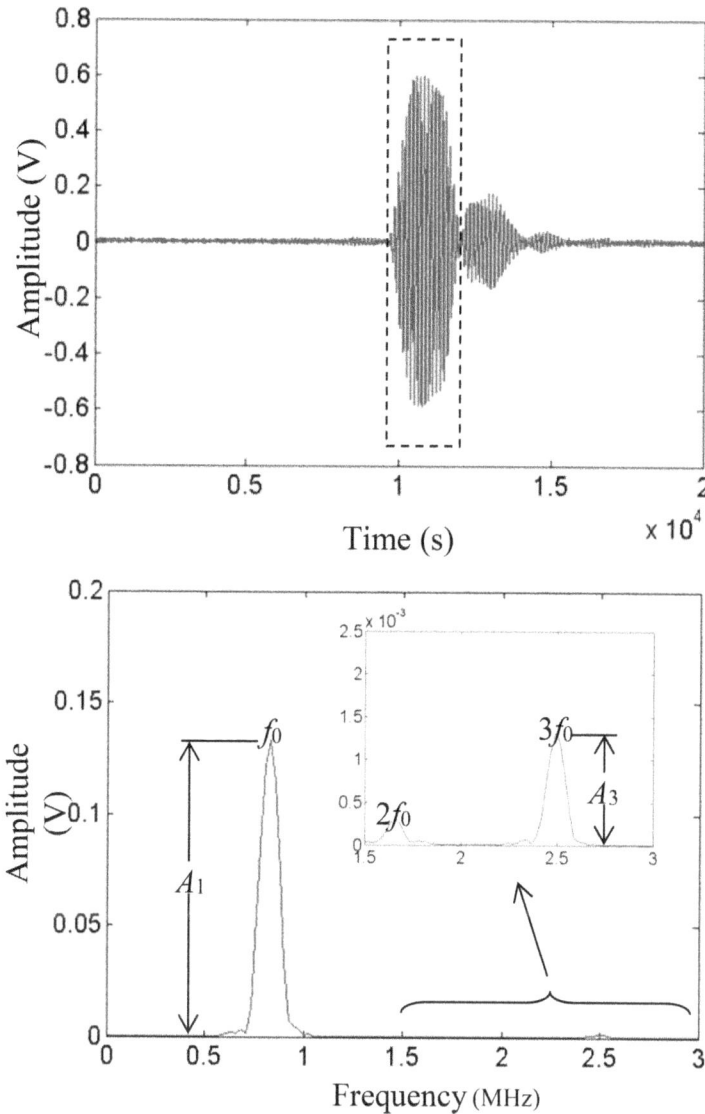

Figure 12.4. Sample A-scan and frequency spectrum for SH0 mode with magnified view of third harmonic. Primary frequency $f_0 = 0.83$ MHz. (Reproduced with permission from [23]. Copyright 2014 Springer Nature.)

(directly changing A_1) and the third order relative nonlinearity coefficient $\gamma' = A_3/A_1^3$ (inspired by example 3.5) was computed. Use of attenuators as described for B.2 in section 11.1 would have been preferred. Figure 12.5 plots γ' as a function of the output level. Working from right to left along the abscissa, γ' is constant from output level 100 down to 65, it then decreases to output level 52, and finally increases until the output level is 40. The nonlinearities in these measurements are due to the material (desired) and the system (undesired). We have seen in

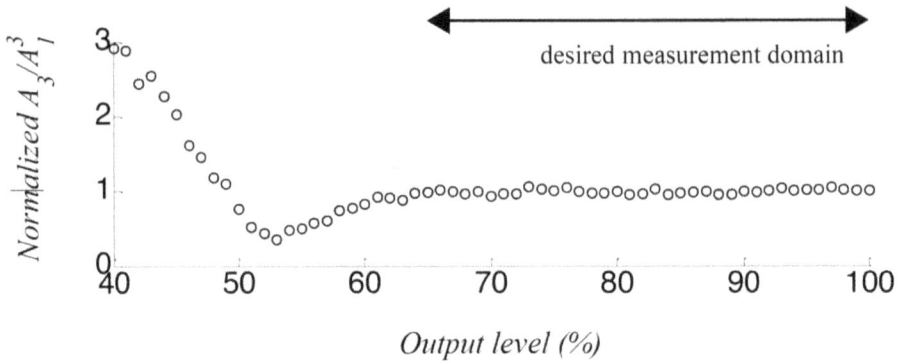

Figure 12.5. Normalized relative nonlinearity coefficient γ' as a function of the amplifier output level. The normalization value for an output level of 100 is $\gamma' = 0.1474$.

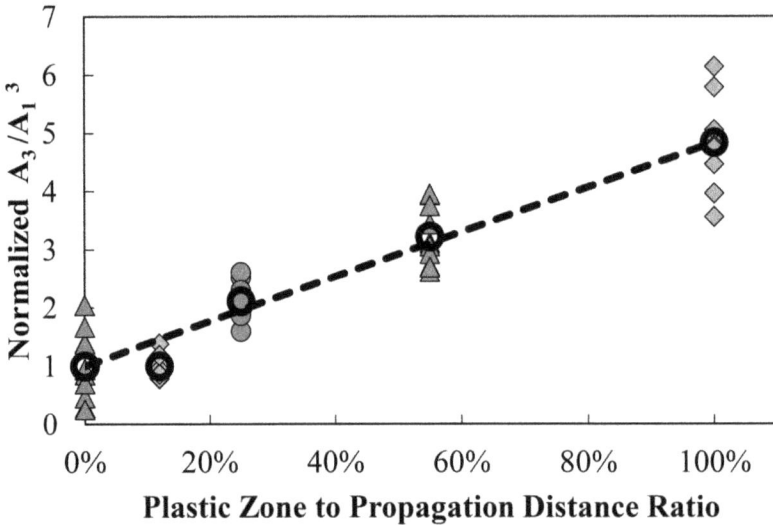

Figure 12.6. Normalized γ' as a function of the plastic zone to propagation distance ratio with the plastic zone always centered between the transmitter and receiver. The normalization value for the pristine material is $\gamma' = 0.1395$ [23]. (Reproduced with permission from [23]. Copyright 2014 Springer Nature.)

example 3.5 that third harmonics from material nonlinearity should be proportional to A_1^3, but this relationship is not anticipated for system nonlinearities. Thus, we seek to make measurements at output levels between 65 and 100 where γ' is constant. For output levels below 52 the results become increasingly dominated by the noise, i.e. the third harmonic amplitude A_3 has a very low signal-to-noise ratio.

The normalized γ' is observed to increase linearly as the ratio of the plastic zone to propagation distance increases, as illustrated in figure 12.6 [23], which shows all ten repetitions and the mean value. When the plastic zone (having a maximum value of $0.073 \ \text{m m}^{-1}$) fully covered the propagation distance the mean value of γ' increased by a factor of 4.8. It is not surprising that the measurements are less sensitive to more localized plastic deformation because the received signal is influenced by all of the material between the transmitter and receiver. Likewise, the normalized γ' is plotted as a function of fatigue life in figure 12.7 [24]. The steadily increasing slope corresponds with an increasing rate of fatigue damage, which is in contrast to low cycle fatigue where much of the plastic deformation damage occurs in the very first cycle. At 80% of the fatigue life (defined by a single test), γ' has increased by a factor of 3.65. Wen *et al* [25] investigated the SH0 third harmonic for assessment of overaging in an aluminum plate. Balachandran and Balasubramaniam [26] applied SH0 third harmonic testing to adhesively bonded lap joints in aluminum plates.

Let's close this section by giving some perspective on self-interaction. Of the three IR points identified for self-interaction of Lamb waves, IR point 1 has the most promise, and a transducer capable of preferentially actuating the in-plane displacement at the surface (such as a macrofiber composite as in [14, 27]) is expected to greatly improve measurements. The other IR points not discussed here include second harmonics at mode cutoffs and quasi-Rayleigh waves. Second harmonics at mode cutoffs would have to be generated by primary waves at mode cutoffs, neither of which would propagate, and thus we will not consider them. Quasi-Rayleigh waves will be discussed in section 12.3. Of the numerous IR points having an SH primary wave that generates symmetric Lamb waves, IR points 4–7 have large group velocity mismatches, which will limit the size of the wave mixing zone. Self-interaction of SH0 waves generates third harmonic SH0 waves that have been shown to have some very useful characteristics.

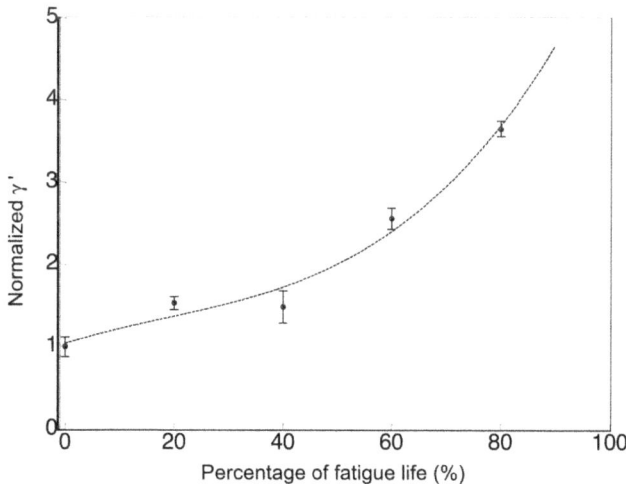

Figure 12.7. Normalized γ' as a function of fatigue life from interrupted tests. The normalization value for the pristine material is $\gamma' = 0.08455$. Symbols represent mean values and error bars are computed from the standard error. A cubic polynomial is fit to the data for visualization purposes. (Reproduced with permission from [24].)

12.2 Mutual interaction

While de Lima and Hamilton [28] modeled the mutual interaction of guided waves in plates in 2003, the first experimental results were not reported until 2017 by Hasanian and Lissenden [29]. In [29], 23 IR points are tabulated based on co-directional, counter-propagating, and non-collinear mixing of Lamb and SH waves that generate combinational harmonics at the sum or difference frequency. For each IR point a 'mode triplet' is defined that is comprised of the interacting primary wave modes and frequencies as well as the secondary wave mode generated. They later tabulate another 22 IR Points and give remarks on some of their interesting attributes [17]. As already evident from chapter 8, neither list is comprehensive.

Finite element simulations were used to demonstrate the value of a simple subtraction-based signal processing step [29]. Using two independent function generators, primary waves A and B are actuated at frequencies f_a and f_b, respectively. Each of waves A and B may generate second and third harmonics by self-interaction, but it is their mutual interaction at $f_a \pm f_b$ that is of interest for second order interactions. Therefore, three tests are performed: test A alone, test B alone, and test A + B together. The mutual interactions received by the transducer are then determined from the signal, S, by subtraction:

$$S_{\text{Diff}} = S(A + B) - [S(A) + S(B)]. \tag{12.1}$$

The motivation for this subtraction can be traced back to (4.118) and (4.119) for the first Piola–Kirchhoff stress. The part of the stress associated with the mutual interaction can be written as

$$\begin{aligned}
\mathbf{S}^{NL}(\mathbf{H}_a, \mathbf{H}_b, 2) + \mathbf{S}^L(\mathbf{H}_{ab}) = {} & \mathbf{S}(\mathbf{H}) \\
& - [\mathbf{S}^L(\mathbf{H}_a) + \mathbf{S}^L(\mathbf{H}_{aa}) + \mathbf{S}^{NL}(\mathbf{H}_a, \mathbf{H}_a, 2)] \\
& - [\mathbf{S}^L(\mathbf{H}_b) + \mathbf{S}^L(\mathbf{H}_{bb}) + \mathbf{S}^{NL}(\mathbf{H}_b, \mathbf{H}_b, 2)].
\end{aligned} \tag{12.2}$$

From this perspective, we are assuming that the entirety of the S_{Diff} signal is associated with nonlinearity. Naturally, the subtraction is not necessary if the secondary waves from mutual interaction propagate in a direction different from the primary waves, or possibly if their polarity is distinctly different. It is unfortunate that the symbols S and $\mathbf{S}$ are used here because they represent completely different things; the former is the received signal while the latter is the stress.

Hasanian and Lissenden [29] used counter-propagating SH0 waves at $f_a = 1.72$ MHz and $f_b = 0.34$ MHz to generate internally resonant S0 waves at the sum frequency of 2.06 MHz in a 1 mm thick aluminum plate. While this mode triplet is not identified in chapter 8, it is IR point 8 in table III of Hasanian and Lissenden [29]. There is no propagating symmetric Lamb wave mode at the wavenumber associated with the difference frequency. Thus, mutual interaction will only be apparent at the sum frequency. Magnetostrictive transducers actuated the primary waves A and B and a 2.25 MHz contact transducer coupled to a 36° acrylic wedge was used to receive the secondary S0 waves. At 2.06 MHz the S0 mode has a larger u_3 component on the surface than the u_1 component. The signals received (with 700 averages and 40 dB amplification) by the angle-beam transducer are shown in figure 12.8 [29]. Signals are received for test A and test B due to the curvature of the wavefronts despite the polarity difference between primary and secondary waves. Thus, the signal difference S_{Diff} is computed. The peaks in the frequency spectra shown in figure 12.9 [29] demonstrate the difference signal is dominated by the sum frequency. The secondary waves are shown to be the S0 mode based on their group velocity and that they leak into fluid placed on the surface of the plate. In addition, a region of the plate was thermally aged and the amplitude of the frequency spectrum at the sum frequency was observed to increase.

In follow-up work, Cho et al [30] extended the finite element simulations to show the evolution of the secondary wave amplitude with propagation distance and expanded the experiments to the assessment of localized fatigue degradation. A machined notch localized the fatigue damage in the plate and results are compared for when the wave mixing zone is in the degraded region and when it is not. Magnetostrictive transducers were again used to actuate the counter-propagating SH0 waves, but a novel polyvinylidene difluorine (PVDF) sensor was used to receive

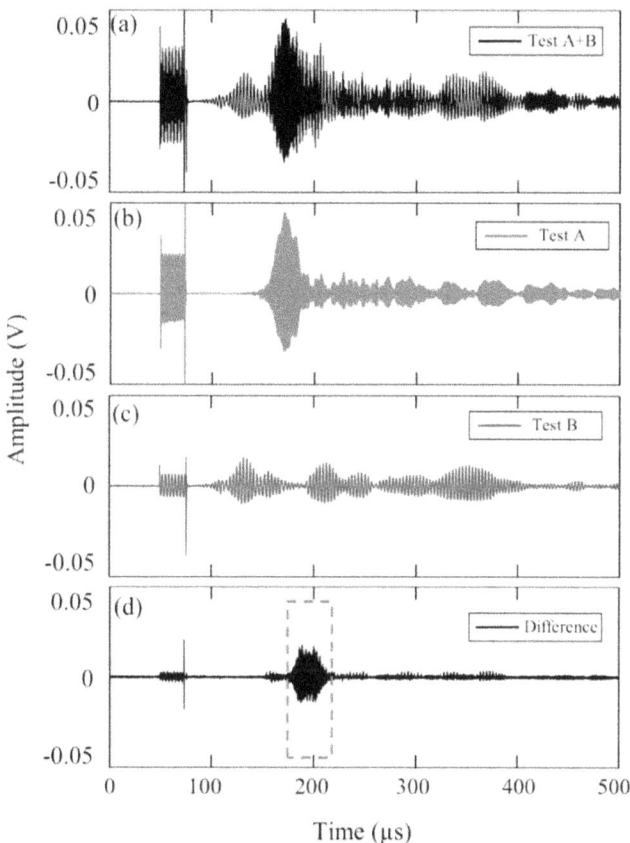

Figure 12.8. Counter-propagating SH0 wave mixing A-scans for (a) waves A + B, (b) waves A, (c) waves B, and (d) the difference signal, S_{Diff}. (Reprinted with permission from [29]. Copyright 2017 AIP Publishing.)

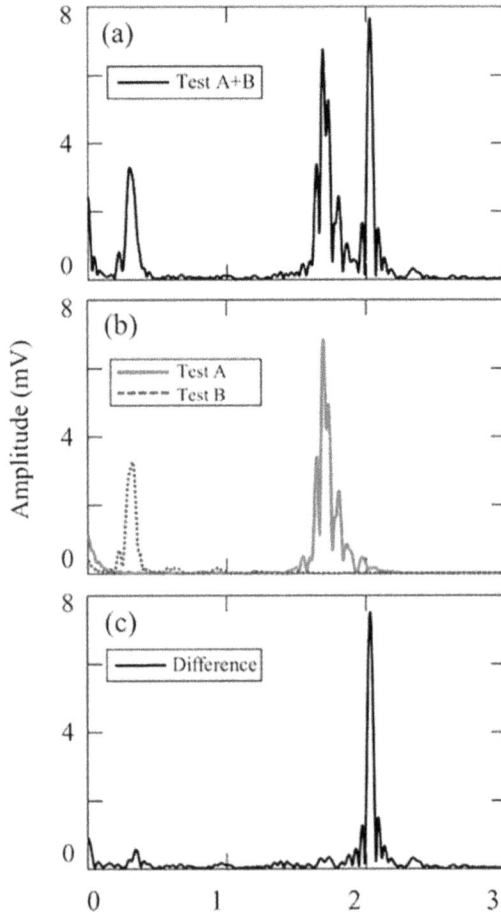

Figure 12.9. Counter-propagating SH0 wave mixing frequency spectra for (a) waves A + B, (b) waves A and waves B, and (c) the difference signal, S_{Diff}. (Reprinted with permission from [29]. Copyright 2017 AIP Publishing.)

both the primary SH0 waves and secondary S0 waves, which have different polarities. The electroactive sensing film is anisotropic because it is stretched during processing. By orienting the stretched direction at 45° to the wave propagation direction, the film is sensitive to both SH particle motion in the X_2 direction and S0 particle motion in the X_1–X_3 plane [31]. It is sufficiently compliant to not appreciably affect the waves it is receiving and can remain permanently affixed for structural health monitoring applications. Moreover, the PVDF film is very broadband, making it ideal for receiving waves over a range of frequencies and enabling determination of the wave amplitude ratio (in volts), $A_{ab}/(A_a A_b)$, even when the waves have different polarities. A 3D laser Doppler vibrometer provides these capabilities for laboratory testing and has the additional benefit of being noncontact.

Shan *et al* [20] investigated co-directional mixing of SH0 waves at different frequencies. Any combination of *fd* products adding up to the *fd* product where the

S0 and SH0 modes intersect (~3.39 MHz-mm for aluminum) will generate internally resonant S0 waves at the sum frequency. At this fd product the $U_1(X_3 = \pm h)$ and $U_3(X_3 = \pm h)$ wavestructure components are quite large as shown in figure 8.3 for IR point 4. Thus, an air-coupled transducer (sensitive to $U_3(X_3 = \pm h)$) can be used to receive the S0 waves, and because it is noncontact, it is easy to scan over the surface to assess the cumulative nature of the secondary waves. In this case the group velocity of the S0 waves is less that the SH0 waves, therefore the mixing zone is limited and after the SH0 and S0 waves separate the amplitude growth decreases as described in the simplified analytical model of Hasanian and Lissenden [17]. When a gel filter is used on the surface the rate of cumulative growth (i.e. the mixing power) of the secondary S0 waves over propagation distances of 20–90 mm correlated well with the number of fatigue cycles as shown in figure 12.10 [20]. The number of fatigue cycles is given as the fatigue level (%) in the figure (0%, 25%, 50%, and 75%).

Li *et al* [32] analyse mutual interactions between co-directional Lamb waves theoretically and also conducted some enlightening experiments. After presenting the model development, they selected the A1 mode at $f_a = 2.58$ MHz and the S0 mode at $f_b = 1.42$ MHz as the primary waves in a 0.95 mm thick aluminum plate, which are synchronized with the S1 mode at the sum frequency of 4.00 MHz. No synchronized mode exists at the difference frequency (1.16 MHz, $k_a - k_b = 2520 - 1620 = 900$ m^{-1}). The mode triplet for the sum frequency is not shown in table 8.7 for same type different nature because it is not well synchronized. Both the authors [32] and table 7.4 indicate that no power flows to symmetric Lamb modes for mixing of symmetric and antisymmetric primary Lamb modes, but rather it flows only to antisymmetric Lamb modes. One point of the experiments is to show that because of the symmetry conditions, the second order interactions do not generate the S1 mode. The other point of the experiments is to show that third order interactions are sufficient to generate four different tertiary modes that are at least approximately synchronized. It is not clear whether the authors searched for primary waves that would give four synchronized third order secondary waves or if it was a serendipitous finding, either way it is marvelous that

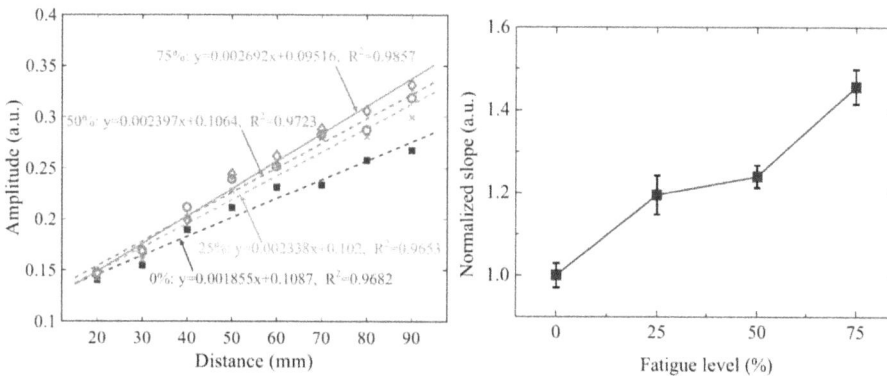

Figure 12.10. Cumulative nature of the secondary S0 waves, computed from S_{Diff}. The rate of accumulation increases with the fatigue level. (Reproduced with permission from [20]. Copyright 2019 Elsevier.)

such a combination exists. The frequencies and wavenumbers of the third order waves are given in table 12.2. The 'combination' wavenumber is based on the wavenumbers of the primary waves. The 'mode' wavenumber is the wavenumber of the third order combinational harmonic mode. The 'difference' wavenumber is the detuning from synchronization. Both the authors [32] and table 7.5 indicate that these third order combinational harmonics have nonzero power flow due to the symmetry conditions.

Angle-beam transducers were used to send and receive. The receiving contact transducer is sufficiently broadband for all combinational harmonics except $2f_b - f_a = 0.26$ MHz, but as shown in figure 12.11 [32] this low frequency is received, albeit with a lower amplitude.

The fact that all four third order combinational harmonic frequency peaks are within two orders of magnitude of the primary frequencies is surprising, as that is typical of second order interactions and third order interactions are generally expected to be smaller. This is despite the receiving transducer being at sub-optimal angles for the third order wave modes since they each have a unique phase velocity (Snell's law is used to determine the optimal angle for each mode and frequency).

Table 12.2. Wave propagation parameters for third order combinational harmonics given the primary waves: A1 mode at $f_a = 2.58$ MHz and S0 mode at $f_b = 1.42$ MHz in a 0.95 mm thick aluminum plate based on [32].

Combination	Frequency (MHz)	Mode	Wavenumber (m^{-1})		
			Combination	Mode	Difference
$2f_b - f_a$	0.26	A0	$2k_b - k_a = 720$	800	−80 (−11.1%)
$2f_a - f_b$	3.74	S1	$2k_a - k_b = 3420$	3500	−80 (−2.3%)
$2f_b + f_a$	5.42	A2	$2k_b + k_a = 5760$	5150	+610 (+5.9%)
$2f_a + f_b$	6.58	S2	$2f_a + f_b = 6600$	6560	+40 (+0.61%)

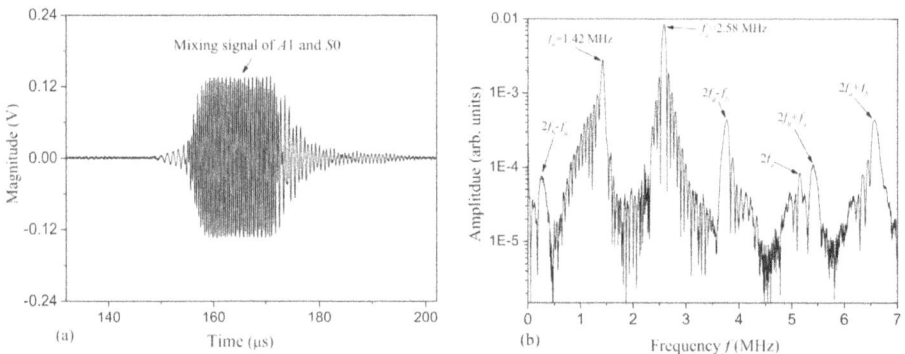

Figure 12.11. Co-directional wave mixing of A1 mode ($f_a = 2.58$ MHz) and S0 mode ($f_b = 1.42$ MHz): A-scan and frequency spectrum. (Reproduced with permission from [32]. Copyright 2018 AIP Publishing.)

The distance between the receiver and nearest transmitter is 520 mm in figure 12.11, but the actual mixing distances are less because of the mismatched group velocities. Nonetheless, the authors show that the $2f_a - f_b$ and $2f_a + f_b$ waves have increasing amplitudes over 360–520 mm. Later, Li *et al* [33] investigated co-directional low-frequency (0.60 and 0.80 MHz) S0 Lamb wave mixing. Both simulations and experiments were conducted. During and after the mutual wave interactions, peaks in the frequency spectrum appear at both the difference and sum frequencies. In addition, the nonlinearity coefficient for both difference and sum frequencies increase with propagation distance between 200 and 600 mm. To supplement the frequency spectra obtained by transforming the A-scan data, bandpass filtering is used to visualize the secondary wave signals in the time domain.

Metya *et al* [34] applied co-directional S0 Lamb waves at frequencies of 0.41 and 0.73 MHz to assess creep damage in 2 mm thick steel plates. The mixing zone was electronically scanned along the length of the plate by using time delays based on the mismatching group velocities of the primary waves and short five cycle tonebursts. Oil coupled angle-beam transducers were used, with the receiver placed at four distinct locations. The relative nonlinearity coefficient computed based on the sum frequency depends on creep strain and position. With a focus on detecting localized material degradation, Sun *et al* [35] study mutual interaction of counter-propagating S0 Lamb waves by simulation and experiment. The primary frequencies used here are not shown in table 8.8 because the low-frequency S0 waves are not sufficiently synchronized. However, synchronism is less important for counter-propagating waves because the wave mixing zone is relatively small by design in this case. The simulations and experiments assess the secondary waves propagating at the sum frequency within the mixing zone and then beyond it. Although one could argue with their terminology referring to these as cumulative waves, their results compare well with analogous model and simulation results for counter-propagating SH waves [17]. In trying to avoid future confusion, we point out that Sun *et al*'s [35] statement referring to [29] that '...the generated Lamb wave propagates in a plane perpendicular to the mixing zone...' is confusing. The primary SH and secondary S0 wave propagation is collinear; it is likely that what they mean is the particle motion is different, i.e. the wave polarizations are orthogonal.

In the last couple of years there are several articles reporting guided wave mixing experimental results. For example:

- Blanloeuil *et al* [36] use non-collinear mixing of SH0 waves that generate S0 secondary waves to interrogate the contact acoustic nonlinearity at an interface.
- By mixing A0 waves at an angle of 145°, Pineda Allen and Ng [37] found that the secondary S0 waves at the sum frequency propagated at 55° and that the nonlinearity coefficient increased with propagation distances of 10–100 mm.
- Mora *et al* [38] used non-collinear mixing of fundamental Lamb wave modes to generate a zero group velocity (ZGV) mode that is detected by a noncontact laser interferometer.
- One-way wave mixing is when two co-directional waves generate secondary waves that back propagate towards the point of origin. Liu *et al* [39] conducted

simulations and experiments mixing co-directional S0 and SH0 waves to generate back-propagating waves at the difference frequency. The back-propagating waves are shown to be sensitive to localized artificial surface corrosion.

- Zhu *et al* [40] mix short tonebursts of co-directional low-frequency S0 waves and demonstrate that the nonlinearity coefficient increases with fatigue cycling. Rather than transform to the frequency domain, they use the phase inversion (or phase-reversal) method due to the short tonebursts.

- Zhu *et al* [41] implement a time shifting scheme to optimize the frequency selection for S0 wave mixing by separating the sum frequency from the second harmonics of the primary waves, four sets of S0 primary wave pairs are actuated by surface-bonded piezoelectric discs.

- Hughes *et al* [42] mixed fundamental edge modes[1] in a plate to assess material nonlinearity, one novelty of this study is that they evaluate different amplitude ratios for the primary waves.

- The quasi-static pulse, which can be thought of as energy at the difference frequency for self-interaction, has been shown to be the S0 mode regardless of whether the primary waves are S0, A0, or SH0 modes [43–45].

- Barely visible impact damage in a 1 mm thick unidirectional composite plate was detected by co-directional mixing of symmetric waves at the primary frequencies 1.95 and 2.25 MHz using a sideband peak counting method [46]. Wave propagation distances are probably significantly smaller than in metal plates due to attenuation and because the target material degradation is highly localized, but this is not reported.

Much of our understanding of nonlinear guided waves in plates translates to pipes, although the symmetry arguments for zero power flow do not, as noted in section 8.3.

12.3 Quasi-Rayleigh waves

We start by noting that chapter 9 discusses pulse-generated finite amplitude progressive surface waves while this brief section focuses on narrowband weakly nonlinear quasi-Rayleigh waves, which are more representative of ongoing efforts. The material is often isotropic and its thickness is many times the wavelength, such that the waves are nondispersive. It appears that the pioneering second harmonic generation tests were reported simultaneously [47, 48]. Barnard *et al* [47] used angle-beam transducers to send and receive surface waves in fatigued Ni and Ti alloy samples. Blackshire *et al* [48] used a 5 MHz surface acoustic wave transducer to send waves and a heterodyne laser to receive. Scanning the detection laser enabled the primary and secondary wave fields on the surface to be constructed. Kim, Jacobs, and Qu have a long-standing collaboration that has applied second harmonic generation of surface waves to monitor the evolution of a broad variety of materials

---

[1] An edge mode could be considered a type of 'feature guided wave' where in this case the feature is the edge of a plate. Other types of feature guided waves include bends, stiffeners, and welds.

where linear methods are insufficient, e.g. [49–55]. We close by mentioning the recent work of Gartsev *et al* [56], who studied the mutual interactions of non-collinear Rayleigh waves to generate secondary bulk longitudinal and shear waves.

12.4 Closure

While there are several internal resonance points for self-interaction of guided waves, in many cases their group velocities are mismatched, limiting their interaction. The S1–S2 and S2–S4 Lamb wave pairs do not suffer from group velocity mismatch, but the S1 and S2 modes are difficult to preferentially actuate. An alternative that is easier to actuate is the S0–S0 Lamb wave pair at low frequency, but this mode pair is not synchronized and because of the lower frequency there is less power flow to the secondary waves. Wave mixing dramatically opens the design space for selecting wave triplets based on their mutual interactions in addition to enabling the separation of material nonlinearity from measurement system non-linearity. Moreover, nondispersive quasi-Rayleigh waves have unique characteristics that can be leveraged for many practical applications.

References

[1] Deng M, Wang P and Lv X 2005 Experimental verification of cumulative growth effect of second harmonics of Lamb wave propagation in an elastic plate *Appl. Phys. Lett.* **86** 124104

[2] Deng M, Wang P and Lv X 2005 Experimental observation of cumulative second-harmonic generation of Lamb-wave propagation in an elastic plate *J. Phys. Appl. Phys.* **38** 344–53

[3] Rose J L 2014 *Ultrasonic Guided Waves in Solid Media* (Cambridge: Cambridge University Press)

[4] Hurley D C and Fortunko C M 1997 Determination of the nonlinear ultrasonic parameter using a Michelson interferometer *Meas. Sci. Technol.* **8** 634–42

[5] Deng M and Pei J 2007 Assessment of accumulated fatigue damage in solid plates using nonlinear Lamb wave approach *Appl. Phys. Lett.* **90** 121902

[6] Bermes C, Kim J-Y, Qu J and Jacobs L J 2007 Experimental characterization of material nonlinearity using Lamb waves *Appl. Phys. Lett.* **90** 021901

[7] Pruell C, Kim J-Y, Qu J and Jacobs L J 2007 Evaluation of plasticity driven material damage using Lamb waves *Appl. Phys. Lett.* **91** 231911

[8] Pruell C, Kim J-Y, Qu J and Jacobs L J 2009 Evaluation of fatigue damage using nonlinear guided waves *Smart Mater. Struct.* **18** 035003

[9] Pruell C, Kim J-Y, Qu J and Jacobs L J 2009 A nonlinear-guided wave technique for evaluating plasticity-driven material damage in a metal plate *NDT E Int.* **42** 199–203

[10] Xiang Y, Deng M, Xuan F-Z and Liu C-J 2012 Effect of precipitate-dislocation interactions on generation of nonlinear Lamb waves in creep-damaged metallic alloys *J. Appl. Phys.* **111** 104905

[11] Liu Y, Chillara V K and Lissenden C J 2013 On selection of primary modes for generation of strong internally resonant second harmonics in plate *J. Sound Vib.* **332** 4517–28

[12] Metya A K, Ghosh M, Parida N and Balasubramaniam K 2015 Effect of tempering temperatures on nonlinear Lamb wave signal of modified 9Cr–1Mo steel *Mater. Charact.* **107** 14–22

[13] Wang R, Wu Q, Yu F, Okabe Y and Xiong K 2019 Nonlinear ultrasonic detection for evaluating fatigue crack in metal plate *Struct. Health Monit.* **18** 869–81

[14] Srivastava A and Lanza di Scalea F 2009 On the existence of antisymmetric or symmetric Lamb waves at nonlinear higher harmonics *J. Sound Vib.* **323** 932–43

[15] Matlack K H, Kim J-Y, Jacobs L J and Qu J 2011 Experimental characterization of efficient second harmonic generation of Lamb wave modes in a nonlinear elastic isotropic plate *J. Appl. Phys.* **109** 014905

[16] Zhu W, Xiang Y, Liu C-J, Deng M and Xuan F-Z 2018 A feasibility study on fatigue damage evaluation using nonlinear Lamb waves with group-velocity mismatching *Ultrasonics* **90** 18–22

[17] Hasanian M and Lissenden C J 2018 Second order ultrasonic guided wave mutual interactions in plate: arbitrary angles, internal resonance, and finite interaction region *J. Appl. Phys.* **124** 164904

[18] Liu Y, Kim J-Y, Jacobs L J, Qu J and Li Z 2012 Experimental investigation of symmetry properties of second harmonic Lamb waves *J. Appl. Phys.* **111** 053511

[19] Achenbach J D and Xu Y 1999 Wave motion in an isotropic elastic layer generated by a time-harmonic point load of arbitrary direction *J. Acoust. Soc. Am.* **106** 83–90

[20] Shan S, Hasanian M, Cho H, Lissenden C J and Cheng L 2019 New nonlinear ultrasonic method for material characterization: codirectional shear horizontal guided wave mixing in plate *Ultrasonics* **96** 64–74

[21] Zuo P, Zhou Y and Fan Z 2016 Numerical and experimental investigation of nonlinear ultrasonic Lamb waves at low frequency *Appl. Phys. Lett.* **109** 021902

[22] Hu X, Ng C-T and Kotousov A 2022 Early damage detection of metallic plates with one side exposed to water using the second harmonic generation of ultrasonic guided waves *Thin-Walled Struct.* **176** 109284

[23] Lissenden C J, Liu Y, Choi G W and Yao X 2014 Effect of localized microstructure evolution on higher harmonic generation of guided waves *J. Nondestruct. Eval.* **33** 178–86

[24] Lissenden C J, Liu Y and Rose J L 2015 Use of non-linear ultrasonic guided waves for early damage detection *Insight: Non-Destr. Test. Cond. Monit* **57** 206–11

[25] Wen F, Shan S and Cheng L 2020 Third harmonic shear horizontal waves for material degradation monitoring *Struct. Health Monit.* **20** 147592172093698

[26] Balachandran A and Balasubramaniam K 2023 Quality assessment of adhesive joints using third harmonics of fundamental shear horizontal wave mode *J. Vib. Control* https://doi.org/10.1177/10775463231182771

[27] Lissenden C J, Lesky A and Soorgee M H 2013 Ultrasonic guided wave based SHM of a steel shell structure under different operating conditions *9th Int. Workshop on Structural Health Monitoring: A Roadmap to Intelligent Structures (Palo Alto)* ed F Chang pp 925–32

[28] de Lima W J N and Hamilton M F 2003 Finite-amplitude waves in isotropic elastic plates *J. Sound Vib.* **265** 819–39

[29] Hasanian M and Lissenden C J 2017 Second order harmonic guided wave mutual interactions in plate: vector analysis, numerical simulation, and experimental results *J. Appl. Phys.* **122** 084901

[30] Cho H, Hasanian M, Shan S and Lissenden C J 2019 Nonlinear guided wave technique for localized damage detection in plates with surface-bonded sensors to receive Lamb waves generated by shear-horizontal wave mixing *NDT E Int.* **102** 35–46

[31] Ren B, Cho H and Lissenden C J 2017 A guided wave sensor enabling simultaneous wavenumber-frequency analysis for both Lamb and shear-horizontal waves *Sensors* **17** 488

[32] Li W, Deng M, Hu N and Xiang Y 2018 Theoretical analysis and experimental observation of frequency mixing response of ultrasonic Lamb waves *J. Appl. Phys.* **124** 044901

[33] Li W, Xu Y, Hu N and Deng M 2020 Numerical and experimental investigations on second-order combined harmonic generation of Lamb wave mixing *AIP Adv.* **10** 045119

[34] Metya A K, Tarafder S and Balasubramaniam K 2018 Nonlinear Lamb wave mixing for assessing localized deformation during creep *NDT E Int.* **98** 89–94

[35] Sun M, Xiang Y, Deng M, Tang B, Zhu W and Xuan F-Z 2019 Experimental and numerical investigations of nonlinear interaction of counter-propagating Lamb waves *Appl. Phys. Lett.* **114** 011902

[36] Blanloeuil P, Rose L R F, Veidt M and Wang C H 2021 Nonlinear mixing of non-collinear guided waves at a contact interface *Ultrasonics* **110** 106222

[37] Pineda Allen J C and Ng C T 2023 Mixing of non-collinear Lamb wave pulses in plates with material nonlinearity *Sensors* **23** 716

[38] Mora P, Chekroun M, Raetz S and Tournat V 2022 Nonlinear generation of a zero group velocity mode in an elastic plate by non-collinear mixing *Ultrasonics* **119** 106589

[39] Liu Y, Zhao Y, Deng M, Shui G and Hu N 2022 One-way Lamb and SH mixing method in thin plates with quadratic nonlinearity: numerical and experimental studies *Ultrasonics* **124** 106761

[40] Zhu H, Ng C T and Kotousov A 2022 Low-frequency Lamb wave mixing for fatigue damage evaluation using phase-reversal approach *Ultrasonics* **124** 106768

[41] Zhu H, Ng C T and Kotousov A 2023 Frequency selection and time shifting for maximizing the performance of low-frequency guided wave mixing *NDT E Int.* **133** 102735

[42] Hughes J M, Kotousov A and Ng C-T 2022 Wave mixing with the fundamental mode of edge waves for evaluation of material nonlinearities *J. Sound Vib.* **527** 116855

[43] Sun X, Shui G, Zhao Y, Liu W, Hu N and Deng M 2020 Evaluation of early stage local plastic damage induced by bending using quasi-static component of Lamb waves *NDT E Int.* **116** 102332

[44] Jiang C, Li W, Deng M and Ng C-T 2021 Static component generation and measurement of nonlinear guided waves with group velocity mismatch *JASA Express Lett.* **1** 055601

[45] Jiang C, Li W, Deng M and Ng C-T 2022 Quasistatic pulse generation of ultrasonic guided waves propagation in composites *J. Sound Vib.* **524** 116764

[46] Li W, Xu Y, Hu N and Deng M 2020 Impact damage detection in composites using a guided wave mixing technique *Meas. Sci. Technol.* **31** 014001

[47] Barnard D J, Brasche L J H, Raulerson D and Degtyar A D 2003 Monitoring fatigue damage accumulation with Rayleigh wave harmonic generation measurements *AIP Conf. Proc.* **657** 1393–400

[48] Blackshire J L, Sathish S, Na J and Frouin J 2003 Nonlinear laser ultrasonic measurements of localized fatigue damage *AIP Conf. Proc.* **657** 1479–88

[49] Herrmann J, Kim J-Y, Jacobs L J, Qu J, Littles J W and Savage M F 2006 Assessment of material damage in a nickel-base superalloy using nonlinear Rayleigh surface waves *J. Appl. Phys.* **99** 124913

[50] Kim J-Y, Jacobs L J, Qu J and Littles J W 2006 Experimental characterization of fatigue damage in a nickel-base superalloy using nonlinear ultrasonic waves *J. Acoust. Soc. Am.* **120** 1266–73

[51] Liu M, Kim J-Y, Jacobs L and Qu J 2011 Experimental study of nonlinear Rayleigh wave propagation in shot-peened aluminum plates—feasibility of measuring residual stress *NDT E Int.* **44** 67–74

[52] Thiele S, Kim J-Y, Qu J and Jacobs L J 2014 Air-coupled detection of nonlinear Rayleigh surface waves to assess material nonlinearity *Ultrasonics* **54** 1470–5

[53] Torello D, Thiele S, Matlack K H, Kim J-Y, Qu J and Jacobs L J 2015 Diffraction, attenuation, and source corrections for nonlinear Rayleigh wave ultrasonic measurements *Ultrasonics* **56** 417–26

[54] Morlock M B, Kim J-Y, Jacobs L J and Qu J 2015 Mixing of two co-directional Rayleigh surface waves in a nonlinear elastic material *J. Acoust. Soc. Am.* **137** 281–92

[55] Kim J-Y, Wall J J, Joo Y-S, Park D-G and Jacobs L J 2020 Device and method for nonlinear ultrasonic measurements on highly irradiated 304 stainless steel specimens in a hot cell environment *Rev. Sci. Instrum.* **91** 025103

[56] Gartsev S, Zuo P, Rjelka M, Mayer A and Köhler B 2022 Nonlinear interaction of Rayleigh waves in isotropic materials: numerical and experimental investigation *Ultrasonics* **122** 106664

Chapter 13

Perspective

In the game of chess, one dreams and schemes ways to checkmate the opponent's king by looking ahead at what could happen many moves in the future. Dreaming and scheming is made much easier by so-called forcing moves, i.e. moves that limit the opponent to preferably just one move, making it much easier to consider what comes next. In this closing chapter we contemplate what might be the forcing moves that help propel nonlinear ultrasonic guided waves for nondestructive evaluation and structural health monitoring into practice.

Image credit: Jessie Lissenden

Let's start with some perspective on the contents of the book. There is a clear focus on guided waves in plates for simplicity with respect to other waveguides, nonlinearity in waveguides having other geometries is similar in many respects, but at best the theoretical analysis is more complicated. The book has a clear theme of analyzing both Lamb and SH waves, i.e. it is not limited to Lamb waves. The aim is not simply to be comprehensive, but rather due to a belief that SH waves are crucial to realistic nonlinear guided wave detection of material nonlinearity. Here is why. While the complicated nature of Lamb wave propagation associated with their dispersion curves and wavestructures is interesting to study and important to understand in order to find key wave modes that are useful, the multiple modes and difficulty in preferential mode generation often make them difficult to utilize. The simplicity of the SH modes and the advent of simple, inexpensive, magneto-strictive transducers to generate finite amplitude SH waves has clear advantages for real applications. Large changes in the relative nonlinearity coefficient for SH0 third harmonics are observed for plastic deformation and fatigue in chapter 12. Furthermore, the differences in polarity between SH and Lamb waves provide advantages for wave mixing applications with respect to isolating material non-linearity from measurement system nonlinearity.

Future nonlinear ultrasonic guided wave applications are envisioned for non-destructive inspection and testing as well as for structural health monitoring. The emergent technology is likely to gradually transition to real products and applica-tions to show value and build confidence. This chapter will share personal opinions and perspective on how measurement methodology and analysis could advance through the following five considerations:

1. Separation of material nonlinearity from measurement system nonlinearity.
2. Link with the structural design that identifies hot spots to be monitored and a plan for inclusion of nonlinear ultrasonic guided waves in the operations management and maintenance planning.
3. Standards for test methods that are broad enough to be applicable to the emerging needs for offline inspection and in-service monitoring.
4. Define specifications needed to build monitoring systems into self-aware smart structures.
5. A solid connection between nonlinear wave propagation characteristics and the material microstructure that dictates its strength and fracture properties.

We are not trying to predict the future; we are hoping to influence it. Let's take these one at a time.

13.1 Separation of material nonlinearity from measurement system nonlinearity

In contrast to bulk waves, there are numerous strategies that can be followed to account for, lessen the impact of, or eliminate the effect of system nonlinearity on nonlinear ultrasonic guided wave measurements.

As previously observed, measuring the weak ultrasonic nonlinearity associated with evolving material microstructure is challenging for numerous reasons. The earlier that evolution can be detected the better to make maintenance decisions and manage operations. Thus, we seek to isolate small changes due to material nonlinearity from background noise and measurement system nonlinearity. If the nonlinearity at the source can be quantified it may be possible to model its presence at the receiver and remove it. Additionally, a narrowband transmitter may be able to filter out second and third harmonics from the signal generator and amplifier. Cumulative secondary guided waves that can be received at different propagation distances enable the cumulative secondary waves from material nonlinearity to be separated from the measurement system nonlinearity. Noncontact sensors (e.g. air-coupled piezoelectrics, laser Doppler vibrometers, electromagnetic acoustic trans-ducers, micro-electro-mechanical system ultrasonic microphone arrays) that can be scanned along the wave propagation path are well-suited for this task in an inspection modality. Likewise, an array of sensors (e.g. PVDF film) permanently bonded to the surface, and which do not affect the wavefield, are well-suited for structural health monitoring. Primary wave modes of one polarity that generate secondary wave modes having a different polarity (e.g. SH primary waves that generate secondary Lamb waves) could isolate the material nonlinearity. Wave mixing enables selection of primary wave frequencies that generate secondary waves at sum and/or difference frequencies sufficiently far removed from measurement system nonlinearities at integer multiples of the primary frequencies. Finally, non-collinear and one-way wave mixing can be arranged such that the secondary waves propagate in a different direction than the primary waves that carry the measure-ment system nonlinearities.

13.2 Link with the structural design that identifies hot spots to be monitored and a plan for inclusion of nonlinear ultrasonic guided waves in the operations management and maintenance planning

The concept of digital twin [1] provides a way to manage the operations of an asset. A digital twin is a virtual representation of a physical entity that is connected to its counterpart by data. The digital twin can be based on the as-built drawings, provide material specifications and pedigrees, be able to simulate the actual operating conditions, and have its condition updated based on inspections and health monitoring data. Ultimately, the digital twin should someday be capable of prognosticating the remaining life based on the current condition and planned operating conditions. The first step in this process is to select critical subdomains, or hot spots, that should be monitored. Next, the specific nonlinear guided wave methodology is designed based on the potential modes of degradation and laboratory tested, before testing is expanded to mockups of the asset. Depending on the asset, the monitoring may be done online or offline. Then probability of detection [2] or receiver operating characteristic curves [3] can be determined and used to quantify the reliability of the diagnostic data acquired from the monitoring system.

13.3 Standards for test methods that are broad enough to be applicable to the emerging needs for offline inspection and in-service monitoring

The first question is who regulates the operation of the assets being monitored and whether there are requirements or standards for managing the structural health. If such requirements or standards exist, then we must work within their framework, or work with them to develop nonlinear ultrasonic guided wave test methods. If no such standards exist, then they need to be developed. Offline inspection and in-service monitoring methodologies are sufficiently different to warrant separate parts of the standard if both inspection and monitoring are applicable to a specific asset. For example, inspection may be conducted with transducers temporarily coupled to the asset, while in-service monitoring is more likely to involve permanently attached transducers. Thus, in the former case standard methods to obtain repeatable coupling is important, while in the latter case standard methods to confirm that the transducer performance remains steady over time is important. In addition, inspection methods must be reproducible by all inspectors, signaling the need for training and certification. It seems likely that each fleet of similar assets will require a set of mockups that can be used for calibration of both inspection and monitoring systems.

13.4 Define specifications needed to build monitoring systems into self-aware smart structures

The basis for fielding nonlinear ultrasonic guided wave-based monitoring systems is shown in figure 1.2. It relies upon early detection of material degradation in order to manage operational life and maintenance actions based on current condition. The underlying premise is that damage will be repaired when it begins to decrease reliability below a prescribed threshold. Thus, some amount of early material degradation is expected to occur and does not require immediate repair. This is a change from current operations in some fields, where all detected fatigue cracks must be immediately repaired, or the component replaced. The justification for the change in philosophy is that the monitoring system is sensitive to incipient damage that precedes macroscale crack growth. Those fields that already practice damage-tolerance and permit operating with known subcritical cracks based on fracture mechanics principles have no need to change philosophy, they will simply benefit from the knowledge acquired earlier in the life.

The nonlinear ultrasonic guided wave system is expected to have maximum impact by implementing structural health monitoring that follows the hierarchy of: detection, localization, classification (type), characterization (sizing), and prognosis [4]. The first two steps in the hierarchy are much easier than the last two steps. There is evidence in this book that nonlinear ultrasonic guided waves can detect material degradation at an early stage. The degradation is at worst localized between the transmitter and receiver. The methods can be improved and have to be proven on particular assets, but there is solid foundational work here. On the other hand, characterization and prognosis remain wide open territory. One example of fatigue prognosis using nonlinear ultrasonics is given by Kulkarni and Achenbach [5]. Progress exploring this territory will be touched upon in the next section 13.5. Presuming that we can work through this entire hierarchy, self-aware smart structures become viable. The self-awareness comes from the nonlinear ultrasonic guided wave sensing system data and analysis thereof through the digital twin introduced previously (section 13.2). The digital twin may become smart enough to limit the operating conditions to avert failures, requiring material state feedback into the structural life analysis software embedded in the digital twin.

13.5 Solid connection between nonlinear wave propagation characteristics and the material microstructure that dictates its strength and fracture properties

It is relatively common for nonlinear ultrasound investigations to correlate non-linear characteristics to the phenomenology of the damage progression process, e.g. second harmonic generation is correlated with the fatigue life, strain to failure, time to creep rupture, time at temperature, radiation exposure. This is inherently valuable, but there is much more to be gained by further investigation. A fundamental tenet of materials science is that there exist key relationships between process-structure-properties. While the second order elastic constants are relatively insensitive to the microstructure (i.e. the Young's moduli and Poisson's ratio do not vary significantly for different aluminum alloys), the third order elastic constants are quite sensitive to the microstructure. Likewise, failure and fracture are thought of as weakest-link phenomena and are known to be very dependent upon the micro-structure. For example, Dowling [6] provides the following data:

Material	Mass density (kg m^{-3})	Young's modulus (GPa)	Typical ultimate strength (MPa)
Aluminum and alloys	2700	70	140–550
Titanium and alloys	4510	120	340–1200
Iron and alloys	7870	212	200–2500
Nickel and alloys	8900	210	340–1400

Third order elastic constants (TOECs) can be computed from first-principles for a material based on its lattice structure, see e.g. [7]. However, this is not the micro-structure of interest to us. We are interested in the lattice structure, but more than that, we are interested in the defected lattice structure (i.e. point defects, line defects, volumetric defects) and morphology. We seek to understand how microstructural features such as dislocation dipoles, cells, forests, as well as precipitates and inclusions, grain boundaries, twins, and persistent slip bands affect the acoustic nonlinearity because these features are also connected to the strength and fracture properties. It is worth pointing out that the nonlinear elastic material model is used to represent the material's dynamic response during elastic wave propagation. It is not used to represent inelastic material responses like plastic deformation and creep.

One modeling approach is to let the nonlinear material model represent the anharmonic response of the defected lattice given the actual morphology. This is a phenological approach where we posit that the TOECs (and fourth order elastic constants) can be determined at different damage states [8]. Suppose high cycle fatigue is the degradation process of interest. Conceptually at least, the TOECs could be determined as a function of fatigue life and then the acoustic nonlinearity coefficient evolution could be predicted and compared with experimental results. The lack of literature demonstrating this approach suggests that it may not be practical to determine TOECs as a function of damage progression. Instead, following the lead of Hikata *et al* [9–11] the effect of defected microstructure has been added directly to the acoustic nonlinearity parameter as reviewed succinctly by Matlack *et al* [12]. The first mechanism modeled is dislocation bowing associated with obstacles pinning each end of the dislocation. As cited in [12], many researchers have extended the original model. The nonlinear force–displacement relation associated with a dislocation dipole has also been shown to contribute to the acoustic nonlinearity. Fatigue damage in wavy-slip metals occurs as vein and channel structures (veins have very high dislocation densities and channels have much lower dislocation densities), that form the basis for persistent slip bands; see for example [13]. The presence of precipitates and their interaction with dislocations adds to the acoustic nonlinearity, as shown in [14]. The commonality between these nonlinear mechanisms is that a localized strain is produced in the lattice. Finally, just as breathing cracks exhibit a contact acoustic nonlinearity, distributions of microcracks can do the same [15]. All of these models are idealized to a specific mechanism and it remains to see whether they can represent the interactions between mechanisms and realistic material microstructures as will be needed to provide a strong connection between the measured acoustic nonlinearity and the material nonlinearity. Multiscale mechanics modeling that crosses both spatial and temporal domains should be capable of connecting nonlinear ultrasonic guided wave signals with the microstructural features that dictate the strength and fracture properties necessary for making remaining life predictions.

In closing, at this point in time:

- physics-informed deep learning is disrupting the entirety of science and engineering;
- metamaterial designs are greatly changing electromagnetic, acoustic, and elastic wave propagation possibilities;
- advances in manufacturing processes are limited by needs for process monitoring and quality assurance testing;

and are likely to provide opportunities for the nondestructive evaluation community in concert with nonlinear ultrasonic guided wave research and development activities. Applying nonlinear ultrasonic guided waves to nondestructive evaluation and structural health monitoring will almost certainly utilize physics-informed deep learning [16] for tasks such as analyzing the sensor data, connecting elastic wave distortion to microstructural features and mechanical properties, and prognostics. Elastic metamaterials can provide unprecedented control over wave propagation to be leveraged by clever researchers. *In situ* monitoring of manufacturing processes often has unique requirements related to the environment and necessary testing speed that require special considerations.

There is much to do, let's get to work.

References

[1] Sjarov M *et al* 2020 The digital twin concept in industry—a review and systematization *25th IEEE Int. Conf. on Emerging Technologies and Factory Automation (ETFA) (Vienna)* (Piscataway, NJ: IEEE) pp 1789–96

[2] Cawley P 2021 A development strategy for structural health monitoring applications *J. Nondestruct. Eval. Diagn. Progn. Eng. Syst.* **4** 041012

[3] Liu C, Dobson J and Cawley P 2017 Efficient generation of receiver operating characteristics for the evaluation of damage detection in practical structural health monitoring applications *Proc. R. Soc. Math. Phys. Eng. Sci.* **473** 20160736

[4] Farrar C R and Worden K 2007 An introduction to structural health monitoring *Philos. Trans. R. Soc. Math. Phys. Eng. Sci.* **365** 303–15

[5] Kulkarni S S and Achenbach J D 2008 Structural health monitoring and damage prognosis in fatigue *Struct. Health Monit.* **7** 37–49

[6] Dowling N E 2013 *Mechanical Behaivior of Materials* 4th edn (London: Pearson)

[7] Liao M *et al* 2021 Elastic3rd: a tool for calculating third-order elastic constants from first-principles calculations *Comput. Phys. Commun.* **261** 107777

[8] Chillara V K and Lissenden C J 2015 On some aspects of material behavior relating microstructure and ultrasonic higher harmonic generation *Int. J. Eng. Sci.* **94** 59–70

[9] Hikata A, Chick B B and Elbaum C 1963 Effect of dislocations on finite amplitude ultrasonic waves in aluminum *Appl. Phys. Lett.* **3** 195–7

[10] Suzuki T, Hikata A and Elbaum C 1964 Anharmonicity due to glide motion of dislocations *J. Appl. Phys.* **35** 2761–6

[11] Hikata A, Chick B B and Elbaum C 1965 Dislocation contribution to the second harmonic generation of ultrasonic waves *J. Appl. Phys.* **36** 229–36

[12] Matlack K H, Kim J-Y, Jacobs L J and Qu J 2015 Review of second harmonic generation measurement techniques for material state determination in metals *J. Nondestruct. Eval.* **34** 273

[13] Cantrell J H 2004 Substructural organization, dislocation plasticity and harmonic generation in cyclically stressed wavy slip metals *Proc. R. Soc. Lond. Ser. Math. Phys. Eng. Sci.* **460** 757–80

[14] Cantrell J H and Yost W T 2000 Determination of precipitate nucleation and growth rates from ultrasonic harmonic generation *Appl. Phys. Lett.* **77** 1952–4

[15] Nazarov V E and Sutin A M 1997 Nonlinear elastic constants of solids with cracks *J. Acoust. Soc. Am.* **102** 3349–54

[16] Willard J, Jia X, Xu S, Steinbach M and Kumar V 2022 Integrating scientific knowledge with machine learning for engineering and environmental systems arXiv: 2003.04919

www.ingramcontent.com/pod-product-compliance
Lightning Source LLC
Chambersburg PA
CBHW082138210326
41599CB00031B/6026